Leukocyte Adhesion Molecules

Proceedings of the First International Conference on:
"Structure, Function and Regulation of Molecules
Involved in Leukocyte Adhesion"

Held in Titisee, West Germany

September 28 – October 2, 1988

Co-Chairmen:
Donald C. Anderson, M.D.
Timothy A. Springer, Ph.D.

T.A. Springer   D.C. Anderson
A.S. Rosenthal   R. Rothlein
Editors

Leukocyte
Adhesion Molecules

Springer-Verlag

New York Berlin Heidelberg
London Paris Tokyo

Timothy A. Springer
Latham Family Professor of Pathology
Harvard Medical School and Center for
    Blood Research
Boston, Massachusetts 02115

Robert Rothlein
Principle Scientist
Department of Immunology
Boehringer Ingelheim Pharmaceuticals Inc.
Ridgefield, Connecticut 06877

Donald C. Anderson
Professor of Pediatrics & Cell Biology
Head
Section of Leukocyte Biology
Baylor College of Medicine and Texas
    Children's Hospital
Houston, Texas 77054

Alan S. Rosenthal
Vice President
Research & Development
Boehringer Ingelheim Pharmaceuticals Inc.
Ridgefield, Connecticut 06877

Art on cover © Boehringer Ingelheim International GmbH.

Library of Congress Cataloging in Publication Data
Leukocyte adhesion molecules.
    Based on a meeting held in Titisee in 1988
    1. Leucocytes—Congresses.   2. Cell adhesion—Con-
gresses.   I. Springer, Timothy A. [DNLM: 1. Cell
Adhesion—congresses.   2. Leukocytes—congresses.
QH 623 L652]
QR185.8.L48L46   1989   616.07'9   89-21917

ISBN-13: 978-1-4612-7927-3   e-ISBN-13: 978-1-4612-3234-6
DOI: 10.1007/978-1-4612-3234-6

© 1990 by Springer-Verlag New York Inc.
Softcover reprint of the hardcover 1st edition 1990

Typeset by Impressions, Inc., Madison, Wisconsin

9 8 7 6 5 4 3 2 1

Dedication

Dr. P. Brian Stewart, Senior Vice President, Research and Development, Boehringer Ingelheim Pharmaceuticals, Inc., and Vice President, Research and Development, Boehringer Ingelheim Corporation, died April 1, 1987.

Dr. Stewart was a graduate of Middlesex Hospital Medical School, University of London, and attended Harvard Graduate School of Business. He served for five years during World War II in the Royal Air Force, as a fighter pilot, for which he won the Distinguished Flying Cross.

Throughout his 30-year career with Boehringer Ingelheim, Dr. Stewart was a dedicated scientist-clinician whose personal and professional standards of excellence were the foundation of Boehringer Ingelheim's Research and Development programs. Prior to joining Boehringer Ingelheim Pharmaceuticals, he was Managing Director from 1962 to 1977 of Boehringer's Pharma Research Ltd. in Montreal. Beginning in 1978, he and his colleagues moved to Ridgefield, Connecticut, to build Boehringer's U.S. Research and Development site. His vision and dedication were responsible for our current research commitment efforts to discover ways to therapeutically regulate clinically relevant aspects of leukocyte adhesion. His colleagues remember him with fondness. We wish to dedicate this scientific meeting to that memory.

Alan S. Rosenthal, M.D.
December 12, 1988

Acknowledgments

The editors wish to offer their sincere thanks to Dr. Ronald Faanes and Ms. Leah Neumaier for their organizational and administrative contributions. It was their enthusiasm and perseverance that accounted for the reality of having a symposium on leukocyte adhesion at Titisee, West Germany and for the successful completion of this book.

We would also like to thank Dr. med Hasso Schroeder for his assistance in arranging for the symposium and details during our stay.

Contents

Participants

Donald C. Anderson, Randall W. Barton, Sarah Bodary, Eric Brown, Eugene C. Butcher, Jill P. Buyon, George Cianciolo, Robert B. Colvin, Angel Corbi, A. Benedict Cosimi, Jürgen Dämmgen, Michael Diamond, Michael Dustin, Ronald B. Faanes, Carl G. Figdor, Alain Fischer, David Frankhouser, William M. Gallatin, Nils Lauge Hansen, John M. Harlan, Martin Hemler, Nancy Hogg, Jeffrey Huth, Douglas Jones, Takashi Kishimoto, Steve Kohl, Richard Krogsrud, Richard Larson, Peter E. Lipsky, Steven D. Marlin, Vincent J. Merluzzi, Leah C. Neumaier, Michael Pierschbacher, Suman Rakhit, Alan B. Rickinson, Alan S. Rosenthal, Robert Rothlein, Hasso Schroeder, Elke Seewaldt-Becker, Stephen Shaw, C. Wayne Smith, Timothy A. Springer, Christian Stratowa, Robert F. Todd III, Craig D. Wegner, Peter C. Wilkinson, Chester Wood, Samuel Wright.

Contributors

Steven B. Abramson, M.D., Associate Professor of Medicine, New York University School of Medicine, 561 First Avenue, New York, NY 10016

Donald C. Anderson, M.D., Professor, Pediatrics, Microbiology and Immunology, Head, Leukocyte Biology Section, Baylor College of Medicine, Texas Children's Clinical Care Center, 8080 North Stadium, Suite #2100, Houston, TX 77054

Ms. Terri Anderson, Technologist, Photomedicine, Harvard Medical School, Massachusetts General Hospital, 100 Blossom Street, The Cox-5 Building, Boston, MA 02115

Randall W. Barton, Ph.D., Senior Principal Scientist, Department of Immunology, Boehringer Ingelheim Pharmaceuticals, Inc., 90 East Ridge Road, Box 368, Ridgefield, CT 06877

Ellen Berg, Ph.D., Postdoctoral Fellow, Department of Pathology, Stanford University School of Medicine, Room #L-235, Stanford, CA 94305

Michael P. Bevilacqua, M.D., Ph.D., Instructor, Vascular Research Division, Department of Pathology, Harvard Medical School, Brigham and Women's Hospital, 75 Francis Street, Boston, MA 02115

Anne Marie Buckle, Ph.D., Macrophage Laboratory, Imperial Cancer Research Fund, Lincoln's Inn Fields, London WC2A 3PX, England

Bean L. Burr, B.S., Department of Pediatrics, Baylor College of Medicine, Texas Children's Clinical Care Center, 8080 North Stadium, Suite #2100, Houston, TX 77054

Jill P. Buyon, M.D., Assistant Professor, Department of Medicine, New York University School of Medicine, Hospital for Joint Diseases, 301 E. 17th Street, New York, NY 10003

Eugene C. Butcher, M.D., Assistant Professor, Department of Pathology, Stanford University School of Medicine, Room L-235, Stanford, CA 94305

Pina M. Cardarelli, Ph.D., Staff Scientist, Immunotec, 11045 Roselle Street, San Diego, CA 92121

Robert B. Colvin, M.D., Associate Professor, Pathology, Director, Immunopathology Unit, Harvard Medical School, Massachusetts General Hospital, 100 Blossom Street, The Cox-5 Building, Boston, MA 02114

David Conti, M.D., Transplant Unit, Harvard Medical School, Massachusetts General Hospital, 100 Blossom Street, The Cox-5 Building, Boston, MA 02114

A. Benedict Cosimi, M.D., Chief, Department of Clinical Transplant Surgery, Massachusetts General Hospital, MGH White 4, Room #5, Boston, MA 02114

Carol Crouse, Ph.D., Harvard Medical School, Dana Farber Cancer Institute, Room #D920, 44 Binney Street, Boston, MA 02115

Robert B. Colvin, M.D., Harvard Medical School, 100 Blossom Street, The Cox-5 Building, Boston, MA 02114

Angel L. Corbi, Ph.D., Postdoctoral Fellow, Center for Blood Research, Harvard Medical School, 800 Huntington Avenue, Boston, MA 02115

Jürgen W. Dämmgen, Ph.D., Director, Department of Pharmacology, Dr. Karl Thomae GmbH, Postfach 720, Birkendorfer Strasse 65, D-7950 Biberach/Riss 1, West Germany

Patricia A. Detmers, Ph.D., Laboratory of Cell Physiology and Immunology, Rockefeller University, 1230 York Avenue, New York, NY 10021

Claire M. Doerschuk, M.D., Assistant Professor, Department of Pathology, University of British Columbia, Pulmonary Research Laboratory, Vancouver, British Columbia V6Z1Y6

Graeme Dougherty, Ph.D., Terry Fox Laboratory for Hematology/Oncology, British Columbia Cancer Research Centre, 601 West 10th Avenue, Vancouver, British Columbia V5Z 1L3, Canada

Mr. Michael L. Dustin, Predoctoral Fellow, Center for Blood Research, Harvard Medical School, 800 Huntington Avenue, Boston, MA 02115

Mariano Elices, Ph.D., Harvard Medical School, Dana Farber Cancer Institute, Room #D920, 44 Binney Street, Boston, MA 02115

Ronald B. Faanes, Ph.D., Director, Department of Immunology, Boehringer Ingelheim Pharmaceuticals, Inc., 90 East Ridge Road, Box 368, Ridgefield, CT 06877

Dr. Carl G. Figdor, Senior Investigator, Division of Immunology, The Netherlands Cancer Institute, Plesmanlaan 121, 1066 CX Amsterdam, The Netherlands

Alain Fischer, M.D., Departement de Pediatrie, Unite d'Immunologie & d'Hematologie, Hopital Necker—Enfants Malades, 149 Rue de Sevres, 75015 Paris, France

Ms. Cheryl Geoffrion, Technologist, Transplantation Unit, Harvard Medical School, Massachusetts General Hospital, 100 Blossom Street, The Cox-5 Building, Boston, MA 02114

Michael A. Gimbrone, Jr., M.D., Professor and Director, Vascular Research Division, Department of Pathology, Harvard Medical School, Brigham and Women's Hospital, 75 Francis Street, Boston, MA 02115

Claude Griscelli, M.D., U 132 Inserm, Hopital Necker—Enfants Malades, 149 Rue de Sevres—75015 Paris, France

Barbara Grospierre, M.D., U 132 Inserm, Hopital Necker—Enfants Malades, 149 Rue de Sevres—75015 Paris, France

Nils Lauge Hansen, M.D., Department of Dermatology, Bispebjerg Hospital, Bispebjerg Bakke 23, 2400 Copenhagen NV, Denmark

John M. Harlan, M.D., Associate Professor of Medicine, Chief, Section of Hematology-Oncology, University of Washington, Harborview Medical Center, Harborview Hall, ZA-34, Seattle, WA 98104

Hal K. Hawkins, M.D., Ph.D., Department of Pathology, Baylor College of Medicine, Texas Children's Clinical Care Center, 8080 North Stadium, Suite #2100, Houston, TX 77054

Martin E. Hemler, Ph.D., Assistant Professor, Department of Pathology, Harvard Medical School, Dana Farber Cancer Institute, 44 Binney Street, Room #D920, Boston, MA 02115

Nancy Hogg, Ph.D., Macrophage Laboratory, Imperial Cancer Research Fund, Lincoln's Inn Fields, London, WC2A 3PX, England

Douglas H. Jones, M.D., NCI, NIH, Building #37, Room #2E16, Bethesda, MD 20892

Mark A. Jutila, Ph.D., Postdoctoral Fellow, Department of Pathology, Stanford University School of Medicine, Room #L-235, Stanford, CA 94305

Charles A. Kennedy, B.S., Scientist II, Department of Immunology, Boehringer Ingelheim Pharmaceuticals, Inc., 90 East Ridge Road, Box 368, Ridgefield, CT 06877

Takashi K. Kishimoto, Ph.D., Department of Pathology, Stanford University School of Medicine, Room #L-235, Stanford, CA 94305

Ms. Sharon Krater, Technician, Department of Pediatrics, Baylor College of Medicine, Texas Children's Clinical Care Center, 8080 North Stadium, Suite #2100, Houston, TX 77054

John F. Ksiazek, B.A., Scientist III, Department of Immunology, Boehringer Ingelheim Pharmaceuticals, Inc., 90 East Ridge Road, Box 368, Ridgefield, CT 06877

Mr. Richard S. Larson, Predoctoral Fellow, Center for Blood Research, Harvard Medical School, 800 Huntington Avenue, Boston, MA 02115

Kathleen Last-Barney, M.S., Scientist III, Department of Immunology, Boehringer Ingelheim Pharmaceuticals, Inc., 90 East Ridge Road, Box 368, Ridgefield, CT 06877

Michael B. Lawrence, Ph.D., Biomedical Engineering Laboratory, Rice University, Houston, TX 77005

Francoise LeDeist, M.D., U 132 Inserm, Hopital Necker—Enfants Malades, 149 Rue de Sevres—75015 Paris, France

David M. Lewinsohn, M.D., Ph.D., Predoctoral Fellow, Department of Pathology, Stanford University School of Medicine, Room #L-235, Stanford, CA 94305

Peter E. Lipsky, M.D., Professor, Department of Internal Medicine and Microbiology, University of Texas Health Science Center, 5323 Harry Hines Blvd., Dallas, TX 75235

Siu Kong Lo, Ph.D., Laboratory of Cell Physiology and Immunology, Rockefeller University, 1230 York Avenue, New York, NY 10021

Benedict R. Lucchesi, M.D., Ph.D., Department of Pharmacology, University of Michigan Medical School, Room #7423, MED SCI Building #1, Ann Arbor, MI 48109-0724

M. William Makgoba, M.D., Ph.D., Division of Molecular Endocrinology, Department of Chemical Pathology, Royal Post-Graduate Medical School, Hammersmith Hospital, DU-Cane Road, London W12-OHS, England

Steven D. Marlin, Ph.D., Senior Scientist, Department of Immunology, Boehringer Ingelheim Pharmaceuticals, Inc., 90 East Ridge Road, Box 368, Ridgefield, CT 06877

Fabienne Mazerolles, Ph.D., U 132 Inserm, Hopital Necker—Enfants Malades, 149 Rue de Sevres—75015 Paris, France

Larry V. McIntyre, Ph.D., Biomedical Engineering Laboratory, Rice University, Houston, TX 77005

Vincent J. Merluzzi, Ph.D., Section Leader, Department of Immunology, Boehringer Ingelheim Pharmaceuticals, Inc., 90 East Ridge Road, Box 368, Ridgefield, CT 06877

Sunil S. Patel, M.D., Ph.D., Medical Science Training Program, Southwestern Medical Center, Dallas, TX 75235

Mark R. Phillips, M.D., New York University School of Medicine, 561 First Avenue, New York, NY 10016

Michael D. Pierschbacher, Ph.D., Vice President and Scientific Director, Telios, 10901 N. Torrey Pines Road, LaJolla, CA 92037

Elizabeth Ralfkiaer, M.D., Department of Pathology, Rigshospitalet, University Hospital of Copenhagen, Copenhagen, Denmark

Robert Rothlein, Ph.D., Senior Principal Scientist, Department of Immunology, Boehringer Ingelheim Pharmaceuticals, Inc., 90 East Ridge Road, Box 368, Ridgefield, CT 06877

Ms. Helen E. Rudloff, Technician, Department of Pediatrics, Baylor College of Medicine, Texas Children's Clinical Care Center, 8080 North Stadium, Suite #2100, Houston, TX 77054

Frank C. Schmalstieg, M.D., Ph.D., Department of Pediatrics, University of Texas Medical Branch, Room #C2-30, Children's Health Center, Galveston, TX 77550

Elke Seewaldt-Becker, Ph.D., Department of Pharmacology, Dr. Karl Thomae GmbH, Postfach 720, Birkendorfer Strasse 65, D-7950 Biberach/Riss 1, West Germany

Steven Shaw, M.D., Senior Investigator, Experimental Immunology Branch, NCI, NIH, Immunology Branch, Bldg. 10, Room #4B17, Bethesda, MD 20892

Yoji Shimizu, Ph.D., Experimental Immunology Branch, NIC, NIH, Immunology Branch, Bldg. 10, Room #4B17, Bethesda, MD 20892

Paul J. Simpson, Ph.D., Senior Scientist, Lilly Research Laboratories, Room #304, Indianapolis, IN 46285

Seth G. Slade, B.S., Research Associate, New York University School of Medicine, Hospital for Joint Diseases, 301 E. 17th Street, New York, NY 10013

C. Wayne Smith, M.D., Baylor College of Medicine, Texas Children's Clinical Care Center, Leukocyte Biology Section, 8080 North Stadium, Suite #2100, Houston, TX 77054

Timothy A. Springer, Ph.D., Associate Professor, Center for Blood Research, Harvard Medical School, 800 Huntington Avenue, Boston, MA 02115

Donald E. Staunton, Ph.D., Postdoctoral Fellow, Center for Blood Research, Harvard Medical School, 800 Huntington Avenue, Boston, MA 02115

Carol D. Stearns, B.A., Scientist IV, Department of Immunology, Boehringer Ingelheim Pharmaceuticals, Inc., 90 East Ridge Road, Box 368, Ridgefield, CT 06877

Yoshikazu Takada, Ph.D., Biochemistry Division, National Cancer Center Research Institute, Tsukiji 5-chrome, Chuo-ku, Tokyo, Japan

Robert F. Todd III, M.D., Ph.D., Associate Professor of Internal Medicine, Department of Internal Medicine, University of Michigan Medical Center, 102 Observatory Street, Ann Arbor, MI 48109-0724

Gunhild Lange Vejlsgaard, M.D., Ph.D., Bispebjerg Hospital, Department of Dermatology, Bispebjerg Bakke 23, 2400 Copenhagen NV, Denmark

Anje A. te Velde, Predoctoral Fellow, Division of Immunology, The Netherlands Cancer Institute, Plesmanlaan 121, 1066 CX Amsterdam, The Netherlands

Jan E. de Vries, Ph.D., Senior Investigator, Dnax Research Institute of Molecular and Cellular Biology, Inc., 901 California Avenue, Palo Alto, CA 94304-1104

Mary C. Wacholtz, M.D., Ph.D., Assistant professor, Department of Internal Medicine, University of Texas Health Science Center, 5323 Harry Hines Blvd., Dallas, TX 75235

Craig D. Wegner, Ph.D., Principal Scientist, Department of Pharmacology, Boehringer Ingelheim Pharmaceuticals, Inc., 90 East Ridge Road, Box 368, Ridgefield, CT 06877

Gerald Weissmann, M.D., Professor of Medicine, New York University School of Medicine, 561 First Avenue, New York, NY 10016

Robert Winchester, M.D., Professor of Medicine, New York University School of Medicine, 561 First Avenue, New York, NY 10016

Robert K. Winn, Ph.D., Research Associate and Professor, Department of Physiology and Biophysics and Surgery, University of Washington, Harborview Medical Center, Seattle, WA 98104

Chester Wood, M.D., Associate Director, Department of Clinical Research, Boehringer Ingelheim Pharmaceuticals, Inc., 90 East Ridge Road, Box 368, Ridgefield, CT 06877

Samuel Wright, Ph.D., Assistant Professor, Laboratory Cell Physiology and Immunology, Rockefeller University, 1230 York Avenue, New York, NY 10021

Introduction

ALAN S. ROSENTHAL

The excitement that prevails today in the field of leukocyte adhesion research has its origin in the expansion of knowledge in the fields of immunology and molecular biology. Those of us who are products of this age of discovery often find it difficult to remain dispassionate and constrain our enthusiasm for the exciting developments of the past few years.

A more scholarly perspective on the events of the past two to three decades is rather revealing. We are not far removed from the equally extraordinary discoveries of Gowans and his colleagues (1) of the circulating behavior of lymphocytes. From this and other in vivo and traditional histological approaches to "inflammation," we can now interpret histopathology in more dynamic terms. The immune system is a fluid or mobile organ system with its multiplicity of specialized T- and B-cell subsets and monocyte-macrophage antigen-presenting cells. The movement of specialized phagocytic cells such as granulocytes and monocytes through vessel walls during inflammation (2) and macrophage-dependent antigen recognition by T lymphocytes during immune recognition (3,4) are some of the complex behavioral events that require direct physical contact between cooperating cells (Fig. 1).

How does one regulate, or better still, orchestrate processes and events which have such enormous potential for host defense as well as host damage?

Immune responsiveness is controlled via MHC-linked Ir genes at the level of antigen processing by the antigen-presenting cells. Such antigen-specific cell interactions involve an antigen independent receptor-ligand interaction mediated by the $LFA_1/ICAM_1$ (5,6). What, however, has more recently come to the fore is the role played by cytokines such as IL-1, TNF, and an interferon as inducers of these specialized cell-recognition structures that regulate cell-type access to inflammatory events (7–9). Members of the integrin family of molecules (10) also play a pivotal role in normal and pathological cell physiology, as evidenced by the papers that are included in this volume.

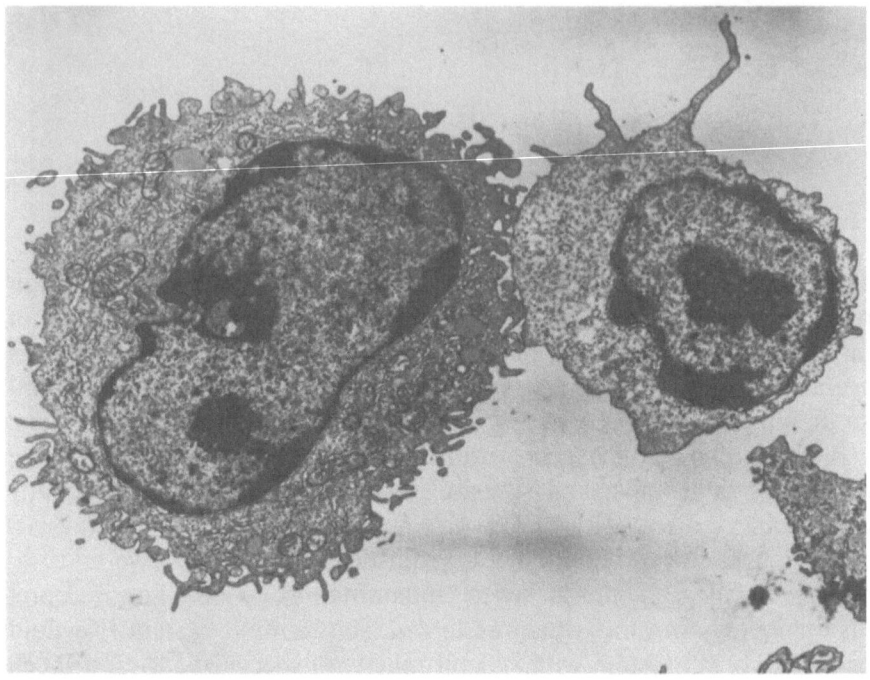

FIGURE 1. Electron micrograph of antigen-specific macrophage-lymphocyte interaction. (From Gallin, J.I. et al., editors, *Inflammation: Basic Principles and Clinical Correlates*, with permission of Raven Press, New York.)

Our meeting at Titisee in the Black Forest was the first international meeting to focus the attention of the scientific and medical community on this rapidly evolving field of leukocyte adhesion and to provide insight into opportunities for therapeutic regulation or modulation of transplantation rejection, reperfusion injury, and chronic immune-based inflammatory disease. In the final analysis, these proceedings are also an extraordinary testimony to the impact of modern immunology and molecular biology on medical progress.

References

1. Gowans JL: The recirculation of lymphocytes from blood to lymph in the rat. *J Physiol* (London) 146:54, 1959.
2. Harlan JM: Leukocyte-endothelial interactions. *Blood* 65:513, 1985.
3. Lipsky PE, Rosenthal AS: Macrophage-lymphocyte interaction. I. Characteristics of the antigen-independent-binding of guinea pig thymocytes and lymphocytes to syngeneic macrophages. *J Exp Med* 138:900, 1973.
4. Lipsky, PE, Rosenthal AS: Macrophage-lymphocyte interaction. II. Antigen-mediated physical interactions between immune guinea pig lymph node lymphocytes and syngeneic macrophages. *J Exp Med* 141:138, 1975.

5. Springer TA, Dustin ML, Kishimoto TK, Marlin SD: The lymphocyte function-associated LFA-1, CD2, and LFA-3 molecules: Cell adhesion receptors of the immune system. *Ann Rev Immunol* 5:223, 1987.
6. Martz E: LFA-1 and other accessory molecules functioning in adhesions of T and B lymphocytes. *Hum Immunol* 18:3, 1987.
7. Dustin ML, Rothlein R, Bahn AK, Dinarello CA, Springer TA: Induction by IL-1 and Interferon-gamma: Tissue distribution, biochemistry, function of a natural adherence molecule (ICAM-1). *J Immunol* 137:245, 1986.
8. Prober JS, Gimbrone MA, Lapierre LA, Mendrick DL, Fiere W, Rothlein, R, Springer TA: Overlapping patterns of activation of human endothelial cells by interleukin-1, tumor necrosis factor and immune interferon. *J Immunol* 137:1893, 1986.
9. Rothlein R, Czajkowski M, O'Neill MM, Marlin SD, Merluzzi VJ: Induction of intercellular adhesion molecule-1 (ICAM-1) on primary and continuous cell lines by pro-inflammatory cytokines: Regulation by pharmacological agents and neutralizing antibodies. *J Immunol* 141:1665, 1988.
10. Hynes RO: Integrins: A family of cell surface receptors. *Cell* 48:549, 1987.

Part 1
Structure and Function of the Integrin Family

1

Leukocyte Integrins

Takashi K. Kishimoto, Richard S. Larson,
Angel L. Corbi, Michael L. Dustin,
Donald E. Staunton, and Timothy A. Springer

Introduction

Cellular adhesion and recognition mechanisms are among the most basic requirements for the evolution of multicellular organisms. During the development of an embryo, cellular adhesion proteins can impart position-specific information that guide cell migration, localization, and the transfer of information between cells. As cells are triggered to differentiate to form tissues or organs, adhesion proteins help to maintain the organization and integrity of the body. The immune system is comprised of a network of cells in which cellular recognition mechanisms have been highly specialized. The function of the immune system is to distinguish self from nonself and to eliminate the latter. Two major protein families, the integrin family and the immunoglobulin superfamily, have evolved to guide cell-extracellular matrix (ECM) and cell-cell interactions for both developmental processes and immune function. The immunoglobulin superfamily, which includes the polymorphic antigen-specific receptors of lymphocytes, has recently been reviewed (1). This review will focus on the molecular biology of the leukocyte integrins, LFA-1, Mac-1, and p150,95 and their role in mediating inflammation.

Three recent developments have underscored the importance of the leukocyte integrins as adhesion receptors of the immune system: (i) The recognition that the leukocyte integrins are evolutionarily related to other integrins, such as the fibronectin receptor and platelet glycoprotein IIb/IIIa, which guide cell localization during embryogenesis and wound healing. The leukocyte integrins provide a similar mechanism in the immune system for guiding leukocyte localization during inflammation. (ii) Identification of ICAM-1, a ligand for LFA-1, which is induced during inflammation and may regulate leukocyte migration and localization. This receptor-ligand pair demonstrates the first known interaction between a member of the integrin family (LFA-1) and a member of the immunoglobulin superfamily (ICAM-1). (iii) Discovery and characterization of immunodeficiency patients who are genetically deficient in their expression of the leukocyte integrins. Leukocytes from these patients fail to mobilize during inflammation, and as a consequence, these patients suffer from recurrent life-threatening bacterial and fungal infections.

Nomenclature

LFA-1 is an acronym for lymphocyte function-associated antigen-1. Mac-1 is an abbreviation for macrophage antigen-1 and is also called Mo-1, OKM-1, and complement receptor type-3 (CR3). In some reports, p150,95 is called complement receptor type-4 (CR4) and Leu M5. The Third International Workshop on Human Leukocyte Differentiation Antigens (2) has designated the α subunits of LFA-1, Mac-1, and p150,95 to be CD11a, b, and c, respectively, and the common β subunit to be CD18.

The LFA-1, Mac-1, and p150,95 family has been called the LFA-1 family, leukocyte adhesion proteins, LeuCAM, and the leukocyte integrins. The homology of this family to other integrin receptors makes the latter name preferable.

Initial Characterization of LFA-1, Mac-1, and p150,95

Mac-1 was first defined by monoclonal antibodies (MAb) as a marker for myeloid cells (3). In contrast, LFA-1 was identified independently of Mac-1 by screening MAb for the ability to inhibit CTL-mediated killing of tumor cell targets (4). Further analysis showed that the LFA-1 MAb prevent the Mg^{2+}-dependent conjugate formation step, rather than the actual killing event (5). Mac-1 and LFA-1 antigen are both high molecular weight $\alpha\beta$ heterodimers (6). Detailed immunochemical and physiochemical comparison of the two antigens showed that the α subunits of LFA-1 and Mac-1 are unique, but that the β subunit is identical in both proteins (7–9). The remarkable structural similarity of LFA-1 and Mac-1 led to the hypothesis that Mac-1 would also function in adhesion. A function for Mac-1 was discovered when Mac-1 MAb were found to inhibit Mg^{2+}-dependent binding of the C3bi fragment of complement by mouse and human myeloid cells, thus defining Mac-1 as the complement receptor type-3 (CR3) (10,11). Analysis with a β subunit-specific MAb led to the identification of a third heterodimeric protein, p150,95, which shares the common β subunit (9).

Expression of all three leukocyte integrins is restricted to immune cells. LFA-1 is expressed by virtually all immune cells (8,12), with the exception of some tissue macrophages (6,13). Mac-1 has a more limited distribution (14): it is found on monocytes, macrophages, granulocytes, large granular lymphocytes, and immature and CD5+ B cells (15). The p150,95 protein has a similar distribution to Mac-1, although it is also expressed on some activated lymphocytes and is a marker for hairy cell leukemia (16,17).

Leukocyte Adhesion Deficiency Disease

Since 1974, a number of investigators have identified a class of immunodeficient patients who suffer from recurrent, life-threatening bacterial and fungal infections, and who have neutrophils deficient in chemotaxis and phagocytosis (18–22). Infected, necrotic lesions in these patients contain few leukocytes, despite the observation that these patients have chronic leukocytosis. One of the earliest reports described a possible actin dysfunction (23). Crowley et al. (24), however, showed that neutrophils from these patients were deficient in a high molecular weight surface protein. They further proposed that defects in chemotaxis and phagocytosis were secondary to a defect in adhesion. In 1984, several groups used MAb to demonstrate that the missing glycoprotein was actually the LFA-1, Mac-1, and p150,95 complex (25–28). In every case studied, expression of all three leukocyte integrins were found to be deficient. More recently, we have shown that LAD is caused by heterogenous defects in the common β subunit (29). Although LAD is a rare disease, the analysis of this disease has greatly increased our understanding of the biology of the leukocyte integrins and their role in inflammatory responses.

LFA-1, Mac-1, and p150,95 Are Members of the Integrin Family

LFA-1, Mac-1, and p150,95 are evolutionarily related to the integrin receptors that mediate cell adhesion to the extracellular matrix during development and wound healing (Fig. 1.1). There are three subfamilies of integrins, each defined by a common β subunit that shares multiple distinct α subunits (30–32). The three β subunits are designated β1, β2, and β3. The β1 and β3 subfamilies include receptors for extracellular matrix (ECM) components. The β1 subunit is shared by at least six VLA antigens (VLA-1–VLA-6) (33), which include the fibronectin receptor (VLA-5). VLA-3 has fibronectin and laminin-binding activity, while VLA-2 has recently been shown to be identical to platelet glycoprotein IaIIa, a collagen-binding receptor. The β2 subunit is shared by the leukocyte integrins, LFA-1, Mac-1, and p150,95. The β3 subunit is shared by the vitronectin receptor and platelet glycoprotein IIbIIIa.

The integrin family is ancient in origin. Homologous structures, termed *position-specific (PS) antigens*, have been implicated in guiding Drosophila development (34,35). The structure, function, and primary sequence of the integrins have been highly conserved in evolution. Some of the integrins involved in cell-matrix interactions recognize a sequence, Arg-Gly-Asp (RGD), which is embedded in numerous, unrelated matrix components (32). The ECM receptor integrins can show exquisite specificity for one matrix component or broad reactivity to multiple ligands. The

	LFA-1	Mac-1	p150,95	VLA-1	VLA-2	VLA-3	VLA-4	FNR(VLA-5)	VLA-6	VLA-7	IIb/IIIa	VNR
Primary Function	Immune cell adherence			Guiding morphogenesis and wound healing							Guiding morphogenesis and wound healing	
General Distribution	Leukocytes			Broad							Platelets	Broad
Structure	$\alpha_1\beta_1$			$\alpha_1\beta_1$							$\alpha_1\beta_1$	
Common β subunit	Yes			Yes							Yes	
Cleavage of α subunit upon reduction	No	No	No	No	No	Yes	No	Yes	Yes	?	Yes	Yes
Ligands	ICAM-1	IC3b Fibrinogen Factor X	(IC3b)	?	Collagen	Laminin Fibronectin Collagen	?	Fibronectin	Laminin	?	Fibronectin Fibrinogen Vitronectin von Willebrand factor	Vitronectin
Recognition Sequence								Arg-Gly-Asp			Arg-Gly-Asp	Arg-Gly-Asp
Interaction with cytoskeleton								Yes				

FIGURE 1.1. The integrin family.

leukocyte integrins are the first integrins known to mediate cell-cell interactions. The name *integrin* denotes that these are membrane receptors which integrate the extracellular environment (ECM or other cells) with the intracellular cytoskeletal network. The evolutionary and functional significance of LFA-1, Mac-1, and p150,95 as integrin proteins will be discussed.

Structural Features of the Leukocyte Integrins

Basic Structure

The leukocyte integrins are $\alpha_1\beta_1$ heterodimers (8), in which the α subunit is noncovalently associated with the β subunit. The α subunits of LFA-1, Mac-1, and p150,95 are 180,000, 170,000, and 150,000 daltons, respectively. The α subunits have been shown to be distinct by MAb-reactivity, antigen-preclearing studies, and tryptic peptide mapping. In contrast, the β subunit, $M_r = 95,000$, has been shown to be identical in all three proteins by the same criteria (7–9). Deglycosylation of the LFA-1, Mac-1, and p150,95 α subunits and the common β subunit reveal polypeptide backbones of 149,000, 137,000, 132,000, and 78,000 daltons, respectively (29,36,37). Heterogeneity in the glycosylation of LFA-1 has been reported. N-linked oligosaccharides of LFA-1 on T cells, but not B cells or macrophages, are sulfated (38). Moreover, sialyation patterns of LFA-1 on B and T cells differ, with LFA-1 on B cells being more acidic (39). Finally, only a subset of leukocyte integrins on neutrophils contain a lacto-N-fucopentaose II oligosaccharide moiety (40). The functional significance of this heterogeneity is unknown.

Biosynthesis

The α subunits of LFA-1, Mac-1, and p150,95 and the common β subunit are synthesized as distinct precursors of 165,000, 160,000, 146,000, and 89,000 daltons, respectively (9,36,37). The newly synthesized precursors contain high-mannose N-linked oligosaccharides (37). Association of the α subunit precursor and the β subunit precursor is required for further processing to complex-type N-linked oligosaccharides (26,41). This oligosaccharide modification occurs in the Golgi apparatus (42) and is evident by a decrease in the electrophoretic mobility of the mature polypeptide and resistance to endoglycosidase H digestion. The mature proteins are then transported to the cell surface or, in some cases, to intracellular granules (43–45).

Primary Structure

β Subunit Structure

β Subunit cDNA

The cDNA encoding the common β subunit was isolated and characterized by us (31) and Law et al. (46) independently. The deduced 769 amino acid sequence (Fig. 1.2) has the characteristic features of an integral membrane protein, with a 677 amino acid extracellular domain containing 6 potential N-glycosylation sites, a 23 amino acid transmembrane domain, and a 46 amino acid cytoplasmic domain. A striking feature of the β subunit is the high cysteine content (7.4% overall). A cysteine-rich (20%) region of 186 amino acid contains a fourfold repeat of an unusual cysteine motif. The high cysteine content is predicted to give the β subunit a rigid tertiary structure.

Northern blot analysis, Southern blot analysis, and peptide sequence data from the β subunit isolated independently from purified LFA-1, Mac-1, and p150,95, confirm immunochemical evidence that a single gene encodes the β subunit of all three leukocyte integrins (31).

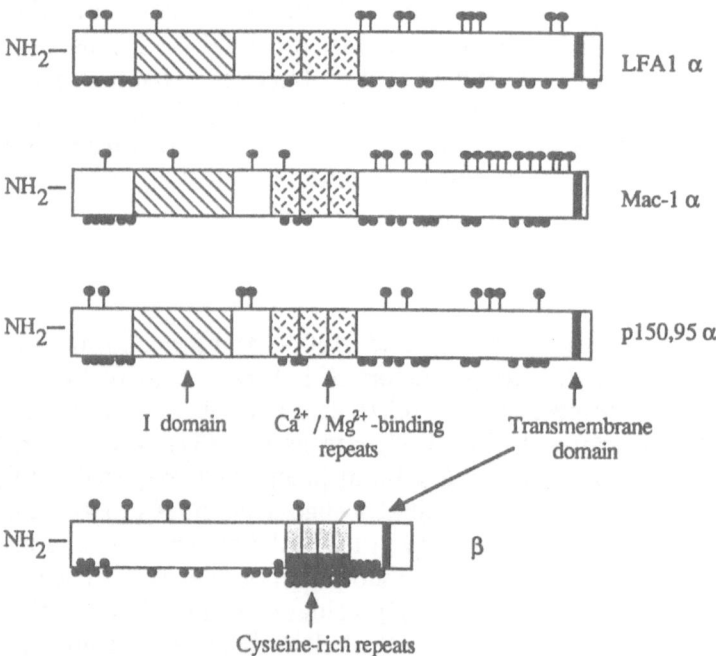

FIGURE 1.2. Schematic representation of primary structure of the LFA-1, Mac-1, and p150,95 α subunits and common β subunit. Black lollipops and circles represent N-linked glycosylation and cysteines, respectively.

Homology to Other Integrin β Subunits

The three β subunits, which define the three integrin subfamilies, share 37 to 45% amino acid identity, with particularly high conservation of the cytoplasmic domain, the transmembrane domain, and a stretch of 241 amino acids in the extracellular domain (70, 47, and 64%, respectively) (31,46–48). All 56 cysteine residues are conserved in each of the three β subunits, including the fourfold repeat of the unusual cysteine motif, first described for the chick β1 subunit (47). This high cysteine content probably gives the β subunits similar tertiary structures and may account for the observed differences in mobility in reducing vs. nonreducing gels (32,49). Interestingly, β1 (47) and β3 (48) have a consensus tyrosine phosphorylation sequence in the cytoplasmic domain, which is not found in the leukocyte β2 subunit (31,46).

α Subunit Structures

α Subunit cDNAs

cDNAs encoding the α subunits of p150,95 (50), Mac-1 (51–53), and more recently LFA-1 (54) have been cloned and characterized. The deduced amino acid sequences (Fig. 1.2) define homologous integral membrane proteins with a long extracellular domain (αX, 1081 amino acids; αM, 1092 amino acids; αL, 1063 amino acids), a hydrophobic transmembrane domain (αX, 26 amino acids; αM, 26 amino acids; αL, 29 amino acids), and a short cytoplasmic domain (αX, 29 amino acids; αM, 19 amino acids; αL, 53 amino acids). Although the polypeptide backbones are of similar size, the differences in apparent molecular weight of the mature polypeptides may be due, in part, to differences in the number of potential N-glycosylation sites (αX, 10 potential sites; αM, 19 potential sites; αL, 12 potential sites).

A striking feature of the α subunits is that each contains three homologous repeats that have putative divalent cation-binding sites that are similar to the Ca^{2+}-binding "EF-hand loop" sequences of calmodulin, troponin C, and parvalbumin. These putative metal-binding sites may account for the Mg^{2+}-dependency of leukocyte integrin-mediated adhesion. The concept that exogenous divalent cations stabilize the interaction of integrin α and β chains (55) has been used in the immunopurification of leukocyte integrins in functional form (162; S.A. Stacker, M.S. Diamond, and T.A. Springer, unpublished). The stabilizing effect of Mg^{2+} and Ca^{2+} on leukocyte integrins is the most direct evidence for binding of divalent cations by these heterodimers.

The Mac-1 and p150,95 α subunits share 63% amino acid identity with each other, but only 35% identity with the LFA-1 α subunit. The transmembrane domain and the three homologous repeats containing putative divalent cation-binding sites are the most highly conserved regions (88 and 87% amino acid identity, respectively, between Mac-1 and p150,95).

Homology to the Extracellular Matrix Receptor Integrins

The α subunits of the three subfamilies of integrins share 25 to 63% amino acid identity (50–54,56–58). The evolutionary relationships among integrin α subunits can be assessed by percent amino acid sequence identity (Fig. 1.3). The integrins may be divided into two functional groups, those which bind ECM ligands (ECM receptors) and those involved in cell-cell interactions and expressed on leukocytes (the leukocyte integrins). The ECM receptor integrins use both $\beta 1$ and $\beta 3$ subunits while the leukocyte integrins use the $\beta 2$ subunit. The α subunits of the leukocyte integrins are more similar to each other ($\bar{x}$ = 47% identity) than to those of the ECM receptor integrins ($\bar{x}$ = 27% identity). Moreover, the ECM receptor integrin α subunits are more related to one another ($\bar{x}$ = 42% identity) than to the leukocyte integrins. The α subunits of the ECM integrins, VLA-3, VLA-5 (fibronectin receptor), vitronectin receptor, and gpIIb/IIIa share a sequence in the extracellular domain which is posttranslationally cleaved, with resulting fragments bridged by a disulfide bond (32). This sequence is not found in the α subunits of leukocyte integrins (50–54). This explains the increased electrophoretic mobility in reducing gels observed for the α subunits of the ECM receptor integrins, but not the leukocyte integrins (32). The regions of highest conservation are the transmembrane domain and the putative metal-binding domains. Like leukocyte integrins, the ECM receptor integrins are dependent upon divalent cations for activity; however, the metal specificity is for Ca^{2+}, Mg^{2+}, or Mn^{2+}. Radioactive calcium has been shown to bind directly to the α subunit of gpIIbIIIa (59) and the fibronectin receptor (60).

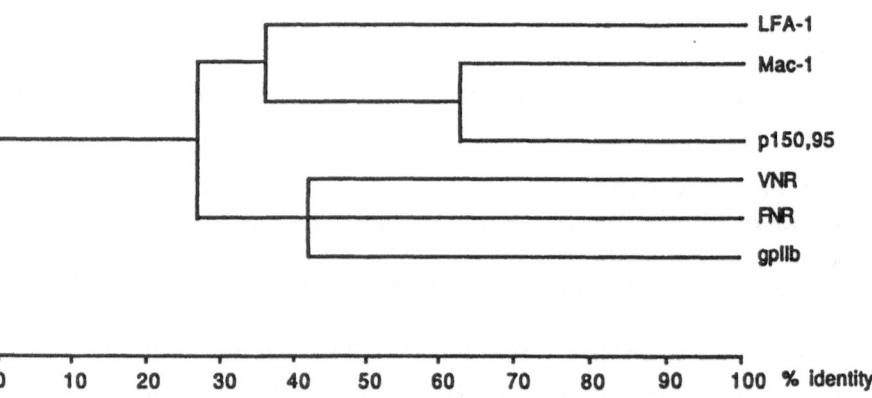

FIGURE 1.3. Evolutionary relationship of integrin α subunits. A pathway of evolution is suggested by the percent identity among α subunits, indicated by the scale below.

I Domain

All three leukocyte integrin α subunits have a 200 amino acid segment in the extracellular domain which has a counterpart in only one of five known ECM receptor integrin α subunit sequences (163). This domain, termed the I domain for inserted/interactive domain, is homologous to the three A domains of von Willebrand factor (vWF), to a domain in the complement cascade proteins C2 and factor B, and to two domains in the cartilage matrix protein (CMP) (51–54).

The relationships among the α subunits suggest the following evolutionary scheme (Fig. 1.3). A primordial α integrin gene duplicated and gave rise to at least two branches of integrin α subunits, the leukocyte integrins and the ECM receptor integrins. The primordial leukocyte integrin α subunit gene, containing an I domain, then duplicated and gave rise to LFA-1 and Mac-1/p150,95 primordial α subunit genes. Further duplication of the Mac-1/p150,95 primordial gene gave rise to the Mac-1 and p150,95 α subunits.

Expression

As a first step to uncover the structural basis for the adhesive activities displayed by LFA-1, Mac-1, and p150,95, their corresponding cDNAs (α and β subunits) have been inserted in the vector CDM8 (61) to obtain transient expression of the three heterodimers. After DEAE-dextran-mediated co-transfection of α and β cDNAs, LFA-1, Mac-1 and p150,95 α/β complexes have been detected on the surface of COS cells by immunofluorescence (164). Immunoprecipitation studies have demonstrated that the α and β subunits of LFA-1, Mac-1, and p150,95 are noncovalently associated. Moreover, the LFA-1 and Mac-1 heterodimers expressed by the transfected cells were functional. COS cells expressing LFA-1 bound to purified ICAM-1 absorbed to plastic, and this binding was inhibited by anti-LFA-1, anti-ICAM-1, and EDTA (Larson and Springer, unpublished). COS cells expressing Mac-1 specifically bound to erythrocytes sensitized with human or mouse C3bi, and this was inhibited with anti-Mac-1α OKM10 MAb and anti-β 60.3 MAb (Corbi and Springer, unpublished).

Relation of Structure to Ligand Binding

The three-dimensional structure of the leukocyte integrins and the individual contributions of the α and β subunits to ligand binding remain to be determined. The ECM receptor integrins recognize the RGD sequence and related sequences such as KQAGD within their ligands (62,63). Since the integrins within a given subfamily share the same β subunit but have distinct ligand-binding specificity, it seems reasonable to propose that the α subunits impart this specificity. The α subunits may influence recognition by changing the conformation of the β subunit to recognize different conformations of the RGD sequence present in different matrix proteins, or by binding to further sites on the matrix protein. Complementation studies with fragments of fibronectin suggest that a second site distinct from RGD is required for binding to the fibronectin receptor (64).

Chemical crosslinking studies show that although radiolabeled RGDS tetrapeptide binds predominately to a highly conserved region of the β subunit of gpIIb/IIIa, there is significant labeling of the α subunit as well (65). The RGD-binding site of the β subunit maps to the N-terminal portion of a highly conserved 241 amino acid segment of the extracellular domain. The RGD-binding site of the α subunit has not been determined. One model is that divalent cations held in the metal-binding domains of the α subunits may help to stabilize interaction with the RGD sequence (50). X-ray crystallography of the Ca^{2+}-binding EF-hand loops shows that amino acids with oxygen-containing side groups form the coordination axes for ligating the metal (66). The putative metal-binding domains of all integrins differ from these classical EF-hand loops in that a glutamic acid in the $-Z$ position of the coordination axes is missing from the former. This may leave the metal free to coordinate in the $-Z$ position with a residue on the ligand. It is tempting to speculate that the metal bound to the receptor may coordinate with the aspartic acid (D) of the RGD recognition sequence. This sequence appears suited for metal binding since GD and DG sequences appear frequently both in the EF-hand loop and in the metal-binding domains of the integrins.

The I domain appears as a functional unit in diverse proteins. The A1 domain of vWF binds glycoprotein Ib and heparin, while both A1 and A3 domains are involved in binding collagen (67). Partial sequence data from the cartilage matrix protein reveals two I domain-like repeats separated by an epidermal growth factor-like sequence (68). The I domain-like region in factor B and C2 is clearly demarcated on the N-terminal side by the cleavage site that gives the active Bb factor, and on the C-terminal side by the serine protease domain (69). Interestingly, both factor B and the homologous protein C2 bind C3b, whereas Mac-1 and p150,95 are receptors for the C3bi fragment of C3. It is tempting to speculate that the I domains of Mac-1 and p150,95 contribute to this specificity. The ECM integrins that recognize RGD do not contain I domains. The I domain may impart distinctive and additional recognition specificities to leukocyte integrins.

Chromosomal Localization

β Subunit Gene on Chromosome 21

The gene encoding the common β subunit was first mapped to chromosome 21 by gene complementation in somatic cell hybrids (70–72). With the availability of the β subunit cDNA, the gene has been further localized to band 21q22 by in situ hybridization to metaphase chromosomes (73). More recently, the β subunit has been shown to be the most distal marker known on the long arm of chromosome 21 (21q22.3) by hybridization to a panel of chromosome 21 deletion mutants and linkage of restriction fragment length polymorphisms (74). The β subunit gene may be a useful marker for analysis of trisomy 21 in Down's syndrome. In addition, band 21q22 has been identified as a breakpoint in chromosomal translocations (t(3;21)(q25;q22) associated with the blast phase of chronic myelogenous leukemia (CML) (75). Hematopoietic progenitor cells in CML show abnormal adhesive interactions with bone-marrow stroma (76). Further studies are required to determine if the β subunit gene is involved in this translocation and contributes to the progression of CML.

α Subunit Gene Cluster on Chromosome 16

The LFA-1 α subunit gene has been mapped to chromosome 16 by gene complementation in somatic cell hybrids (70). Subsequent Southern blot analysis of DNA from LFA-1 α hybrid cells containing human chromosome 16, has shown that the genes encoding Mac-1 and p150,95 are on the same chromosome (73). Furthermore, in situ hybridization data show that all three α subunit genes are clustered between bands p11–p13.1 on chromosome 16, defining a gene cluster involved in cell adhesion (73). The close proximity of the α subunit genes provides further support for their evolution by gene duplication.

Inversions (inv(16)(p13;q22)) and translocations (t(16;16)(p13;22)) involving this region of chromosome 16 are frequently observed in patients with acute myelomonocytic leukemia (77). Further investigation is required to determine if the genes encoding the leukocyte integrin α subunits are involved in these arrangements.

The Leukocyte Integrins in Inflammation

The Role of Leukocytes in the Inflammatory Process

The acute inflammatory reaction provides a rapid host-defense response to contain and eliminate infectious agents in extravascular tissues. The peripheral blood leukocytes constitute the recruitable force that infiltrates the infected tissues. Neutrophils and monocytes migrate in response to chemotactic factors released at the infection site. Chemotactic stimulation causes leukocyte cell polarization, granule release, and increased adhesiveness. Diapedesis requires the ability of leukocytes to bind vascular endothelial cells, cross the basement membrane, and enter the infected tissues. Regulation of the inflammatory response must be exquisite, so that leukocytes only enter the infected area and do not damage healthy tissues. Until recently, the molecular mechanisms that regulate and mediate leukocyte extravasation have been largely unknown. The role of the leukocyte integrins in this process has been elucidated by the characterization of human leukocyte adhesion deficiency disease.

Insights from the Study of Leukocyte Adhesion Deficiency

The clinical hallmarks of leukocyte adhesion deficiency (LAD) are recurrent necrotic and indolent infections of soft tissues, such as the skin, mucous membranes, and intestinal tract (18). The infectious microbes include a wide spectrum of fungi and bacteria, but most commonly Staphylococcal or gram-negative enteric bacteria. A key feature of LAD is that infected skin lesions are largely devoid of granulocytes, despite chronic peripheral blood leukocytosis (5–20 times normal levels). This failure of leukocytes to mobilize is also observed in Rebuck skin window assays. Transfused normal leukocytes are capable of mobilization to the infected tissues (78). Adhesion-independent responses, such as cell polarization in response to chemotactic factors, fMLP-binding, and respiratory burst from soluble stimuli, are normal. These observations demonstrate a direct involvement of the leukocyte integrins in the extravasation of leukocytes during an inflammatory response.

In vitro, neutrophils, and monocytes from LAD patients show profound defects in adhesion-related functions (18). Mac-1 and p150,95 are deficient both on the cell surface and in intracellular granules. Chemoattractants induce granule mobilization, but not upregulation of leukocyte integrin expression or homotypic cell aggregation. Neutrophils and monocytes fail to adhere and spread on artificial substrates, such as glass and plastic, and on endothelial cell monolayers. As a consequence, leukocytes have impaired directed motility in response to chemoattractants. C3bi-coated particles fail to induce phagocytosis or subsequent respiratory burst. Antibody-dependent cellular cytotoxicity mediated by granulocytes and monocytes is also abnormal (79,80).

Functional Studies

LFA-1 Function

LFA-1 serves to mediate cellular adhesion events in a wide spectrum of both antigen-dependent and antigen-independent interactions of immune cells (81,82). The role of LFA-1 in conjugate formation during CTL- and NK-mediated cytolysis has been extensively reviewed (81,82) and will not be discussed here. Lymphocyte localization to lymphoid organs, sites of inflammation, and grafts is dependent upon specific interaction with the vascular endothelium. T lymphocyte adherence to endothelial cells (83–85), fibroblasts (86), epidermal keratinocytes (87), synovial cells (88), and hepatocytes (89) is inhibitable partially or totally by LFA-1 MAb. Typically adherence is greatly increased by stimulation of the nonhematopoietic cells with cytokines such as TNF, LPS, IFN-γ, and IL-1 and by activation of the lymphocytes.

T lymphocyte and lymphoblast adherence to cultured endothelial cells has both an LFA-1-dependent and an LFA-1-independent component (84,85). Both pathways are increased upon stimulation of the endothelial monolayer with cytokines. LFA-1⁻ lymphoblasts from LAD patients exhibit the LFA-1-independent pathway, but not the LFA-1-dependent pathway (85). The impaired primary responses of LAD lymphocytes to specific antigen, but near normal secondary responses (90,91) may reflect selective involvement of LFA-1-independent and LFA-1-dependent pathways. Leukocyte adhesion deficiency patients display essentially normal delayed type hypersensitivity responses. Inflammatory tissues, which are devoid of granulocytes, do contain lymphocytes. The nature of the LFA-1-independent pathway of adhesion to endothelium is unknown, but does not appear to be the CD2/LFA-3 pathway. In the case of lymphocyte binding to high endothelial venules of lymphoid organs, adherence is partially inhibitable by LFA-1 MAb (92); however, organ-specificity is mediated by lymphocyte-homing receptors (93).

An in vitro model of LFA-1 dependent adhesion is the homotypic aggregation of lymphocytes and lymphoid cell lines in response to phorbol ester stimulation (94–96). Time lapse videomicroscopy (96) shows that phorbol ester activated peripheral blood lymphocytes are motile and show uropod formation and extensive membrane ruffling. Contact between pseudopodia of adjacent cells leads to adhesion, followed by mass aggregation. This aggregation event is both inhibitable and reversible by LFA-1 MAb. Moreover, cells from patients that are genetically deficient in LFA-1 expression do not aggregate (96). Although phorbol ester-induced aggregation is an in vitro phenomenon, it has been a useful model to study the cell biology of LFA-1 function and probably correlates with adhesion events in vivo. The requirements for aggregation are remarkably similar to the adhesion phase of CTL conjugate formation. Both events are Mg^{2+}-, energy-, and temperature-dependent and require an intact cytoskeleton (96–98). Aggregation-like cluster formation is also observed shortly after T helper cell activation by antigen presentation (99). Homotypic aggregation of activated lymphocytes may be an important step in extravasation of lymphocytes.

Improved isolation procedures have allowed functional studies on purified LFA-1 (162). LFA-1 purified in the presence of Mg^{2+} remains associated in an αβ complex and is functional since it mediates adhesion of several human cell lines when reconstituted into planar membranes or absorbed directly to plastic. This binding is blocked by pretreatment of cells with ICAM-1 MAb, pretreatment of the monolayer with LFA-1 MAb, and removal of divalent cations. The major difference between LFA-1-ICAM-1 mediated cell-cell adhesion and adhesion of cells to purified LFA-1 is that the former requires high temperature and metabolic activity, while the latter does not. In contrast, binding of cells to ICAM-1 monolayers is temperature and energy dependent (100). This suggests that the energy and temperature requirements are met on the LFA-1 bearing-cell side of the interaction, consistent with cellular regulation of LFA-1 avidity for its ligand.

Mac-1 Function

A functional role for Mac-1 was first demonstrated by the ability of Mac-1 MAb to inhibit monocyte and granulocyte binding of C3bi-coated erythrocytes (10). Thus, Mac-1 is equivalent to the complement receptor type 3. Mac-1 can mediate phagocytosis and lysis of C3bi-coated erythrocytes (101) and contributes to elevated NK activity against C3bi-coated target cells (102). Mac-1 has also been implicated in the ability of macrophages to bind *Leishmania promastigotes* (103,104), *E. coli* (105), and *Histoplasma capsulatum* (106).

Recent evidence suggests that Mac-1, like LFA-1, may play a more general role in mediating adhesive interactions of myeloid cells. Activated neutrophils, like phorbol ester-stimulated lymphocytes, form homotypic aggregates in vitro. Neutrophil aggregation is inhibitable by Mac-1 MAb, and not by LFA-1 MAb (107). Neutrophil and monocyte chemotaxis (107,108) and adherence to glass and plastic (107,108) and to endothelial (109) and epithelial (110) monolayers also involve Mac-1. Differential MAb blocking of CR3 activity and general adhesion suggest that Mac-1 may be a multifunctional receptor (107,108).

p150,95 Function

A role for p150,95 as a complement receptor was first discovered when both Mac-1 and p150,95 were found to specifically elute from a C3bi affinity column (111,112). The physiological relevance of this finding was tested with p150,95 on intact cells. The contribution of p150,95 to C3bi-binding by neutrophils and monocytes could only be assessed after antibody blockade of CR1 and Mac-1, which are expressed in 10-fold excess of p150,95. The remaining C3bi-binding activity on these antibody-treated cells could be blocked with the p150,95 MAb. These results suggest that p150,95, like Mac-1, may have some C3bi-binding activity, and it has been designated the CR4 (113). However, p150,95 expressed in COS cells does not bind C3bi (J Garcia-Aguillar, AL Corbi, and TA Springer, unpublished).

Like Mac-1, p150,95 probably has a broad role as a general adhesion protein. Anderson et al. (107) showed that p150,95 MAb could partially inhibit the adhesion of neutrophils to substrates. Mac-1, however, appears to have a more important role. In contrast, Figdor and his colleagues reported that p150,95 is a major component of peripheral blood monocyte adhesion to substrates and endothelial cells, phagocytosis of latex particles, and chemotaxis (114,115). These results were somewhat unexpected since p150,95 is expressed only at low levels on blood monocytes but high levels on tissue macrophages (16).

Recently, p150,95 expression has been reported on some activated lymphocytes and lymphocytic cell lines (17,116). The p150,95 MAb was found to inhibit conjugate formation by cytotoxic T lymphocytes (CTL) expressing comparable amounts of p150,95 and LFA-1 equally well as LFA-1 MAb (116). The inhibitory effects of the p150,95 MAb and LFA-1 MAb are additive. One group (117) reported no effect of the p150,95 MAb on CTL activity. This discrepancy may reflect differences in p150,95 expression by CTL clones.

Functional Redundance Among the Leukocyte Integrins

All three leukocyte integrins appear to function as general adhesion proteins for immune cell function. MAb directed to the α subunits have provided powerful tools to dissect the functions of the individual proteins. While the relative importance of LFA-1, Mac-1, and p150,95 may very with different systems, there is clearly some redundancy, particularly with neutrophil and monocyte functions. Comparative studies have shown that all three leukocyte integrins contribute to neutrophil and monocyte adhesion to endothelial cells and artificial substrates (107,115). These results are consistent with the finding that MAb to the common β subunit are often the most potent inhibitors of adhesion-related functions. Moreover, none of the LAD patients examined to date have been found to have a selective deficiency in the expression of only one of the leukocyte integrins (reviewed in 18–20), suggesting that the immune system might be able to compensate for the loss of one but not all three proteins.

Animal Models

While the study of human LAD has provided tremendous insight into the role of the leukocyte integrins in inflammation, the next stage of investigation—experimental manipulation in vivo—requires a suitable animal model. One possibility is a canine LAD model (118), which appears completely analogous to human LAD. A serious limitation is the difficulty and expense of maintaining a stable colony.

A second approach has been to mimic the LAD state by in vivo administration of MAb directed against the leukocyte integrins. An anti-β subunit MAb given intravenously to rabbits inhibits leukocyte extravasation and leukocyte-dependent plasma leakage in response to intradermal administration of chemotactic agents fMLP, LTB4, and C5a (119). Intravital microscopy of the rabbit tenuissimus muscle revealed that the MAb inhibited adherence of leukocytes to the vascular endothelium, but not rolling of leukocytes along the venule wall (119). In another system, mice given an intravenous injection of Mac-1 MAb fail to mobilize monocytes in response to thioglycollate-induced inflammation of the peritoneum (120). These results suggest that of the three leukocyte integrins, Mac-1 may play the major role in myeloid cell extravasation.

Potential Therapeutic Value of Leukocyte Integrin MAb

Inflammation is an integral part of the host defense to infection and injury of extravascular tissue. However, an inappropriate or uncontrolled inflammatory response may contribute to the pathogenesis of chronic disease states, such as arthritis, and acute ischemic shock followed by reperfusion. Leukocytes are thought to be a major culprit in promoting tissue damage following ischemia-reperfusion injury, perhaps by generating reactive oxygen metabolites, proteases, and phospholipases. Animals depleted of peripheral blood leukocytes show significantly reduced damage from myocardial ischemia and reperfusion.

Two recent studies have investigated the use of leukocyte integrin MAb to reduce ischemia-reperfusion injury. In a dog model of myocardial infarction, arterial flow is interrupted, then re-established. Reperfusion injury, measured as infarct size as a percentage of area at risk, was reduced twofold by in vivo administration of Mac-1 MAb (121). Histological studies show that Mac-1 MAb-treated dogs had fewer neutrophils in the myocardium. A rabbit model of hemorrhagic shock and resuscitation showed similar protective effects of an anti-β subunit MAb against liver and gastrointestinal injury but not lung injury (122). All MAb-treated animals survived five days, compared to a 29% survival rate among control animals. These results suggest potential therapeutic value for leukocyte integrin MAb in controlling tissue and organ injury following myocardial infarction, hemmorhagic shock, and other trauma which cause ischemia and are followed by re-establishment of normal circulatory flow. Of particular relevance is treatment of myocardial infarction with streptokinase or tissue plasminogen activator (TPA), which dissolve clots and allow circulation to be re-established in affected areas of the heart. Administration of leukocyte integrin MAb or ICAM-1 MAb together with anti-clotting agents promises to significantly reduce the amount of damaged cardiac tissue.

Regulation of Expression and Functional Activity

De Novo Synthesis

Cell surface leukocyte integrin expression is regulated both by de novo biosynthesis and by upregulation of preformed material. Change in biosynthetic rate is a relatively slow and inefficient means of responding to rapid changes in the environment, such as in an infection, and is primarily associated with immune cell differentiation. LFA-1 is expressed during the differentiation of hematopoietic stem cells. In the B cell and myeloid lineages, LFA-1 expression is first detected in the cytoplasmic μ chain-positive pre-B cells and late myeloblasts, respectively (123). Mac-1 expression is associated with committed granulocyte and monocyte precursors in the bone marrow (124). These observations concur with the in vitro differentiation of the promyeloblastic cell line HL-60. Undifferentiated HL-60 expresses only LFA-1. Differentiation to the monocytic lineage, induced with phorbol esters, or to the granulocytic lineage, induced with retinoic acid, results in expression of both Mac-1 and p150,95 (17,50,51). Peripheral blood monocytes express high levels of Mac-1 and low levels of p150,95. Upon extravasation and maturation to tissue macrophages, however, the pattern of expression is reversed to low levels of Mac-1 and high levels of p150,95 (16). Work in the murine system shows that many peripheral monocytes lose expression of LFA-1 after differentiation to tissue macrophages (8,13). These results suggest some role for p150,95 in tissue macrophage function.

Mobilization of an Intracellular Pool

In contrast to the slow time-course for upregulation by de novo synthesis, Mac-1 (26,43,44,125) and p150,95 (44,117,126) expression on phagocytic cells can be dramatically upregulated in a matter of minutes. Most of the Mac-1 and p150,95 proteins are stored in intracellular pools of neutrophils (43–45) and monocytes (44). Electron microscopy shows that leukocyte integrins are associated with peroxidase negative granules (44,45). This latent pool is mobilized in response to chemotactic factors, including fMLP, C5a, and leukotriene B4, and results in up to a 10-fold increase in surface expression of Mac-1 and p150,95. LFA-1 is upregulated twofold in monocytes and none in neutrophils. Upregulation presumably aids in the rapid mobilization of monocytes and neutrophils to inflammatory sites.

Receptor Activation

The regulation of the functional activity of leukocyte integrins is more complex than changing the gross level of surface expression. Peripheral blood lymphocytes express LFA-1, but do not spontaneously adhere to each other. Phorbol ester stimulation induces a rapid homotypic aggregation event. Aggregation is inhibitable by MAb against LFA-1 (94–96) and ICAM-1, an LFA-1 ligand (127). Aggregation is not associated with quantitative changes in the surface expression of either LFA-1 or ICAM-1. These results suggest that cell activation induces some change in either LFA-1 or ICAM-1 molecules.

Mac-1 functional activity appears to be similarly regulated on neutrophils. Neutrophils also form homotypic aggregates upon stimulation. Buyon et al. (128) were able to dissociate upregulation of the latent pool of Mac-1 from the homotypic aggregation event. Mac-1 MAb can effectively block neutrophil aggregation. Freshly isolated neutrophils were pretreated with Mac-1 MAb to coat surface Mac-1, washed, and then activated. Precoating the surface Mac-1 did not affect upregulation and expression of the latent pool; however, it did effectively inhibit neutrophil aggregation. Furthermore, upregulation and aggregation were dissociated kinetically and pharmacologically. In contrast, Vedder and Harlan (129) found that treating neutrophils with the anion channel blocking agent, DIDS (4,4'-diisothiocyanostilbene-2,2'-disulfonic acid), inhibited both Mac-1 upregulation and neutrophil aggregation. Interestingly, DIDS did not inhibit Mac-1-dependent binding of neutrophils to endothelial cells.

Phorbol esters have a biphasic effect on the C3bi-binding activity of Mac-1, which is independent of upregulation (130). First, a rapid increase in C3bi-binding activity is observed. After 60 min, however, binding activity is reduced below resting levels, despite the increased expression of Mac-1. Interferon-γ depresses C3bi binding activity without affecting Mac-1 expression. Binding activity is restored by adherence to fibronectin-coated surfaces (131). These results suggest that qualitative changes in Mac-1 are important for functional activity. The quantitative changes as a result of upregulation of intracellular pools may augment the response.

Activation Epitopes

The molecular mechanisms that regulate functional activity are not clear. the simplest explanation is a conformational change which unmasks the ligand binding site. However, most MAb which block adhesion of activated cells also bind to leukocyte integrins on resting cells. Recently, one MAb against a novel epitope of LFA-1 has been shown to induce homotypic aggregation of lymphoid cells, which is inhibitable by other LFA-1 MAb (132). This and several other MAb appear to define activation epitopes on LFA-1 (133,134).

Interaction with Other Receptors

Interaction of the leukocyte integrins with other membrane proteins may also influence binding activity. Pytowski et al. (135) recently raised a MAb against a 157,000 dalton cell surface protein on neutrophils, which specifically inhibits the C3bi-binding activity of Mac-1. The antigen is distinct from the leukocyte integrins and is expressed on cells from LAD patients. Another report suggests that a subset of Mac-1 on monocyte cell surfaces is associated with the Fc receptor and shows no lateral diffusion in the membrane (136). Antibody against the nondiffusable Mac-1 blocks IgG-mediated phagocytosis.

Receptor Clustering

Binding activity may also be regulated by crosslinking of receptors. Phorbol ester-induced C3bi-binding activity of Mac-1 correlates with aggregation of Mac-1 in small clusters on the cell surface. Prolonged exposure to PMA reverses C3bi-binding activity and results in dispersal of the Mac-1 clusters (137).

Phosphorylation

Receptor activity is likely to be regulated by intracellular events. Phorbol esters, which induce homotypic aggregation, are known to activate protein kinase C. Phosphorylation of the LFA-1 β subunit is induced by phorbol ester stimulation of peripheral blood lymphocytes (138). Neutrophils loaded with thiophosphate, which renders phosphorylated proteins resistant to phosphatases, show enhanced C3bi-binding activity upon phorbol ester stimulation. Moreover, thiophosphate-loaded neutrophils do not display the characteristic deactivation of C3bi-binding activity after prolonged exposure to phorbol esters (130). These results suggest that phosphorylation may play an important role in regulating leukocyte integrin function.

Interaction with the Cytoskeleton

Cell activation may also induce association of the leukocyte integrins with cytoskeletal elements. LFA-1 dependent adhesion is disrupted by cytochalasin B (96). Furthermore LFA-1 and actin filaments co-localize to the site of contact between NK cells and target cells (139). More interestingly the fibronectin receptor, an evolutionarily related integrin, has been shown to interact directly with talin, a cytoskeletal protein (140). LFA-1 and talin also co-localize in activated, but not resting, lymphocytes (141), and redistribute to sites of adhesion with specific antigen-bearing cells (142). Redistribution of LFA-1 and talin is one of the earlier events, preceding redistribution of the T-cell antigen receptor and reorientation of the Golgi and microtubule organizing center. Redistribution of LFA-1 may be triggered by initial antigen receptor engagement, a mechanism earlier proposed to regulate adhesion strengthening via LFA-1 based on experiments with phorbol esters (81,96). An unglycosylated intracellular protein of 86,000 daltons has been found in association with cell surface LFA-1 (143). The functional significance is unknown, although this protein may be part of a cytoskeletal linkage. Association and disassociation of the leukocyte integrins with the cytoskeleton would provide a mechanism for rapid adhesion and deadhesion events, as required for CTL-mediated cytolysis and other immune functions.

Ligand Molecules for the Leukocyte Integrins

LFA-1 Ligands

ICAM-1

Identification and Characterization of an LFA-1 Ligand

Phorbol ester-induced homotypic aggregation of lymphoid cells is an LFA-1-dependent phenomenon. Rothlein and Springer (96) showed that LFA-1⁻ cells from LAD patients do not form homotypic aggregates; however, LFA-1⁺ control cells can form aggregates with LFA-1⁻ cells from LAD patients. These results suggested that LFA-1 binds a ligand distinct from itself, and that cells from LAD patients should express the ligand. Rothlein et al. (127) raised MAb against LFA-1⁻ cells from LAD patients and screened them for the ability to inhibit LFA-1-dependent homotypic aggregation, in the hopes of identifying the putative LFA-1 ligand. One MAb (RR1/1) defined a 76,000–114,000 dalton heavily glycosylated molecule, called *intercellular adhesion molecule-1* (ICAM-1), which fit these criteria.

The receptor-ligand relationship of LFA-1 and ICAM-1 was formally proven with purified ICAM-1 incorporated into planar lipid membranes (100) and by gene transfection (144). All of these systems can mediate LFA-1-dependent adhesion of lymphoblasts, which is inhibitable by pretreatment of the lymphoblast with LFA-1 MAb or pretreatment of the ICAM-1 surface with ICAM-1 MAb. Moreover, LFA-1⁻ cells from LAD patients do not bind to purified ICAM-1 (100). Confirming studies with ICAM-1 cross-linked to artificial substrates extended these findings and demonstrated that three independent MAbs defined ICAM-1, one of which had been used to map it to chromosome 19 (145).

Reciprocal studies have been done using purified LFA-1 (162). Binding of some cell types to purified LFA-1 in planar membranes was inhibited by ICAM-1 MAb. Furthermore, purified LFA-1 binds to purified ICAM-1 when ICAM-1 is presented in planar membranes, confirming that LFA-1 dependent adhesion can be accounted for by direct interaction of LFA-1 with ICAM-1 and other ligands (see below).

ICAM-1 is a widely distributed molecule whose expression is highly regulated. Basal ICAM-1 expression on nonhematopoietic cells is normally low, but can be upregulated by a variety of cytokines, including interleukin-1, tumor necrosis factor, and interferon-γ (85–87,146,147). ICAM-1 expression is prominent on cytokine-activated endothelial cells during inflammation (86,148). Increased ICAM-1 expression correlates directly with increased LFA-1-dependent adhesion of lymphoblasts to induced cells (85–87). These results suggest that ICAM-1 provides dynamic "position-specific" information to guide lymphocyte and leukocyte localization during the course of the immune response. ICAM-1 is expressed only weakly on resting peripheral blood leukocytes, but expression is increased upon cell activation. A molecule identical to ICAM-1 (145) was defined as a B-cell activation molecule (149). The LFA-1/ ICAM-1 adhesion pathway has been implicated in lymphocyte adhesion to endothelial cells (85), fibroblasts (86), epidermal keratinocytes (87), synovial cells (88), and other lymphoid cells (127,150).

ICAM-1 has been reported to localize to uropods of the T cell line HSB-2 (151). By indirect immunofluorescence, ICAM-1 expressed in COS cells (Staunton et al., unpublished) and endothelial cells (166) demonstrates a punctate staining. This localized cell membrane distribution may facilitate initial interaction with LFA-1. Spreading of cells on surfaces coated with ECM components or adhesion ligands is an indication of cytoskeletal association. A Reid-Sternberg cell line expressing high levels of ICAM-1 spreads dramatically on LFA-1 bearing planar membranes. This raised the possibility that at least in some cells, ICAM-1 is associated with the cytoskeleton. A truncated form of ICAM-1, which completely lacks a cytoplasmic tail was generated by oligonucleotide mutagenesis (Staunton et al., unpublished). The truncated protein immunoprecipitated from transfected COS cells demonstrates a mobility on SDS-PAGE which is consistent with its predicted 3 kD reduction in molecular weight. Resistance to cleavage by a phosphatidyl inositol (PI) specific phospholipase C suggests that it does not acquire a PI membrane anchor. This truncated form of ICAM-1 still demonstrates its characteristic cell surface localization. Thus, if localization is through cytoskeletal interaction, this must occur in the ICAM-1 transmembrane region, or by association of the extracellular domain with an anchored protein.

Gene Cloning

ICAM-1, which has recently been cloned and sequenced (144,152), has been shown to be a member of the immunoglobulin superfamily with closest relation to the neural cell adhesion molecule (NCAM) and myelin-associated glycoprotein (MAG), another neural adhesion protein. All three proteins contain five immunoglobulin domains. Interestingly, NCAM has been shown to participate in homophilic (like-like) interactions (153), while ICAM-1 participates in heterophilic interactions. LFA-1 is the only known member of the integrin superfamily that binds to a member of the immunoglobulin superfamily. Members of the other two integrin subfamilies bind a conserved Arg-Gly-Asp (RGD) sequence in unrelated ECM proteins. ICAM-1 does not have an RGD sequence (144,152), and neither does ICAM-2 (see below) (165), suggesting that leukocyte integrin binding specificity has diverged from that of other integrins. Preliminary results from mutational studies reveal that at least two sites in ICAM-1 are involved in binding LFA-1 (Staunton et al., unpublished).

Other Ligands

There is evidence that ICAM-1 is not the only ligand for LFA-1. Phorbol ester-induced homotypic aggregation of the SKW3 T cell line is inhibited by LFA-1 MAb, but not by ICAM-1 MAb (150). Similarly some heterotypic cell interactions, such as CTL adhesion to B lymphoblastoid cell lines is inhibitable by LFA-1 MAb but not ICAM-1 MAb (145). LFA-1-dependent T-cell adherence to endothelial cells has both an ICAM-1-dependent and an ICAM-1-independent component (85). The ICAM-1-dependent pathway is inducible with cytokines, such as TNF, while the ICAM-1-independent pathway is unaffected. These results suggest that multiple ligands may serve to mediate distinct adhesion requirements of lymphocytes during different stages of the immune response. Cell binding to purified LFA-1 in glass supported planar membranes or immobilized directly on plastic also suggests the presence of other ligands. The most striking example of this is the T-cell line SKW3 which expresses virtually no ICAM-1 yet binds strongly to purified LFA-1. This binding is blocked by LFA-1 MAb but not by any single ICAM-1 MAb or combination of ICAM-1 MAb (162). Other cell lines show different degrees of inhibition with ICAM-1 MAb which is dependent on the LFA-1 density of purified LFA-1 on the monolayer. With these cell lines, ICAM-1 MAb are inhibitory at low LFA-1 density but not at high LFA-1 density, while LFA-1 MAb are inhibitory at all densities. These results suggest that both ICAM-1 and the alternative LFA-1 ligand are present on these cells, and that interaction is stronger with ICAM-1 than with the alternative ligand. The only nonhematopoietic cells we have encountered which express ICAM-1 in the absence of other LFA-1 ligands are epithelial cells from several tissues and dermal fibroblasts (86,87). The existence of multiple ligands for a single integrin is not unprecedented. Other integrins, such as platelet glycoprotein IIb/IIIa, have as many as four distinct ligands.

Recently, a second LFA-1 ligand, designated ICAM-2, was cloned based on its functional properties (165). A cDNA expression library was screened for ability to confer on COS cells the ability to bind to purified LFA-1 coated on Petri dishes. Screening was in the presence of ICAM-1 MAb. ICAM-2 has two Ig-like domains, in contrast to ICAM-1 which has five, and these are 35% identical to the first two domains of ICAM-1. ICAM-1 and ICAM-2 are much more similar to one another than to other members of the Ig superfamily, and thus represent an Ig subfamily specialized to interact with LFA-1. The functional cDNA isolation approach should have wide application for identifying other adhesion counterstructures.

Mac-1 and p150,95 Ligands

C3bi

Both Mac-1 and p150,95 bind the C3bi fragment of complement (111). A peptide fragment of C3bi which contains an RGD sequence binds to macrophages in a Mac-1 MAb-inhibitable fashion (154). However, it is not clear whether the RGD sequence itself is critical for binding.

Other Ligands

The ability of Mac-1 to mediate a number of cell-cell interactions in which C3bi is not involved, such as neutrophil aggregation, suggests putative cell surface ligand(s) for Mac-1. The epitopes on Mac-1 involved in C3bi-binding activity and general adhesion have been distinguished by MAb-inhibition studies (27,107,108,155,156). Some MAb show differential effects on neutrophil homotypic aggregation and adhesion to endothelial monolayers as compared to C3bi-binding, while other MAb block both.

Molecular Basis of Leukocyte Adhesion Deficiency Disease

Early Studies

Every LAD patient analyzed to date has been found to be deficient in the expression of all three leukocyte integrins (18). The simplest hypothesis is that a defect in the common β subunit could account for LAD. This hypothesis was tested in biosynthesis and human X mouse lymphocyte hybridization experiments.

Biosynthesis experiments utilized EBV-transformed B lymphocyte and mitogen-stimulated T lymphocyte cell lines, which from healthy individuals synthesize the LFA-1 $\alpha(\alpha L)$ subunit and the common β subunit, and express the LFA-1 $\alpha\beta$ complex on cell surfaces. Early studies showed that patient cell lines synthesize apparently normal αL subunit precursor, but that the αL precursor does not undergo carbohydrate processing, does not associate in an $\alpha\beta$ complex, and neither subunit is expressed on the cell surface (26,157). In these studies, the use of available anti-β subunit monoclonal antibodies (MAb) did not allow for the immunoprecipitation of β subunit precursors from either control cells or LAD patient cells.

In human X mouse lymphocyte hybrids, human LFA-1 α and β subunits from normal cells were shown to associate with mouse LFA-1 subunits to form interspecies $\alpha\beta$ complexes. Surface expression of the α, but not the β, subunit of patient cells can be rescued by the formation of interspecies complexes (70). These studies showed that the LFA-1 α subunit in genetically deficient cells is competent for surface expression in the presence of an appropriate mouse β subunit. Taken together, these results suggest that leukocyte adhesion deficiency is secondary to a defect in the common β subunit.

Identification of Heterogenous Mutations in the Common β Subunit

The acquisition of two molecular probes, the β subunit cDNA (31,46) and a rabbit antiserum which recognizes the precursor form of the β subunit (29,158) allowed analysis of the β subunit from LAD patients. Five phenotypes of β subunit expression and structure were identified by Kishimoto et al. (29). One class of mutations resulted in no detectable mRNA or protein precursor. Southern analysis of genomic DNA from these patients showed no gross deletions of the β subunit gene. A second class of mutation is represented by a moderately deficient patient, whose cells synthesized trace amounts of the β subunit precursor and low levels of mRNA message. Dimanche et al. (159) studied two patients with no apparent β subunit precursor synthesis which may fall into one of these two classes; however, further analysis at the RNA level is required.

Two other classes of mutations affect the structure of the common β subunit. One patient synthesizes an aberrantly large β subunit precursor. However after endoglycosidase H digestion, the protein backbone appears about normal in size. One hypothesis is that a point mutation causes an amino acid change which creates a novel consensus N-glycosylation site (Asn-X-Ser/Thr). Four moderately deficient patients, who are all related, synthesize an aberrantly small precursor, which is degraded. The pedigree analysis of 14 members of this kindred shows that inheritance of the aberrant precursor correlates with the expected disease state and surface expression of LFA-1. Endoglycosidase H digestion of N-linked carbohydrates from the precursor shows that the defect is in the protein backbone rather than glycosylation.

Finally three unrelated patients studied by us (29), a group of four patients studied by Dana et al. (158), and one patient studied by Dimanche et al. (159) synthesize both a normal size β subunit precursor and a normal size α subunit precursor. Neither subunit is processed or transported to the cell surface. Although it is likely that there is a point mutation in the β subunit, we cannot exclude the possibility of α subunit mutations.

Molecular Basis of the Severe and Moderate Deficiency Phenotypes

Heterogeneity in the defect causing LAD disease was first observed in the extent of the leukocyte integrin deficiency at the cell surface. Patients are classified as severely deficient (<0.5% normal levels of expression) and moderately deficient (3–10% normal levels of expression) (160). Survival of LAD patients is greatly influenced by the extent of deficiency of the leukocyte integrins. Severely deficient patients often die in the first 2 years of childhood, while moderately deficient patients are less prone to life-threatening infections and can survive to adulthood. However, even those patients who survive to adulthood can suddenly die of complications from severe infections.

The molecular basis for this heterogeneity is unclear. In the case of deficiency of β subunit mRNA and protein precursor, the extent of deficiency correlates with moderate and severe phenotype (29). Two severely deficient patients had no detectable β subunit mRNA expression or protein precursor synthesis, while one moderately deficient patient had low levels of mRNA expression and precursor synthesis (29). Apparently, this low level of expression is sufficient to account for the moderate phenotype and less severe clinical complications. However, among 13 patients synthesizing normal quantities of mRNA and protein precursor, it has been unclear why some are of the moderate and some of the severe deficiency phenotype (29,158,159) (Wardlaw and Springer, unpublished).

We have analyzed four related patients with the moderate deficiency phenotype (161). The predominant form of the β subunit precursor synthesized by these patients is several thousand daltons smaller than normal and is degraded before transport to the Golgi apparatus. However, ^{125}I cell surface labeling shows that the small amount of LFA-1 that reaches the surface contains a normal-sized β subunit. At the RNA level, S1 nuclease protection studies define a 90 nt deletion in the β subunit message from these patients. The Taq polymerase chain reaction (PCR) was adapted to amplify the aberrant mRNA. Sequence analysis shows an in-frame 90 nt deletion in the region encoding the extracellular domain. This 90 bp region is also shown to be encoded on a separate exon in both the normal and patient genome. Sequence analysis of genomic DNA from these patients shows a single G to C substitution in the sequence of the 5′ splice site, suggesting aberrant RNA splicing. A small amount of normally spliced message, detected by S1 nuclease protection analysis and Taq PCR, encodes a normal size β subunit and accounts for the low levels (3% of normal) of cell surface expression of the leukocyte integrins observed in these patients, hence the moderate deficiency phenotype. The 30 amino acids encoded by the deletion region shares 63% amino acid identity with the corresponding region of the fibronectin receptor. The high conservation suggests some functional significance of this region, perhaps in $\alpha\beta$ subunit association.

Future Avenues of Research

The first leukocyte integrin to be functionally characterized was LFA-1, in 1981 (4). Rapid progress in recent years in the study of leukocyte adhesion reflects an ever-increasing appreciation of the diverse functions of leukocyte integrins. The leukocyte integrins, like other members of the integrin superfamily, play a dynamic role in transmitting positional information and linking the extracellular environment with intracellular processes. The interaction of LFA-1 and ICAM-1 represents one of the few defined heterotypic receptor-ligand relationships of cell surface molecules. LFA-1-dependent adhesive interactions are exquisitely regulated at both the level of receptor and ligand. We have an understanding of the role of leukocyte integrins in vivo in both immune and nonimmune inflammatory responses. The recent advances hold promise for many new avenues of research.

Precise structure-function relationships of the leukocyte integrins and ICAM-1 can now be directly tested by in vitro mutagenesis of the appropriate cDNAs. Domain-swapping among the leukocyte integrin α subunits may allow dissection of LFA-1, Mac-1, and p150,95 functional specificity. Domains that are defined as functionally important can be synthesized and tested both in vitro and in vivo. Similar mutagenesis approaches may reveal how integrin receptor activation occurs, what kinds of signals are transmitted to the cell by leukocyte integrins, and the nature of cytoskeletal interaction. In addition, the acquisition of the leukocyte integrin and ICAM-1 cDNAs allows analysis of the basis for gene regulation by a variety of cytokines.

There is also substantial evidence that other ligands for LFA-1, Mac-1, and p150,95 exist. Rational strategies must be designed to identify these ligands and to assess their contributions in different phases of the immune response. Multiple ligands may provide quite distinct signals and positional information to leukocytes.

The answers to these questions of structure, function, and regulation of receptor activity have profound therapeutic implications, as suggested by recent animal models of ischemia and reperfusion injury. The control of pathologic tissue damage by neutrophils during inflammation can be approached by inhibition of receptor activation, modulation of ICAM-1 expression, or by using MAb or fragments of leukocyte integrins or their ligands to block acute adhesive events.

References

1. Williams AF, Barclay AN: The immunoglobulin superfamily-domains for cell surface recognition. *Ann Rev Immunol* 6:381–405, 1988.
2. McMichael AJ (ed): *Leukocyte Typing III: White Cell Differentiation Antigens.* Oxford, Oxford University Press, 1987.

3. Springer T, Galfre G, Secher DS, Milstein C: Mac 1: A macrophage differentiation antigen identified by monoclonal antibody. *Eur J Immunol* 9:301–306, 1979.

4. Davignon D, Martz E, Reynolds T, Kürzinger K, Springer TA: Lymphocyte function-associated antigen 1 (LFA-1): A surface antigen distinct from Lyt-2,3 that participates in T lymphocyte-mediated killing. *Proc Natl Acad Sci USA* 78:4535–4539, 1981.

5. Davignon D, Martz E, Reynolds T, Kürzinger K, Springer TA: Monoclonal antibody to a novel lymphocyte function-associated antigen (LFA-1): Mechanism of blocking of T lymphocyte-mediated killing and effects on other T and B lymphocyte functions. *J Immunol* 127:590–595, 1981.

6. Kurzinger K, Reynolds T, Germain RN, Davignon D, Martz E, Springer TA: A novel lymphocyte function-associated antigen (LFA-1): Cellular distribution, quantitative expression, and structure. *J Immunol* 127:596–602, 1981.

7. Trowbridge IS, Omary MB: Human cell surface glycoprotein related to cell proliferation is the receptor for transferrin. *Proc Natl Acad Sci USA* 78:3039–3043, 1981.

8. Kürzinger K, Springer TA: Purification and structural characterization of LFA-1, a lymphocyte function-associated antigen, and Mac-1, a related macrophage differentiation antigen. *J Biol Chem* 257:12412–12418, 1982.

9. Sanchez-Madrid F, Nagy J, Robbins E, Simon P, Springer TA: A human leukocyte differentiation antigen family with distinct alpha subunits and a common beta subunit: The lymphocyte function-associated antigen (LFA-1), the C3bi complement receptor (OKM1/Mac-1), and the p150,95 molecule. *J Exp Med* 158:1785–1803, 1983.

10. Beller DI, Springer TA, Schreiber RD: Anti-Mac-1 selectively inhibits the mouse and human type three complement receptor. *J Exp Med* 156:1000–1009, 1982.

11. Wright SD, Rao PE, Van Voorhis WC, Craigmyle LS, Iida K, Talle MA, Westberg EF, Goldstein G, Silverstein SC: Identification of the C3bi receptor of human monocytes and macrophages with monoclonal antibodies. *Proc Natl Acad Sci USA* 80:5699–5703, 1983.

12. Krensky AM, Sanchez-Madrid F, Robbins E, Nagy J, Springer TA, Burakoff SJ: The functional significance, distribution, and structure of LFA-1, LFA-2, and LFA-3: Cell surface antigens associated with CTL-target interactions. *J Immunol* 131:611–616, 1983.

13. Strassmann G, Springer TA, Haskill SJ, Miraglia CC, Lanier LL, Adams DO: Antigens associated with the activation of murine mononuclear phagocytes in vivo: Differential expression of lymphocyte function-associated antigen in the several stages of development. *Cell Immunol* 94:265–275, 1985.

14. Springer TA, Unkeless JC: Analysis of macrophage differentiation and function with monoclonal antibodies, in Adams DO, Hanna, Jr MG (eds): *Contemporary Topics in Immunobiology, Vol. 14.* New York, Plenum Press, p1, 1984.

15. de la Hera A, Alvarez-Mon M, Sanchez-Madrid F, Martinez-A C, Durantez A: Co-expression of MAC-1 and p150,95 on CD5+ B cells. Structural and functional characterization in a human chronic lymphocytic leukemia. *Eur J Immunol* 18:1131–1134, 1988.

16. Schwarting R, Stein H, Wang CY: The monoclonal antibodies anti S-HCL 1 (anti Leu 14) and anti S-HCL 3 (anti Leu M5) allow the diagnosis of hairy cell leukemia. *Blood* 65:974–983, 1985.
17. Miller LJ, Schwarting R, Springer TA: Regulated expression of the Mac-1, LFA-1, p150,95 glycoprotein family during leukocyte differentiation. *J Immunol* 137:2891–2900, 1986.
18. Anderson DC, Springer TA: Leukocyte adhesion deficiency: An inherited defect in the Mac-1, LFA-1, and p150,95 glycoproteins. *Ann Rev Med* 38:175–194, 1987.
19. Todd RF, Freyer DR: The CD11/Cd18 leukocyte glycoprotein deficiency. *Hem/Onc Clinics NA* 2:13–31, 1988.
20. Fischer A, Lisowska-Grospierre B, Anderson DC, Springer TA: The leukocyte adhesion deficiency: Molecular basis and functional consequences. *Immunodef Rev* 1:39–54, 1988.
21. Anderson DC, Smith CW, Springer TA: Disorders of leukocyte motility and adherence, in Stanbury JB, Wyngaarden JB, Frederickson DS (eds): *Metabolic Basis of Inherited Disease* 6th ed. In press, 1989.
22. Kishimoto TK, Springer TA: Human leukocyte adhesion deficiency: Molecular basis for a defective immune response to infections of the skin, *Curr Prob Derm* 18:in press, 1989.
23. Boxer LA, Hedley-Whyte ET, Stossel TP. Neutrophil actin dysfunction and abnormal neutrophil behavior. *N Engl J Med* 291:1093–1099, 1974.
24. Crowley CA, Curnutte JT, Rosin RE, Andre-Schwartz J, Gallin JI, Klempner M, Snyderman R, Southwick FS, Stossel TP, Babior BM: An inherited abnormality of neutrophil adhesion: Its genetic transmission and its association with a missing protein. *New Eng J Med* 302:1163–1168, 1980.
25. Anderson DC, Schmalstieg FC, Arnaout MA, Kohl S, Tosi MF, Dana N, Buffone GJ, Hughes BJ, Brinkley BR, Dickey WD, Abramson JS, Springer T, Boxer LA, Hollers JM, Smith CW: Abnormalities of polymorphonuclear leukocyte function associated with a heritable deficiency of high molecular weight surface glycoproteins (GP138): Common relationship to diminished cell adherence. *J Clin Invest* 74:536–551, 1984.
26. Springer TA, Thompson WS, Miller LJ, Schmalstieg FC, Anderson DC: Inherited deficiency of the Mac-1, LFA-1, p150,95 glycoprotein family and its molecular basis. *J Exp Med* 160:1901–1918, 1984.
27. Beatty PG, Harlan JM, Rosen H, Hansen JA, Ochs HD, Price TD, Taylor RF, Klebanoff SJ. 1984. Absence of monoclonal-antibody-defined protein complex in boy with abnormal leucocyte function. *Lancet* I:535–537.
28. Dana N, Todd RF, Pitt J, Springer TA, Arnaout MA. 1984. Deficiency of a surface membrane glycoprotein (Mo1) in man. *J Clin Invest* 73:153–159.
29. Kishimoto TK, Hollander N, Roberts TM, Anderson DC, Springer TA: Heterogenous mutations in the beta subunit common to the LFA-1, Mac-1, and p150,95 glycoproteins cause leukocyte adhesion deficiency. *Cell* 50:193–202, 1987.
30. Hynes RO: Integrins: A family of cell surface receptors. *Cell* 48:549–554, 1987.
31. Kishimoto TK, O'Connor K, Lee A, Roberts TM, Springer TA. Cloning of the beta subunit of the leukocyte adhesion proteins: Homology to an extracellular matrix receptor defines a novel supergene family. *Cell* 48:681–690, 1987.

32. Ruoslahti E, Pierschbacher MD: New perspectives in cell adhesion: RGD and integrins. *Science* 238:491–497, 1987.
33. Hemler ME: Adhesive protein receptors on hematopoietic cells. *Immunol Today* 9:109–113. 1988.
34. Bogaert T, Brown N, Wilcox M: The drosophila PS2 antigen is an invertebrate integrin that, like the fibronectin receptor, becomes localized to muscle attachments. *Cell* 51:929–940, 1987.
35. Wilcox M, Leptin M: Tissue-specific modulation of a set of related cell surface antigens in Drosophila. *Nature* 316:351–354, 1985.
36. Miller LJ, Springer TA. Biosynthesis and glycosylation of p150,95 and related leukocyte adhesion proteins. *J Immunol* 139:842–847, 1987.
37. Sastre L, Kishimoto TK, Gee C, Roberts T, Springer TA: The mouse leukocyte adhesion proteins Mac-1 and LFA-1: Studies on mRNA translation and protein glycosylation with emphasis on Mac-1. *J Immunol* 137:1060–1065, 1986.
38. Dahms NM, Hart GW: Lymphocyte function-associated antigen 1 (LFA-1) contains sulfated N-linked oligosaccharides. *J Immunol* 134:3978–3986, 1985.
39. Takeda A: Sialyation patterns of lymphocyte function-associated antigen 1 (LFA-1) differ between T and B lymphocytes. *Eur J Immunol* 17:281–286, 1987.
40. Skubitz KM, Snook RW II: Monoclonal antibodies that recognize lacto-N-fucopentaose III (CD15) react with the adhesion-promoting glycoprotein family (LFA-1/HMAC-1/GP 150,95) and CR1 on human neutrophils. *J Immunol* 139:1631–1639, 1987.
41. Ho M-K, Springer TA: Biosynthesis and assembly of the alpha and beta subunits of Mac-1, a macrophage glycoprotein associated with complement receptor function. *J Biol Chem* 258:2766–2769, 1983.
42. Kornfeld R, Kornfeld S: Assembly of asparagine-linked oligosaccharides. *Ann Rev Biochem* 54:631–664, 1985.
43. Todd RF III, Arnaout MA, Rosin RE, Crowley CA, Peters WA, and Babior BM: Subcellular localization of the large subunit of Mo1 (Mo1 alpha, formerly gp 110), a surface glycoprotein associated with neutrophil adhesion. *J Clin Invest* 74:1280–1290, 1984.
44. Miller LJ, Bainton DF, Borregaard N, Springer TA: Stimulated mobilization of monocyte Mac-1 and p150,95 adhesion proteins from an intracellular vesicular compartment to the cell surface. *J Clin Invest* 80:535–544, 1987.
45. Bainton DF, Miller LJ, Kishimoto TK, Springer TA: Leukocyte adhesion receptors are stored in peroxidase-negative granules of human neutrophils. *J Exp Med* 166:1641–1653, 1987.
46. Law SKA, Gagnon J, Hildreth JEK, Wells CE, Willis AC, Wong, AJ: The primary structure of the beta subunit of the cell surface adhesion glycoproteins LFA-1, CR3 and p150,95 and its relationship to the fibronectin receptor. *EMBO J* 6:915–919, 1987.
47. Tamkun JW, DeSimone DW, Fonda D, Patel RS, Buck C, Horwitz AF, Hynes RO: Structure of integrin, a glycoprotein involved in the transmembrane linkage between fibronectin and actin. *Cell* 46:271–282, 1986.
48. Fitzgerald LA, Steiner Jr B, Rall SC, Lo S, Phillips DR: Protein sequence of endothelial glycoprotein IIIa derived from a cDNA clone. Identity with

platelet glycoprotein IIIa and similarity to "integrin". *J Biol Chem* 262:3936–3939, 1987.

49. Leptin M: The fibronectin receptor family. *Nature* 321:728, 1986.
50. Corbi AL, Miller LJ, O'Connor K, Larson RS, Springer TA: cDNA cloning and complete primary structure of the alpha subunit of a leukocyte adhesion glycoprotein, p150,95. *EMBO J* 6:4023–4028, 1987.
51. Corbi AL, Kishimoto TK, Miller LJ, Springer TA: The human leukocyte adhesion glycoprotein Mac-1 (Complement receptor type 3, CD11b) alpha subunit: Cloning, primary structure, and relation to the integrins, von Willebrand factor and factor B. *J Biol Chem* 263:12403–12411, 1988.
52. Pytela R: Amino acid sequence of the murine Mac-1 alpha chain reveals homology with the integrin family and an additional domain related to von Willebrand factor. *EMBO J* 7:1371–1378, 1988.
53. Arnaout MA, Gupta SK, Pierce MW, Tenen DG: Amino acid sequence of the alpha subunit of human leukocyte adhesion receptor Mo1 (complement receptor type 3). *J Cell Biol* 106:2153–2158, 1988.
54. Larson RS, Corbi AL, Berman L, Springer TA: Primary structure of the LFA-1 alpha subunit: An integrin with an embedded domain defining a protein superfamily. *J Cell Biol* 108:703–712, 1989.
55. Jennings LK, Phillips DR: Purification of glycoproteins IIb and III from human platelet plasma membranes and characterization of a calcium-dependent glycoprotein IIb-III complex. *J Biol Chem* 257:10458–10466, 1982.
56. Argraves WS, Suzuki S, Arai H, Thompson K, Pierschbacher MD, Ruoslahti E: Amino acid sequence of the human fibronectin receptor. *J Cell Biol* 105:1183–1190, 1987.
57. Poncz M, Eisman R, Heidenreich R, Silver SM, Vilaire G, Surrey S, Schwartz E, Bennett JS: Structure of the platelet membrane glycoprotein IIb: homology to the alpha subunits of the vitronectin and fibronectin membrane receptors. *J Biol Chem* 262:8476–8482, 1987.
58. Suzuki S, Argraves WS, Arai H, Languino LR, Pierschbacher MD, Ruoslahti E: Amino acid sequence of the vitronectin receptor alpha subunit and comparative expression of adhesion receptor mRNAs. *J Biol Chem* 262:14080–14085, 1987.
59. Fujimura K, Phillips DR: Binding of $^{45}Ca^{2+}$ to glycoprotein IIb from human platelet plasma membranes. *Thrombo Haemostasis* 50:251a, 1983.
60. Gailit J, Ruoslahti E: Regulation of the fibronectin receptor affinity by divalent cations. *J Biol Chem* 263:12927–12932, 1988.
61. Seed B: An LFA-3 cDNA encodes a phospholipid-linked membrane protein homologous to its receptor CD2. *Nature* 329:840–842, 1987.
62. Ruoslahti E, Pierschbacher MD: Arg-Gly-Asp: A versatile cell recognition signal. *Cell* 44:517–518, 1986.
63. Ruggeri ZM, Houghten RA, Russell SR, Zimmerman TS: Inhibition of platelet function with synthetic peptides designed to be high-affinity antagonists of fibrinogen binding to platelets. *Proc Natl Acad Sci USA* 83:5708–5712, 1986.
64. Obara M, Kang MS, Yamada KM: Site-directed mutagenesis of the cell-binding domain of human fibronectin: Separable, synergistic sites mediate adhesive function. *Cell* 53:649–657, 1988.

65. D'Souza SE, Ginsberg MH, Burke TA, Lam SC-T, Plow EF: Localization of an Arg-Gly-Asp recognition site within an integrin adhesion receptor. *Science* 242:91–93, 1988.
66. Szebenyi DME, Obendorf SK, Moffat K: Structure of vitamin D-dependent calcium-binding protein from bovine intestine. *Nature* 294:327–332, 1981.
67. Girma J-P, Meyer D, Verweij CL, Pannekoek H, Sixma JJ: Structure-function relationship of human Von Willebrand factor. *Blood* 70:605–611, 1987.
68. Argraves WS, Deak F, Sparks KJ, Kiss I, Goetinck PF: Structural features of cartilage matrix protein deduced from cDNA. *Proc Natl Acad Sci USA* 84:464–468, 1987.
69. Bently DR: Primary structure of human complement component 2. *Biochem J* 239:339–345, 1986.
70. Marlin SD, Morton CC, Anderson DC, Springer TA: LFA-1 immunodeficiency disease: Definition of the genetic defect and chromosomal mapping of alpha and beta subunits of the lymphocyte function-associated antigen 1 (LFA-1) by complementation in hybrid cells. *J Exp Med* 164:855–867, 1986.
71. Suomalainen HA, Lundqvist C, Gahmberg CG, Schröder J: Assignment of gene(s) coding for antigen defined by monoclonal antibody 2B2. *Som Cell Gen* 9:745–756, 1983.
72. Akao Y, Utsumi KR, Naito K, Ueda R, Takahashi T, Yamada K: Chromosomal assignments of genes coding for human leukocyte common antigen, T-200, and lymphocyte function-associated antigen 1, LFA-1 beta subunit. *Somat Cell Mol Genet* 13:273–278, 1987.
73. Corbi AL, Larson RS, Kishimoto TK, Springer TA, Morton CC: Chromosomal location of the genes encoding the leukocyte adhesion receptors LFA-1, Mac-1 and p150,95. Identification of a gene cluster involved in cell adhesion. *J Exp Med* 167:1597–1607, 1988.
74. Gardner K, Watkins P, Munke M, Drabkin H, Jones C, Patterson D: Partial physical map of chromosome 21. *Somat Cell Mol Genet* 14:623–638, 1988.
75. Rubin CM, Larson RA, Bitter MA, Carrino JJ, Le Beau M, Diaz MO, Rowley JD: Metallothionein gene cluster is split by chromosome 16 rearrangements in myelomonocytic leukaemia. *Blood* 70:1338–1343, 1987.
76. Gordan MY, Dowling CR, Riley GP, Goldman JM, Greaves MF: Altered adhesive interactions with marrow stroma of haematopoietic progenitor cells in chronic myeloid leukemia. *Nature* 328:342–344, 1987.
77. Le Beau MM, Diaz MO, Karin M, Rowley JD: Metallothionein gene cluster is split by chromosome 16 rearrangements in myelomonocytic leukaemia. *Nature* 313:709–711, 1985.
78. Bowen TJ, Ochs HD, Altman LC, Price TH, Van Epps DE, Brautigan DL, Rosin RE, Perkins WD, Babior Bernard M, Klebanoff SJ, Wedgewood RJ: Severe recurrent bacterial infections associated with defective adherence and chemotaxis in two patients with neutrophils deficient in a cell-associated glycoprotein. *J Pediatr* 101:932–940, 1982.
79. Kohl S, Springer TA, Schmalstieg FC, Loo LS, Anderson DC: Defective natural killer cytotoxicity and polymorphonuclear leukocyte antibody-dependent cellular cytotoxicity in patients with LFA-1/OKM-1 deficiency. *J Immunol* 133:2972–2978, 1984.
80. Kohl S, Loo LS, Schmalsteig FC, Anderson DC: The genetic deficiency of leukocyte surface glycoproteins Mac-1, LFA-1, p150,95 in humans is asso-

ciated with defective antibody dependent cellular cytotoxicity in vitro and defective protection against herpes simplex virus infection in vivo. *J Immunol* 137:1688–1694, 1986.

81. Springer TA, Dustin ML, Kishimoto TK, Marlin SD: The lymphocyte function-associated LFA-1, CD2, and LFA-3 molecules: Cell adhesion receptors of the immune system. *Ann Rev Immunol* 5:223–252, 1987.

82. Martz E: LFA-1 and other accessory molecules functioning in adhesions of T and B lymphocytes. *Human Immunol* 18:3–37, 1986.

83. Mentzer SJ, Burakoff SJ, Faller DV: Adhesion of T lymphocytes to human endothelial cells is regulated by the LFA-1 membrane molecule. *J Cell Physiol* 126:285–290, 1986.

84. Haskard D, Cavender D, Beatty P, Springer T, Ziff M: T lymphocyte adhesion to endothelial cells: Mechanisms demonstrated by anti-LFA-1 monoclonal antibodies. *J Immunol* 137:2901–2906, 1986.

85. Dustin ML, Springer TA: Lymphocyte function associated antigen-1 (LFA-1) interaction with intercellular adhesion molecule-1 (ICAM-1) is one of at least three mechanisms for lymphocyte adhesion to cultured endothelial cells. *J Cell Biol* 107:321–331, 1988.

86. Dustin ML, Rothlein R, Bhan AK, Dinarello CA, Springer TA: Induction by IL-1 and interferon, tissue distribution, biochemistry, and function of a natural adherence molecule (ICAM-1). *J Immunol* 137:245–254, 1986.

87. Dustin ML, Singer KH, Tuck DT, Springer TA: 1988. Adhesion of T lymphoblasts to epidermal keratinocytes is regulated by interferon gamma and is mediated by intercellular adhesion molecule-1 (ICAM-1). *J Exp Med* 167:1323–1340, 1988.

88. Mentzer SJ, Rothlein R, Springer TA, Faller DV: 1988. Intercellular adhesion molecule-1 (ICAM-1) is involved in the cytolytic T lymphocyte interaction with human synovial cells. *J Cell Physiol* 137:173–178, 1988.

89. Roos E, Roossien FF: Involvement of leukocyte function-associated antigen-1 (LFA-1) in the invasion of hepatocyte cultures by lymphoma and T-cell hybridoma cells. *J Cell Biol* 105:553–559, 1987.

90. Krensky AM, Mentzer SJ, Clayberger C, Anderson DC, Schmalstieg FC, Burakoff SJ, Springer TA: Heritable lymphocyte function-associated antigen-1 deficiency: Abnormalities of cytotoxicity and proliferation associated with abnormal expression of LFA-1. *J Immunol* 135:3102–3108, 1985.

91. Mentzer SJ, Bierer BE, Anderson DC, Springer TA, Burakoff SJ: Abnormal cytolytic activity of lymphocyte function-associated antigen-1-deficient human cytolytic T lymphocyte clones. *J Clin Invest* 78:1387–1391, 1986.

92. Hamann A, Westrich DJ, Duijevstijn A, Butcher EC, Baisch H, Harder R, Thiele HG: Evidence for an accessory role of LFA-1 in lymphocyte-high endothelium interaction during homing. *J Immunol* 140:693–699, 1988.

93. Butcher EC: The regulation of lymphocyte traffic. *Curr Topics Microbiol Immunol* 128:85–122, 1986.

94. Patarroyo M, Beatty PG, Fabre JW, Gahmberg CG: Identification of a cell surface protein complex mediating phorbol ester-induced adhesion (binding) among human mononuclear leukocytes. *Scand J Immunol* 22:171–182, 1985.

95. Mentzer SJ, Gromkowski SH, Krensky AM, Burakoff SJ, Martz E: LFA-1 membrane molecule in the regulation of homotypic adhesions of human B lymphocytes. *J Immunol* 135:9–11, 1985.

96. Rothlein R, Springer TA: The requirement for lymphocyte function-associated antigen 1 in homotypic leukocyte adhesion stimulated by phorbol ester. *J Exp Med* 163:1132–1149, 1986.
97. Martz E: Mechanism of specific tumor cell lysis by alloimmune T-lymphocytes: Resolution and characterization of discrete steps in the cellular interaction. *Contemp Topics Immunobiol* 7:301–361, 1977.
98. Patarroyo M, Jondal M, Gordon J, Klein E: Characterization of the Phorbol 12,13-dibutyrate(P(Bu2)) induced binding between human lymphocytes. *Cell Immunol* 81:373–383, 1983.
99. Werdelin O: Antigen specific physical interaction between macrophages and T lymphocytes, in Unanue ER, Rosenthal AS (eds): *Macrophage Regulation of Immunity*. New York, Academic Press, p. 213–230, 1980.
100. Marlin SD, Springer TA: Purified intercellular adhesion molecule-1 (ICAM-1) is a ligand for lymphocyte function-associated antigen 1 (LFA-1). *Cell* 51:813–819, 1987.
101. Rothlein R, Springer TA: Complement receptor type three-dependent degradation of opsonized erythrocytes by mouse macrophages. *J Immunol* 135:2668–2672, 1985.
102. Ramos OF, Kai C, Yefenof E, Klein E: The elevated natural killer sensitivity of targets carrying surface-attached C3 fragments require the availability of the iC3b receptor (CR3) on the effectors. *J Immunol* 140:1239–1243, 1988.
103. Mosser DM, Edelson PJ: The mouse macrophage receptor for C3bi (CR3) is a major mechanism in the phagocytosis of Leishmania promastigotes. *J Immunol* 135:2785–2789, 1985.
104. Russell DG, Wright SD: Complement receptor type 3 (CR3) binds to an arg-gly-asp-containing region of the major surface glycoprotein, gp63, of Leishmania promastigotes. *J Exp Med* 168:279–292, 1988.
105. Wright SD, Jong MTC: Adhesion-promoting receptors on human macrophages recognize *Escherichia coli* by binding to lipopolysaccharide. *J Exp Med* 164:1876–1988, 1986.
106. Bullock WE, Wright SD: Role of the adherence-promoting receptors, CR3, LFA-1 and p150,95, in binding of *Histoplasma capsulatum* by human macrophages. *J Exp Med* 165:195–210, 1987.
107. Anderson DC, Miller LJ, Schmalstieg FC, Rothlein R, Springer TA: Contributions of the Mac-1 glycoprotein family to adherence-dependent granulocyte functions: Structure-function assessments employing subunit-specific monoclonal antibodies. *J Immunol* 137:15–27, 1986.
108. Dana N, Styrt B, Griffin J, Todd III RF, Klempner M, Arnaout MA: Two functional domains in the phagocyte membrane glycoprotein Mo1 identified with monoclonal antibodies. *J Immunol* 137:3259–3263, 1986.
109. Wallis, WJ, Hickstein DD, Schwartz BR, June CC, Ochs HD, Beatty PG, Klebanoff SJ, Harlan JM: Monoclonal antibody-defined functional epitopes on the adhesion-promoting glycoprotein complex (CDw18) of human neutrophils. *Blood* 67:1007–1013, 1986.
110. Simon RH, DeHart PD, Todd III RF: Neutrophil-induced injury of rat pulmonary alveolar epithelial cells. *J Clin Invest* 78:1375–1386, 1986.
111. Micklem KJ, Sim RB: Isolation of complement-fragment-iC3b-binding proteins by affinity chromatography. *Biochem J* 231:233–236, 1985.

112. Malhotra V, Hogg N, Sim RB: Ligand binding by the p150,95 antigen of U937 monocytic cells: properties in common with complement receptor type 3 (CR3). *Eur J Immunol* 16:1117–1123, 1986.

113. Myones BL, Dalzell JG, Hogg N, Ross GD: Neutrophil and monocyte cell surface p150,95 has iC3b-receptor (CR4) activity resembling CR3. *J Clin Invest* 82:640–651, 1988.

114. Keizer GD, te Velde AA, Schwarting R, Figdor CG, de Vries JE: Role of p150,95 in adhesion, migration, chemotaxis and phagocytosis of human monocytes. *Eur J Immunol* 17:1317–1322, 1987.

115. te Velde AA, Keizer GD, Figdor CG: Differential function of LFA-1 family molecules (CD11 and CD18) in adhesion of human monocytes to melanoma and endothelial cells. *Immunology* 61:261–267, 1987.

116. Keizer GD, Borst J, Visser W, Schwarting R, de Vries JE, Figdor CG: Membrane glycoprotein p150,95 of human cytotoxic T cell clones is involved in conjugate formation with target cells. *J Immunol* 138:3130–3136, 1987.

117. Lanier LL, Arnaout MA, Schwarting R, Warner NL, Ross GD: p150/95, third member of the LFA-1/CR3 polypeptide family identified by anti-Leu M5 monoclonal antibody. *Eur J Immunol* 15:713–718, 1985.

118. Giger U, Boxer LA, Simpson PJ, Lucchesi BR, Todd III RF: Deficiency of leukocyte surface glycoproteins Mo1, LFA-1, and Leu M5 in a dog with recurrent bacterial infections: An animal model. *Blood* 69:1622–1630, 1987.

119. Arfors K-E, Lundberg C, Lindborm L, Lundberg K, Beatty PG, Harlan JM: A monoclonal antibody to the membrane glycoprotein complex CD18 inhibits polymorphonuclear leukocyte accumulation and plasma leakage in vivo. *Blood* 69:338–340, 1987.

120. Rosen H, Gordon S: Monoclonal antibody to the murine type 3 complement receptor inhibits adhesion of myelomonocytic cells in vitro and inflammatory cell recruitment in vivo. *J Exp Med* 166:1685–1701, 1987.

121. Simpson PJ, Todd III RF, Fantone JC, Mickelson JK, Griffin JD, Lucchesi BR, Adams MD, Hoff P, Lee K, Rogers CE: Reduction of experimental canine myocardial reperfusion injury by a monoclonal antibody (anti-Mo1, anti-CD11b) that inhibits leukocyte adhesion. *J Clin Invest* 81:624–629, 1988.

122. Vedder NB, Winn RK, Rice CL, Chi EY, Artors K-E, Harlan JM: A monoclonal antibody to the adherence-promoting leukocyte glycoprotein, CD18, reduces organ injury and improves survival from hemorrhagic shock and resuscitation in rabbits. *J Clin Invest* 81:939–944, 1988.

123. Campana D, Sheridan B, Tidman N, Hoffbrand AV, Janossy G: Human leukocyte function-associated antigens on lympho-hemopoietic precursor cells. *Eur J Immunol* 16:537–542, 1986.

124. Miller BA, Antognetti G, Springer T: Identification of cell surface antigens present on murine hematopoietic stem cells. *J Immunol* 134:3286–3290, 1985.

125. Berger M, O'Shea J, Cross AS, Folks TM, Chused TM, Brown EJ, Frank MM: Human neutrophils increase expression of C3bi as well as C3b receptors upon activation. *J Clin Invest* 74:1566–1571, 1984.

126. Springer TA, Miller LJ, Anderson DC. p150,95, the third member of the Mac-1, LFA-1 human leukocyte adhesion glycoprotein family. *J Immunol* 136:240–245, 1986.

127. Rothlein R, Dustin ML, Marlin SD, Springer TA: A human intercellular adhesion molecule (ICAM-1) distinct from LFA-1. *J Immunol* 137:1270–1274, 1986.
128. Buyon JP, Abramson SB, Philips MR, Slade SG, Ross GD, Weissman G, Winchester RJ: Dissociation between increased surface expression of Gp165/95 and homotypic neutrophil aggregation. *J Immunol* 140:3156–3160, 1988.
129. Vedder NB, Harlan JM: Increased surface expression of CD11b/CD18 (Mac-1) is not required for stimulated neutrophil adherence to cultured endothelium. *J Clin Invest* 81:676–682, 1988.
130. Wright SD, Meyer BC: Phorbol esters cause sequential activation and deactivation of complement receptors on polymorphonuclear leukocytes. *J Immunol* 136:1759–1764, 1986.
131. Wright SD, Detmers PA, Jong MTC, Meyer BC: Interferon-gamma depresses binding of ligand by C3b and C3bi receptors on cultured human monocytes, an effect reversed by fibronectin. *J Exp Med* 163:1245–1259, 1986.
132. Keizer GD, Visser W, Vliem M, Figdor CG: A monoclonal antibody (NKI-L16) directed against a unique epitope on the alpha-chain of human leukocyte function-associated antigen 1 induces homotypic cell-cell interactions. *J Immunol* 140:1393–1400, 1988.
133. Pircher H, Groscurth P, Baumhütter S, Aguet M, Zinkernagel RM, Hengartner H: A monoclonal antibody against altered LFA-1 induces proliferation and lymphokine release of cloned T cells. *Eur J Immunol* 16:172–181, 1986.
134. Morimoto C, Rudd CE, Letvin NL, Schlossman SF: A novel epitope of the LFA-1 antigen which can distinguish killer effector and suppressor cells in human CD8 cells. *Nature* 330:479–482, 1987.
135. Pytowski B, Easton TG, Valinsky JE, Calderon T, Sun T, Christman JK, Wright SD, Michl J: A monoclonal antibody to a human neutrophil-specific plasma membrane antigen: effect of the antibody on C3bi-mediated adherence by neutrophils and expression of antigen during myelopoiesis. *J Exp Med* 167:421–439, 1988.
136. Brown EJ, Bohnsack JF, Gresham HD: Mechanism of inhibition of immunoglobulin G-mediated phagocytosis by monoclonal antibodies that recognize the Mac-1 antigen. *J Clin Invest* 81:365–375, 1988.
137. Detmers PA, Wright SD, Olsen E, Kimball B, Cohn ZA: Aggregation of complement receptors on human neutrophils in the absence of ligand. *J Cell Biol* 105:1137–1145, 1987.
138. Hara T, Fu SM: Phosphorylation of alpha,beta subunits of 180/100-Kd polypeptides (LFA-1) and related antigens, in Reinherz EL, Haynes BF, Nadler LM, Bernstein ID (eds): *Leukocyte Typing II. Vol. 3 Human Myeloid and Hematopoietic Cells.* New York, Springer-Verlag, p77, 1986.
139. Carpen O, Keiser G, Saksela E: LFA-1 and actin filaments co-distribute at the contact area in lytic NK-cell conjugates (abstract). *6th Int Cong Immunol, Toronto* 315:572, 1986.
140. Horwitz A, Duggan K, Buck C, Beckerle MC, Burridge K: Interaction of plasma membrane fibronectin receptor with talin—a transmembrane linkage. *Nature* 320:531–533, 1986.
141. Burn P, Kupfer A, Singer SJ: Dynamic membrane-cytoskeleton interactions: Specific association of integrin and talin arises in vivo after phorbol ester

treatment of peripheral blood lymphocytes. *Proc Natl Acad Sci USA* 85:497–501, 1988.

142. Kupfer A, Singer SJ: Cell biology of cytotoxic and helper T cell functions. Immunofluorescence microscopic studies of single cells and cell couples. *Ann Rev Immunol* 7:309–338, 1989.

143. van Agthoven A, Pierres M, Goridis C: Identification of a previously unrecognized polypeptide associated with lymphocyte function associated antigen one (LFA-1). *Molec Immunol* 22:1349–1358, 1985.

144. Simmons D, Makgoba MW, Seed B: ICAM, an adhesion ligand of LFA-1, is homologous to the neural cell adhesion molecule NCAM. *Nature* 331:624–627, 1988.

145. Makgoba MW, Sanders ME, Luce GEG, Dustin ML, Springer TA, Clark EA, Mannoni P, Shaw S: ICAM-1: Definition by multiple antibodies of a ligand for LFA-1 dependent adhesion of B, T and myeloid cell. *Nature* 331:86–88, 1988.

146. Pober JS, Gimbrone Jr MA, Lapierre LA, Mendrick DL, Fiers W, Rothlein R, Springer TA: Overlapping patterns of activation of human endothelial cells by interleukin 1, tumor necrosis factor and immune interferon. *J Immunol* 137:1893–1896, 1986.

147. Pober JS, Lapierre LA, Stolpen AH, Brock TA, Springer TA, Fiers W, Bevilacqua MP, Mendrick DL, Gimbrone Jr MA: Activation of cultured human endothelial cells by recombinant lymphotoxin: Comparison with tumor necrosis factor and interleukin 1 species. *J Immunol* 138:3319–3324, 1987.

148. Cotran RS, Pober JS, Gimbrone JR MA, Springer TA, Wiebke AA, Gaspari AA, Rosenberg SA, Lotze MT: Endothelial activation during interleukin 2 (IL 2) immunotherapy: A possible mechanism for the vascular leak syndrome. *J Immunol* 140:1883–1888, 1988.

149. Clark EA, Ledbetter JA, Holly RC, Dinndorf PA, Shu G: Polypeptides on human B lymphocytes associated with cell activation. *Hum Immunol* 16:100–113, 1986.

150. Makgoba MW, Sanders ME, Ginther Luce GE, Gugel EA, Dustin ML, Springer TA, Shaw S: Functional evidence that intercellular adhesion molecule-1 (ICAM-1) is a ligand for LFA-1 in cytotoxic T cell recognition. *Eur J Immunol* 18:637–640, 1988.

151. Dougherty GJ, Murdoch S, Hogg N: The function of human intercellular adhesion molecule-1 (ICAM-1) in the generation of an immune response. *Eur J Immunol* 18:35–39, 1988.

152. Staunton DE, Marlin SD, Stratowa C, Dustin ML, Springer TA: Primary structure of intercellular adhesion molecule 1 (ICAM-1) demonstrates interaction between members of the immunoglobulin and integrin supergene families. *Cell* 52:925–933, 1988.

153. Edelman GM: Cell adhesion molecules in the regulation of animal form and tissue pattern. *Ann Rev Cell Biol* 2:81–116, 1986.

154. Wright SD, Reddy PA, Jong MTC, Erickson BW: C3bi receptor (complement receptor type 3) recognizes a region of complement protein C3 containing the sequence Arg-Gly-Asp. *Proc Natl Acad Sci USA* 84:1965–1968, 1987.

155. Sanchez-Madrid F, Simon P, Thompson S, Springer TA: Mapping of antigenic and functional epitopes on the alpha and beta subunits of two related glycoproteins involved in cell interactions, LFA-1 and Mac-1. *J Exp Med* 158:586–602, 1983.

156. Beatty PG, Ledbetter JA, Martin PJ, Price TH, Hansen JA: Definition of a common leukocyte cell-surface antigen (Lp95–150) associated with diverse cell-mediated immune functions. *J Immunol* 131:2913–2918, 1983.
157. Lisowska-Grospierre B, Bohler MCh, Fischer A, Mawas C, Springer TA, Griscelli C: Defective membrane expression of the LFA-1 complex may be secondary to the absence of the beta chain in a child with recurrent bacterial infection. *Eur J Immunol* 16:205–208, 1986.
158. Dana N, Clayton LK, Tennen DG, Pierce MW, Lachmann PJ, Law SA, Arnaout MA: Leukocytes from four patients with complete or partial Leu-CAM deficiency contain the common beta-subunit precursor and beta-subunit messenger RNA. *J Clin Invest* 79:1010–1015, 1987.
159. Dimanche MT, Le Deist F, Fischer A, Arnaout MA, Griscelli C, Lisowska-Grospierre B: LFA-1 beta-chain synthesis and degradation in patients with leukocyte adhesive protein deficiency. *Eur J Immunol* 17:417–419, 1987.
160. Anderson DC, Schmalstieg FC, Finegold MJ, Hughes BJ, Rothlein R, Miller LJ, Kohl S, Tosi MF, Jacobs RL, Waldrop TC, Goldman AS, Shearer WT, Springer TA: The severe and moderate phenotypes of heritable Mac-1, LFA-1 deficiency: Their quantitative definition and relation to leukocyte dysfunction and clinical features. *J Inf Dis* 152:668–689, 1985.
161. Kishimoto TK, O'Connor K, Springer TA: Leukocyte adhesion deficiency: Aberrant splicing of a conserved integrin sequence causes a moderate deficiency phenotype. *J Biol Chem* 264:3588–3589, 1989.
162. Dustin ML, Springer TA: Linkage between the T cell antigen receptor and LFA-1 regulates adhesion and deadhesion, *Nature*, in press, 1989.
163. Takada Y, Hemler ME: The primary structure of VLA-2/collagen receptor/alpha-2 subunit (GPIa): Homology to other integrins and presence of possible collagen-binding domain. *J Cell Biol* in press, 1989.
164. Larson RS, Hibbs ML, Corbi AL, Luther E, Garcia-Aguilar J, Springer TA: The subunit specificity of the CD11a/18, CD11b, CD11c panels of antibodies. in Knapp N et al. (eds): *Leukocyte Typing IV* Oxford, Oxford University Press, in press, 1989.
165. Stauton DE, Dustin ML, Springer TA: Functional cloning of ICAM-2, a cell adhesion ligand for LFA-1 homologous to ICAM-1. *Nature*, 339:61–64, 1989.
166. Smith CW, Lawrence MB, Marlin SD, Rothlein R, McIntire LV, Anderson DC: The role of ICAM-1 in the adherence of human neutrophils to human endothelial cells in vitro. Springer TA et al. (eds): *Leukocyte Adhesion Molecules.* New York, Springer-Verlag, chap 13, 1989.

2

Structure and Functions for the Adhesion Receptors VLA-2 and VLA-4: Comparisons to other Members of the Integrin Superfamily

MARTIN E. HEMLER, YOSHIKAZU TAKADA, MARIANO ELICES, AND CAROL CROUSE

Introduction

Recent understanding of cell adhesion mechanisms has been greatly enhanced by the elucidation of the integrin superfamily of cell adhesion molecules (1). These adhesion receptors include at least 11 different cell-matrix and cell-cell adhesion receptor heterodimers with major structural similarities (Table 2.1). They are subdivided into three families known as (i) VLA proteins (2,3), (ii) cytoadhesins (4), and (iii) LFA-1, Mac-1, p150,95 proteins (5). Among these receptors, the a subunits of a fibronectin receptor (VLA-5) (6,7), vitronectin receptor (8), platelet gpIIa (7,9), Mac-1 (10–12), and p150,95 (13) have been sequenced and show 20 to 60% conservation of amino acids, consistent with their structural relatedness.

After discovery of VLA-1 and VLA-2 on activated T cells, evidence for four more VLA proteins (VLA-3 through VLA-6) has been obtained, and their role in cell adhesion was established (14). All are heterodimers with a unique α subunit associated with the same VLA β_1 subunit (schematically shown in Fig. 2.1). N-terminal amino acid sequence information ((15) and Fig. 2.7 below) confirms that the α subunits are distinct, but structurally similar.

Among the VLA proteins, VLA-2 is an α/β subunit cell surface heterodimer strongly implicated as receptor for collagen because (i) it (VLA-2) is identical (16) to 150K/110K M_r (160k/130K nonreduced) structures implicated in human fibrosarcoma cell (17) and platelet (18–20) attach-

TABLE 2.1. Integrin Family of Cell Adhesion Receptors

Receptor	Subunits	Cleaved α	Distribution	Function	"R-G-D" Role
VLA-1	$\alpha^1 \beta_1$	No	Widespread	(COLL, LM)*	No
VLA-2	$\alpha^2 \beta_1$	No	Widespread	Adheres to COLL (LM?)	No
VLA-3	$\alpha^3 \beta_1$	Yes	Widespread	Adheres to FN, LM, COLL	(No)
VLA-4	$\alpha^4 \beta_1$	No	Lymphocyte, monocyte	Cell-Cell Adhesion?	No
VLA-5	$\alpha^5 \beta_1$	Yes	Widespread	Adheres to FN	Yes
VLA-6	$\alpha^6 \beta_1$	Yes	Widespread	Adheres to LM	No
LFA-1	$\alpha^L \beta_2$	No	Lymphoid, myeloid	Leukocyte adhesion, receptor for ICAM-1, ICAM-2	No
MAC-1	$\alpha^M \beta_2$	No	Lymphoid, myeloid	Leukocyte adhesion, C3bi receptor	Yes?
p150,95	$\alpha^X \beta_2$	No	Lymphoid, myeloid	Leukocyte adhesion, C3bi receptor	?
VNR	$\alpha^V \beta_3$	Yes	Widespread	Adheres to VN, vWF, FB, TSP, OP	Yes
IIB/IIA	$\alpha^{IIB} \beta_3$	Yes	Platelets	Adheres to FN, FB, vWF, VN	Yes

*Collagen, COLL; laminin, LM; fibronectin, FN; vitronectin, VN; von Willebrand factor, VWF; fibrinogen, FB; thrombospondin, TSP; osteoportin, OP

ment to collagen and (ii) a patient deficient in platelet protein Ia (alpha subunit of VLA-2 (21)) also lacked responsiveness to collagen (22).

In contrast to VLA-2, no specific ligand has been identified for VLA-4. Unlike the other VLA proteins, VLA-4 expression is most prevalent on hematopoietic cells (lymphocytes and monocytes) which are often unattached to extracellular matrix, so it is not yet obvious that VLA-4 would have a cell matrix adhesion function.

To further establish structure-function correlations among VLA/integrin sequences, the complete cDNA sequences for the VLA α^2 and α^4 subunits were obtained, and then compared to five other known integrin α subunit sequences. Also, a possible cell-cell adhesion role for VLA-4 has been established.

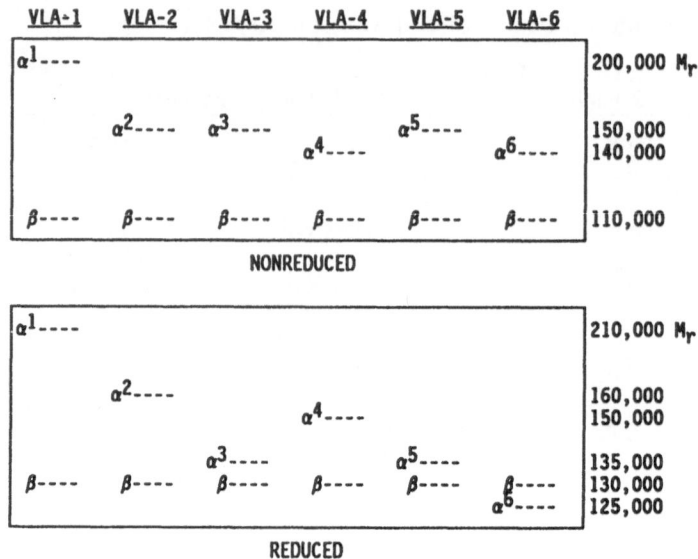

FIGURE 2.1. Schematic diagram of the sizes of VLA protein subunits in reducing and nonreducing conditions

Results and Conclusions

Primary Structure of VLA α^2 and α^4 Subunits

The complete amino acid sequences for the α^2 and α^4 subunits (of VLA-2 and VLA-4, respectively) have been obtained (23,24). RNA hybridization analyses (Fig. 2.2) showed that cDNA clones for α^2, α^3, and α^4 each hybridize to mRNA of the appropriate size (5–8 kb) and specificity. Fibroblasts are known to express VLA-2 and VLA-3 but not VLA-4, whereas T-leukemic cell lines are known to express VLA-4 but not VLA-2 or VLA-3 (25). Both the α^2 and α^4 clones had N-terminal sequences consistent with direct N-terminal amino acid sequencing results, although originally there has been some confusion regarding the α^2 N-terminal sequence (see Fig. 2.7 below). The α^4 subunit has 999 amino acids in its primary structure, and 12 potential N-glycosylation sites, whereas the α^2 subunit has 1,152 amino acids and 10 potential N-glycosylation sites. In each case, the predicted sizes (plus N-glycosylation) are consistent with the estimated protein size from SDS-PAGE. Both the α^2 and α^4 sequences have seven repeating domains in the N-terminal half of the molecule, three potential divalent cation binding sites, and a typical hydrophobic transmembrane region followed by a short cytoplasmic tail. In all of these respects, both of these structures resemble the other integrin α subunits for which complete sequences are known. Also, the majority of cysteines

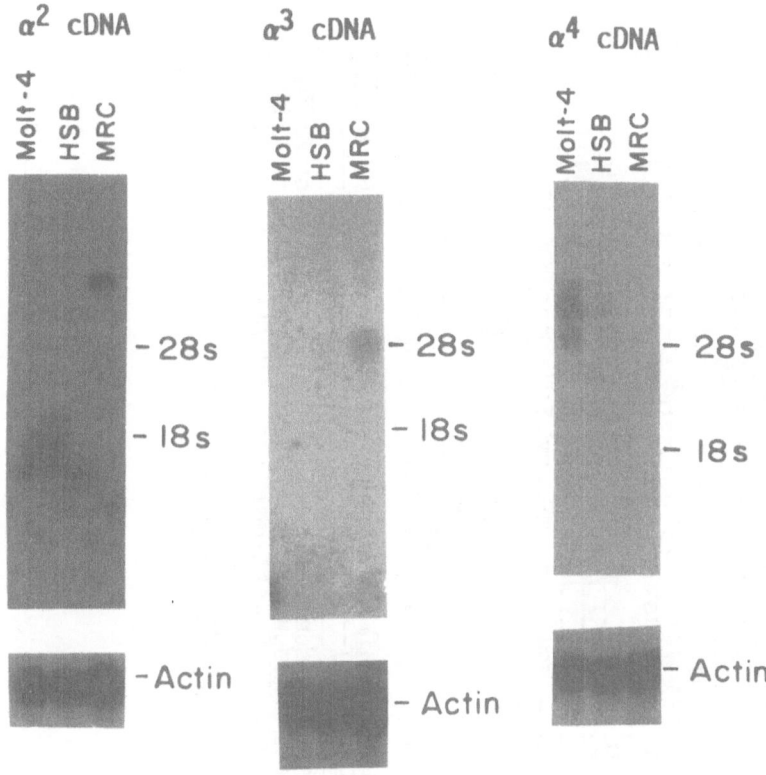

FIGURE 2.2. RNA hybridization analysis of α^2, α^3, and α^4 clones.

(17/20 for α^2 and 19/24 for α^4) are conserved among at least three of the six other known human integrin sequences.

Presence of an "I-domain" in the α^2 Subunit

The greater length of the α^2 amino acid sequence is largely due to a 191 amino acid insert (amino acids 159–349) not found in the α^4, α^5, vitro-nectin receptor or IIb/IIIa sequences. However, a similar inserted se-quence is found in approximately the same relative position in the α subunits of Mac-1 (10–12) and p150,95 (13). This inserted domain has been named the "I-domain," and has notable homology (Fig. 2.3) to collagen-binding domains in cartilage matrix protein (26) and von Wil-lebrands factor (vWF) (27,28). When average linkages between I-domains were calculated (Fig. 2.4) based on percent similarity (using the data in Fig. 2.3, and allowing for conservative amino acid substitutions), it could be seen that the α^2 I-domain was most related to α^M and p150 (48% similarity) and then with CMP-1 and CMP-2 (40% similarity). The vWF

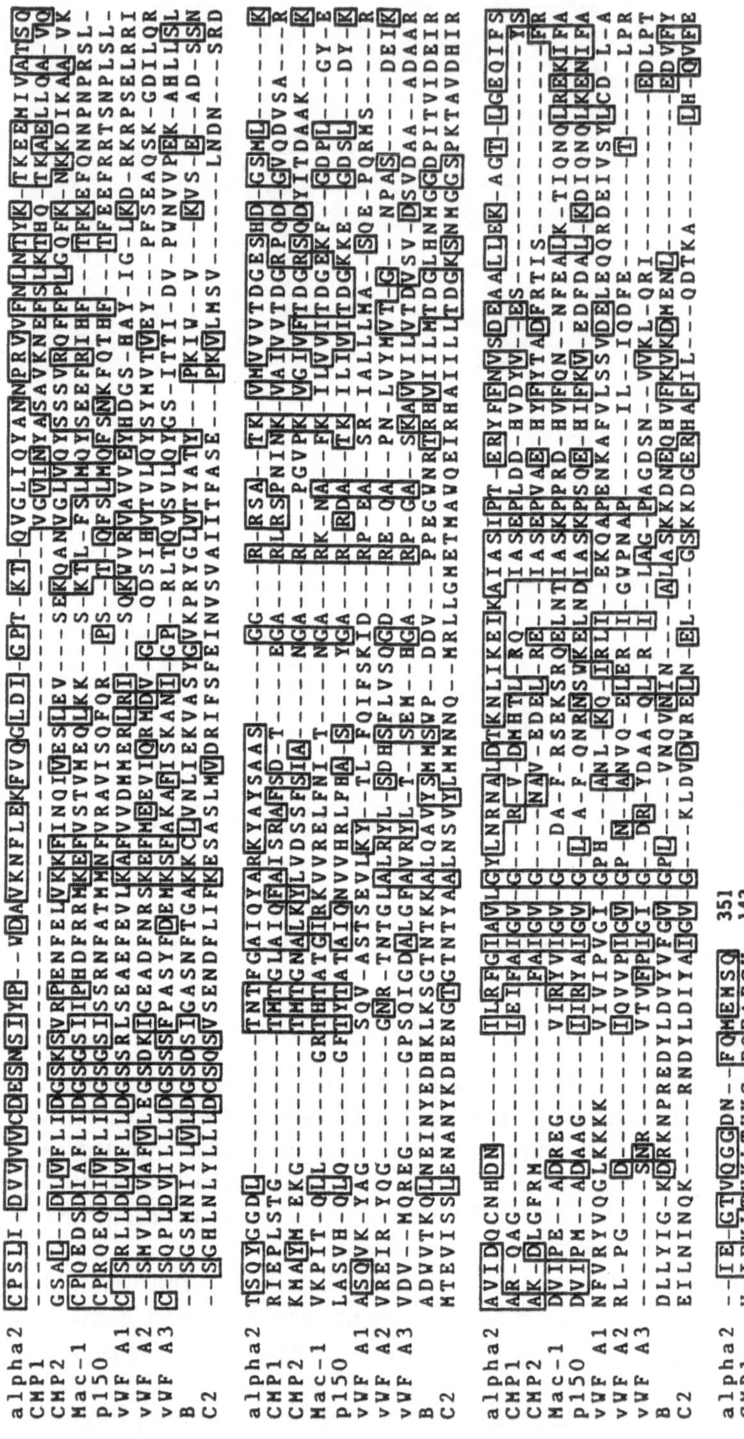

FIGURE 2.3. Comparison of α^2 I-domain with similar sequences in other proteins.

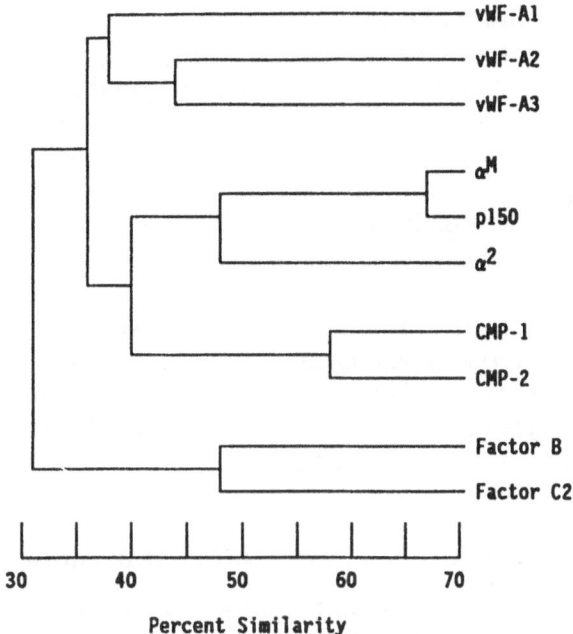

FIGURE 2.4. Linkages between the α^2 I-domain and similar domains in other proteins.

A domains were less related (36%) and complement factors B (29) and C2 (30) were least related (31%) and thus they were grouped separately from all of the others. If percent similarities were calculated utilizing only amino acid identity (not allowing for conservative substitutions), then essentially the same clustering pattern was obtained, except that the α^2 I-domain was grouped first with CMP-1 and CMP-2, and then with α^M and p150 (results not shown). Thus, it is not certain if the α^2 I-domain is more related to α^M and p150 or to CMP-1 and CMP-2, but clearly it is somewhere in between those two sets of domains. Homology between the I-domain of α^2 and collagen-binding domains in other proteins is consistent with the known collagen-binding function of α^2. However, the functional significance of homology to domains in Mac-1, p150, complement factor B, and C2 is not clear, since interaction of these proteins with collagen-like substrates has not been established.

Potential Protease Cleavage Sites in the VLA-4 α^4 Subunit

For three of the integrin α subunits (α^5, α^V, IIb), the C-terminal 15% of the amino acid sequence is cleaved, but then remains attached to the rest of the α chain by a disulfide linkage (6,8,31). The α^4 sequence contains a Lys-Arg (at positions 852–853) which is in the same region as cleavage sites in other integrin α subunits (Line 2, Fig. 2.5). However, there is little additional homology to other cleavage sites in the immediately surrounding residues and no evidence yet that this site is cleaved since the size of the α^4 subunit protein is appropriate for its sequence length. Also, the α^4 subunit does not diminish in size upon reduction (2), suggesting that there are no disulfide-linked cleavage fragments. However, at another site in α^4 (residues 564–583) there is a sequence which somewhat resembles the protease cleavage sites in other integrins (Line 1, Fig. 2.5). Variable cleavage at this site would be consistent with the previously observed splitting of α^4 into 80,000 and 70,000 fragments (2,25). However, unlike for the other cleaved integrin α subunits, cleavage of α^4 is variable and incomplete and there is no evidence for disulfide linkage of the 80,000 and 70,000 M_r cleaved α^4 fragments.

Relative Similarities Between Integrin α Subunits

When the two new VLA α subunit sequences (α^4 and α^2) were compared to the five other previously published integrin α sequences, a surprising pattern was observed. As shown (Fig. 2.6) α^2, α^4, and α^5 did not group together despite being members of the same family and sharing the same β_1 subunit. Rather, α^2 was more similar to the two other subunits which contain a I-domain, and α^5 was more similar to the other subunits which undergo protease cleavage and have disulfide-linked C-terminal fragments. The α^4 sequence, which has no I-domain and has anomalous, incomplete protease cleavage, was a little more similar to the other cleaved subunits, than to the I-domain subunits.

To further emphasize the alignments in Fig. 2.6, patterns of exclusive residue-sharing among sets of α subunits were examined. For example,

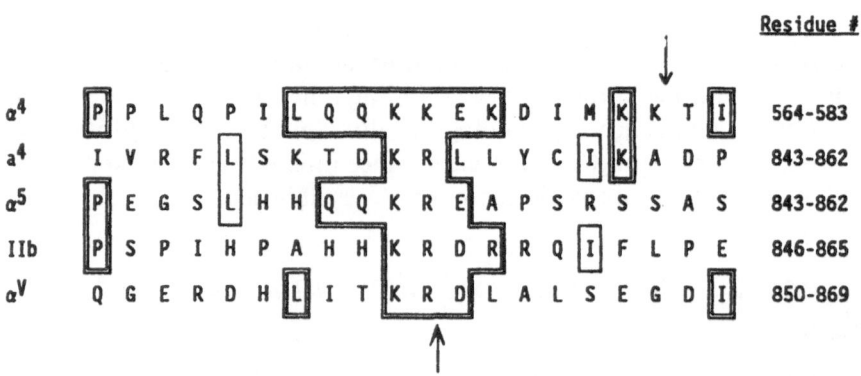

FIGURE 2.5. Potential cleavage sites in α^4 sequence and homology to other integrin α subunit cleavage sites.

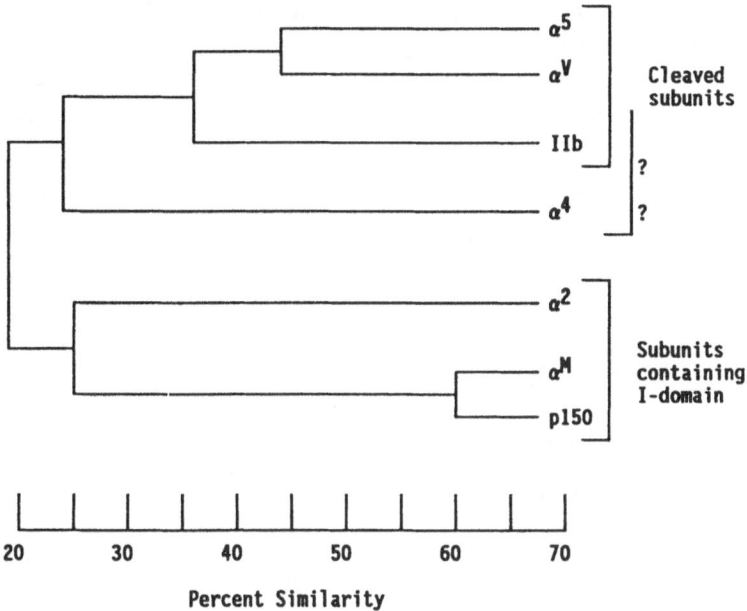

FIGURE 2.6. Linkage map of relative similarities between integrin α subunits.

a computer-generated list was compiled of all amino acid positions in which exactly three (out of seven) residues were shared. From that list it was determined that α^5, α^V, and IIb most often shared amino acids (at 70 positions) and after that, α^2, αM, and p150 exclusively shared residues at 42 positions (Table 2.2, Part A). The designation of these two groups by this method agrees with the results shown in Fig. 2.6. Residues were shared among other sets of three α subunits less frequently (at 29 positions or less). Notably, the VLA α subunits (α^2, α^4, α^5) shared residues at only five positions, and thus this group was only 15th-most prevalent in the list of possible combinations of three α subunits. It should be emphasized that the residues shared by the I-domain subunits (or by the cleaved subunits) are distributed throughout the sequence, and not limited to one region (e.g., the I-domain). In fact, the I-domain amino acids were excluded from the data used for Table 2.2. Because there are so few "β_1-specific" amino acids, it is clear that the VLA α subunits do not depend on large numbers of conserved amino acids for β_1 association.

To address the question of α^4 similarities to other integrin subunits, positions were enumerated in which four amino acids were exclusively shared (Table 2.2, Part B). The set of four sequences which most often exclusively shared residues was α^4, α^5, α^V, IIb (occurring 29 times). The next time α^4 appeared in the list (in third place) it was grouped with the I-domain subunits α^2, αM, and p150 (occurring 22 times). This evidence moderately favors grouping of α^4 with the "cleaved" subunits rather than

TABLE 2.2. Comparison of Subsets of α Subunit Which Exclusively Share Residues at the Most Positions

Subsets of		Subunits	No. of Sites with Shared Residues	Common Features
A.	1	$\to \alpha^5, \alpha^V, IIb$	70	Cleaved subunits
	2	$\to \alpha^2, \alpha^M, p150$	42	I-domain subunits
	3	$\alpha^4, \alpha^M, p150$	29	—
	4	$\alpha^M, IIb, p150$	21	—
	5	a^4, α^5, α^V	20	—
	6	$\alpha^2, \alpha^5, \alpha^V$	18	—
	7	$\alpha^5, \alpha^M, p150$	17	—
	8	$\alpha^V, \alpha^M, p150$	16	—
	9	α^4, α^V, IIb	10	—
	10	α^2, α^V, IIb	10	—
	11	α^2, α^4, IIb	8	—
	12	$\alpha^4, \alpha^V, \alpha^M$	7	—
	13	$\alpha^5, IIb, p150$	7	—
	14	α^2, α^5, IIb	6	—
	15	$\to \alpha^2, \alpha^4, \alpha^5$	5	β_1-associated
B.	1	$\to \alpha^4, \alpha^5, \alpha^V, IIb$	29	—
	2	$\alpha^2, \alpha^5, \alpha^V, IIb$	24	—
	3	$\to \alpha^4, \alpha^2, \alpha^5, \alpha^M, p150$	22	—
	4	$\alpha^4, \alpha^2, \alpha^5, IIb$	10	—

grouping it with the I-domain subunits, which is consistent with the results shown in Figure 2.6, which were obtained by a different method.

Co-purification of a Protein Co-migrating with α^2 Provides Possible Evidence for Another VLA Heterodimer.

Highly purified α^2 subunit from platelets (purified using the MAb 12F1) yielded the N-terminal amino acid sequence YNVGLPEAKIFSGPS (Fig. 2.7, line B). The identical sequence was encoded by α^2 cDNA (Fig. 2.7, line A), thus affirming the α^2 identity of the cDNA clone. Surprisingly, the sequence YNVGLPEAKIFSGPS only partly resembled the previously published "α^2" N-terminal sequence FNLDTXEDNVFRGP (15), with homology in only 5/14 positions (Fig. 2.7, line D). To resolve this discrepancy, VLA proteins were isolated from platelets (using the anti-β MAb A-1A5), and then α^2 plus any comigrating α subunits were further purified using preparative reducing SDS-PAGE. Upon N-terminal amino acid sequencing of purified 160,000 M_r "α^2-like" protein, a mixture of the YNVGLPEAKIFSGPS sequence and the FNLTDXEDNVFRGP sequence was obtained in approximately equivalent molar ratios (Fig. 2.7, line C). Thus, it appears that the latter sequence belongs to a protein of

Residue #:	1	2	3	4	5	6	7	8	9	10	11	12	13	14	15
A. α^2 (cDNA)	Y	N	V	G	L	P	E	A	K	I	F	S	G	P	S
B. α^2 (protein)	Y	N	V	G	L	P	E	A	K	I	F	S	G	P	S
C. $\alpha^2 + \alpha^?$	F/Y	N	L/V	D/G	L	P	E	D/A	N/K	I/V	F	-/S	-	-	-
D. $\alpha^?$ (originally published as α^2)	F	N	L	D	T	X	E	D	N	V	F	R	G	P	-

FIGURE 2.7. Comparison of α^2 erroneous and corrected N-terminal sequences.

160,000 M_r which is distinct from α^2, though it closely resembles α^2 in size. Currently, the nature of this potentially new subunit is not clear and additional independent evidence confirming its existence has not been obtained.

A Cell-Cell Adhesion Function for VLA-4?

A MAb (called L25) which blocked cytolytic function by killer T cells also immunoprecipitated a cell surface antigen complex which remarkably resembled VLA-4 (32). When L25 and anti-VLA-4 MAb were analyzed in side-by-side immunoprecipitation and preclearing experiments, they recognized identical structures (24). Thus further blocking experiments were carried out, testing the effect of anti-VLA-4 MAb on a clone of cytolytic T cells. As shown in Fig. 2.8, four different concentrations of L25 effectively blocked cytolytic T cell activity down to 30 to 40% of control levels. Another anti-VLA-4 MAb (B-5G10) also blocked cytolysis, but was a little less effective. When L25 was pre-incubated with target cells rather than with killer cells, it had no blocking effect. In another control experiment, anti-HLA Class II MAb (LB3.1) effectively blocked killing, as expected since the cytolytic T cells were directed against class II targets. This result confirms the previous L25 result (32), and suggests that VLA-4 could have a cell-cell adhesion role during the association of killer T cells and targets. Obviously, the function of VLA-4 is not likely to be limited to cytolytic T cells, since many non-T cells can also express this antigen (25).

Summary

Although it is appropriate to continue to subdivide the integrin super-family into three subfamilies based on β subunit utilization, there is now reason to consider other subdivisions based on new structural and func-

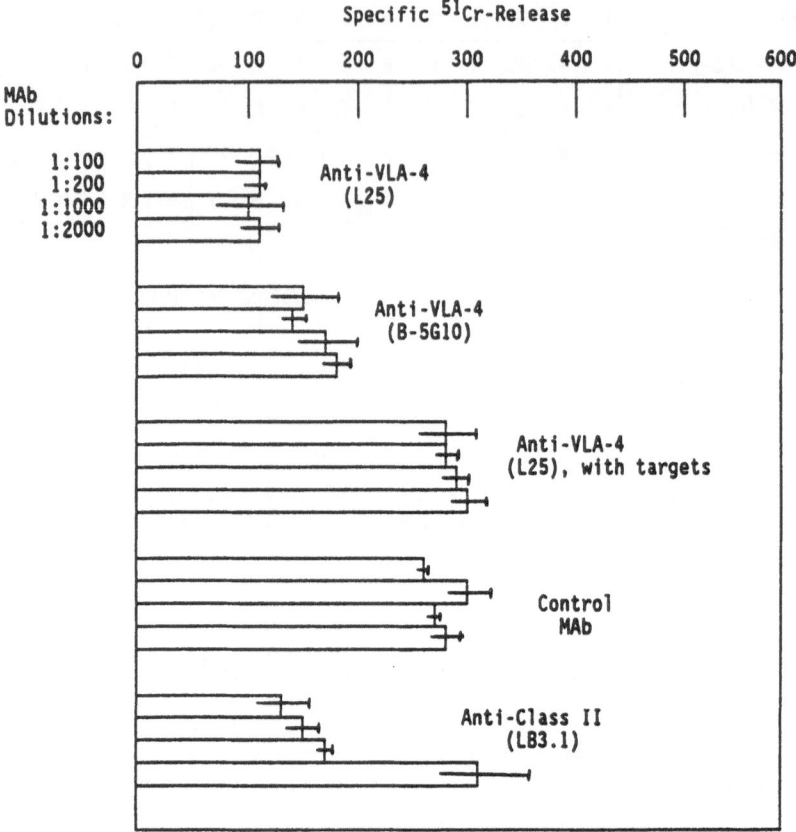

FIGURE 2.8. Inhibition of T cell cytolysis by anti-VLA-4 antibodies.

TABLE 2.3. Grouping of Integrins According to Structural and Functional Similarities

		Structure			Adhesion Function	
Integrin	Subunits	Protease Cleavage Site	Divalent Cation Sites	C-domain	Cell-cell	Cell-matrix
VLA-6	α^6,β_1	Yes	?	?	No?	Yes
VLA-3	α^3,β_1	Yes	?	No	?	Yes
VNR	$\alpha^V\beta_3$	Yes	4	No	No	Yes
IIb/IIIa	IIb,β_3	Yes	4	No	No	Yes
VLA-5	$\alpha^5\beta_1$	Yes	4	No	No	Yes
VLA-4	$\alpha^4\beta_1$	(Yes)	3	No	Yes	?
VLA-2	$\alpha^2\beta_1$	No	3	Yes	No	Yes
Mac-1	$\alpha^M\beta^2$	No	3	Yes	Yes?	No?
p150,95	$\alpha^X\beta_2$	No	3	Yes	Yes?	No?
LFA-1	$\alpha^L\beta_2$	No	?	?	Yes	No?
VLA-1	$\alpha^1\beta_1$	No	?	?	?	?

tional information. In fact, sequence information and biochemical studies suggest that most integrin α subunits can be divided into (i) those which are cleaved, but have no I-domain and (ii) those which have an I-domain but are not cleaved (Table 2.3). Also, it is perhaps significant that the cleaved subunits have four potential divalent cation sites whereas the I-domain subunits have only three. The α subunit of VLA-4 is anomalous in that it has only three divalent cation sites, does not have a I-domain and undergoes an unusual, incomplete cleavage. Information for VLA-6, VLA-3, and LFA-1 is incomplete, but it is predicted that they will align with their respective groups as shown in Table 2.3. Little information is yet available about VLA-1 structure, but because the α^1 subunit does not appear to be cleaved (2,33) it will perhaps belong with the lower group in Table 2.3. Information regarding adhesion functions is incomplete, but a preliminary analysis suggests that cell-matrix adhesion functions can be found in either group, whereas cell-cell adhesion functions (except for VLA-4) are found in the lower group shown in Table 2.3.

References

1. Hynes RO: Integrins: A family of cell surface receptors. *Cell* 48:549–554, 1987.
2. Hemler ME, Huang C, Schwarz L: The VLA protein family. Characterization of five distinct cell surface heterodimers each with a common 130,000 molecular weight beta subunit. *J Biol Chem* 262:3300–3309, 1987.
3. Hemler ME, Crouse C, Takada Y, Sonnenberg A: Multiple very late antigen (VLA) heterodimers on platelets: Evidence for distinct VLA-2, VLA-5 (fibronectin receptor) and VLA-6 structures. *J Biol Chem* 263:7660–7665, 1988.
4. Ginsberg MH, Loftus JC, Plow EF: Cytoadhesins, integrins and platelets. *Thrombos Hemostas* 59:1, 1988.
5. Springer TA, Dustin ML, Kishimoto TK, Marlin SD: The lymphocyte function-associated LFA-1, CD2, and LFA-3 molecules: Cell adhesion receptors of the immune system. *Ann Rev Immunol* 5:223–252, 1987.
6. Argraves WS, Suzuki S, Arai H, Thompson K, Pierschbacher MD, Ruoslahti E: Amino acid sequence of the human fibronectin receptor. *J Cell Biol* 105:1183–1190, 1987.
7. Fitzgerald LA, Poncz M, Steiner B, Rall Jr SC, Bennett JS, Phillips DR: Comparison of cDNA-derived protein sequences of the human fibronectin and vitronectin receptor alpha-subunits and platelet glycoprotein IIb. *Biochemistry* 26:8158–8165, 1987.
8. Suzuki S, Argraves WS, Arai H, Languino LR, Pierschbacher M, Ruoslahti E: Amino acid sequence of the vitronectin receptor alpha subunit and comparative expression of adhesion receptor mRNAs. *J Biol Chem* 262:14080–14085, 1987.
9. Poncz M, Eisman R, Heidenreich R, Silver SM, Vilaire G, Surrey S, Schwartz E, Bennett JS: Structure of the platelet membrane glycoprotein IIb: Homology to the alpha subunits of the vitronectin and fibronectin membrane receptors. *J Biol Chem* 262:8476–8482, 1987.

10. Pytela R: Amino acid sequence of the murine Mac-1 a chain reveals homology with the integrin family and an additional domain related to von Willebrand factor. *EMBO J* 7:1371–1378, 1988.

11. Arnaout MA, Gupta SK, Pierce MW, Tenen DG: Amino acid sequence of the alpha subunit of human leukocyte adhesion receptor Mo1 (complement receptor type 3). *J Cell Biol* 106:2153–2158, 1988.

12. Corbi AL, Kishimoto TK, Miller LJ, Springer TA: The human leukocyte adhesion glycoprotein Mac-1 (Complement receptor type 3, CD11b) a subunit. Cloning, primary structure, and relation to the integrins, von Willebrand factor and factor B. *J Biol Chem* 263:12403–12411, 1988.

13. Corbi AL, Miller LJ, O'Connor K, Larson RS, Springer TA: cDNA cloning and complete primary structure of the alpha subunit of a leukocyte adhesion glycoprotein, p150,95. *EMBO J* 6:4023–4028, 1987.

14. Hemler ME: Adhesive protein receptors on hematopoietic cells. *Immunol Tod* 41:109–113, 1988.

15. Takada Y, Strominger JL, Hemler ME: The very late antigen family of heterodimers is part of a superfamily of molecules involved in adhesion and embryogenesis. *Proc Natl Acad Sci USA* 84:3239–3243, 1987.

16. Takada Y, Wayner EA, Carter WG, Hemler ME: The extracellular matrix receptors, ECMRII and ECMRI, for collagen and fibronectin correspond to VLA-2 and VLA-3 in the VLA family of heterodimers. *J Cell Biochem* 37:385–393, 1988.

17. Wayner EA, Carter WG: Identification of multiple cell adhesion receptors for collagen and fibronectin in human fibrosarcoma cells possessing unique alpha and common beta subunits. *J Cell Biol* 105:1873–1884, 1987.

18. Kunicki TJ, Nugent DJ, Staats SJ, Orchekowski RP, Wayner EA, Carter WG: The human fibroblast class II extracellular matrix receptor mediates platelet adhesion to collagen and is identical to the platelet glycoprotein Ia-IIa complex. *J Biol Chem* 263:4516–4519, 1988.

19. Santoro SA: Identification of a 160,000 dalton platelet membrane protein that mediates the initial divalent cation-dependent adhesion of platelets to collagen. *Cell* 46:913–920, 1986.

20. Santoro SA, Rajpara SM, Staatz WD, Woods VL: Isolation and characterization of a platelet surface collagen binding complex related to VLA-2. *Biochem Biophys Res Commun* 153:217–223, 1988.

21. Pischel KD, Bluestein HG, Woods VL: Platelet glycoproteins Ia, Ic, and IIa are physicochemically indistinguishable from the very late activation antigens adhesion-related proteins of lymphocytes and other cell types. *J Clin Invest* 81:505–513, 1988.

22. Nieuwenhuis HK, Akkerman JWN, Houdijk WPM, Sixma JJ: Human blood platelets showing no response to collagen fail to express surface glycoprotein Ia. *Nature* 318:470–472, 1985.

23. Takada Y, Hemler ME: The primary structure of the VLA-2/collagen receptor α^2 subunit (platelet gp Ia): Homology to other integrins and the presence of a possible collagen-binding domain. *J Cell Biol* in press, 1989.

24. Takada Y, Elices MJ, Crouse C, Hemler ME: The primary structure of the α^4 subunit of VLA-4: Homology to other integrins and a possible cell-cell adhesion function. *EMBO J* in press, 1989.

25. Hemler ME, Huang C, Takada Y, Schwarz L, Strominger JL, Clabby ML: Characterization of the cell surface heterodimer VLA-4 and related peptides. *J Biol Chem* 262:11478–11485, 1987.
26. Argraves WS, Deak F, Sparks KJ, Kiss I, Goetinck PF: Structural features of cartilage matrix protein deduced from cDNA. *Proc Natl Acad Sci USA* 84:464–468, 1987.
27. Girma J-P, Meyer CL, Verweij CL, Pannekoek H, Sixma JJ: Structure-function relationship of human Von Willebrand factor. *Blood* 70:605–611, 1987.
28. Shelton-Inloes BB, Titani K, Sadler JE: cDNA sequences for human von Willebrand factor reveal five types of repeated domains and five possible protein sequence polymorphisms. *Biochemistry* 25:3164–3171, 1986.
29. Mole JE, Anderson JK, Davison EA, Woods DE: Complete primary structure for the zymogen of human complement factor B. *J Biol Chem* 259:3407–3412, 1984.
30. Bentley DR: Primary structure of human complement component C2: Homology to two unrelated protein families. *Biochem J* 239:339–345, 1986.
31. Frelinger AL, Lam SC-T, Plow EF, Smith MA, Loftus JC, Ginsberg MH: Occupancy of an adhesive glycoprotein receptor modulates expression of an antigenic site involved in cell adhesion. *J Biol Chem* 263:12397–12402, 1988.
32. Clayberger C, Krensky AM, McIntyre BW, Koller TD, Parham P, Brodsky F, Linn DJ, Evans EL: Identification and characterization of two novel lymphocyte function-associated antigens, L24 and L25. *J Immunol* 138:1510–1514, 1987.
33. Hemler ME, Jacobson JG, Strominger JL: Biochemical characterization of VLA-1 and VLA-2. Cell surface heterodimers on activated T cells. *J Biol Chem* 260:15246–15252, 1985.

3

Fibronectin Receptor Expression on Thymocytes

MICHAEL D. PIERSCHBACHER AND
PINA M. CARDARELLI

Introduction

Entrance of precursor T cells into the thymus and their subsequent exit into the circulation are prerequisite events for development of mature T-lymphocyte characteristics (1). Thymic epithelial cells may play an integral role in thymocyte differentiation, both through direct contact and by secretion of thymic hormones (2). Thus, positional and developmental cues for maturing thymocytes may come from neighboring cell surfaces (3) or from the extracellular matrix surrounding these cells. However, the full complement of molecules involved in this differentiation process has not been identified. The influence of cell adhesion on cell growth and differentiation is well established for many cell systems (4–10) yet the interaction of lymphocytes with adhesive extracellular matrix proteins, with a few exceptions, has been largely unexplored (11–16).

Cell Interaction with Extracellular Matrix

The extracellular matrix is composed of a number of molecules, some of which have been shown to support the attachment of various types of cells in vitro. Among these are fibronectin (17–21), vitronectin (22–25), the collagens (26–28), and laminin (29–31). The cell-binding site found within the fibronectin molecule has been reproduced by using synthetic peptides carrying the amino acid sequence arginine-glycine-aspartic acid (RGD) (32–35) and the cell attachment-promoting activity of other adhesive molecules such as vitronectin, fibrinogen, von Willebrand factor, thrombospondin, bone sialoprotein, collagen, and tenacin have also been localized to this RGD sequence (36–38).

The discovery that the sequence in the fibronectin molecule that is primarily recognized by cells consists of the three amino acid sequence arginine-glycine-aspartic acid (RGD) (39,32,33,34,36,40) has enabled us to identify the structure at the cell surface that serves as the fibronectin receptor (41). The findings implicating RGD sequences as the cell attachment determinants of some extracellular matrix proteins (37,38) have led also to the identification of a number of other RGD-directed receptors. Thus, a receptor recognizing the RGD sequence in vitronectin (42) and another binding collagen type I through its RGD sequences have been isolated (43). Moreover, an RGD-directed receptor present on platelets that binds fibronectin, vitronectin, fibrinogen, and von Willebrand factor has also been purified. This receptor is the IIb/IIIa protein complex (44). Obviously, the latter receptor does not show the ligand specificity shown by the first two receptors, and some possible explanations for this have been discussed (37,38). We feel that the most likely explanation is that some receptors recognize the RGD sequence in only one or a few restricted environments whereas the binding site of the IIb/IIIa protein accepts the RGD sequence in a large variety of conformations. In fact, recent evidence indicating that receptor selectivity can be achieved with short, conformationally restricted synthetic peptides would seem to confirm this hypothesis (45).

The complete primary structure of the fibronectin receptor has been deduced from cDNA, and a number of physical properties of the receptor can be determined from this sequence (46,47). The protein exists at the cell surface as a heterodimeric complex (though the larger polypeptide is enzymically processed) having both polypeptide chains inserted into the membrane. Each chain extends 30 to 40 residues into the cytoplasmic space, and at least one of the cytoplasmic peptides appears to interact with the cytoskeleton (48). The larger of the two polypeptides, that which we call the α subunit, contains a number of regions that are structurally similar to calmodulin and that apparently mediate the binding of calcium to the receptor. The presence of such divalent cations is required for the receptor to bind ligand. The β subunit is somewhat smaller and has a structure compacted by numerous intrachain disulfide bonds. The cytoplasmic domain of the β subunit contains a potentially phosphorylated tyrosine (38,49).

Partial sequences from other RGD-directed receptors as well as from other adhesion receptors, the ligands for which remain unknown, have revealed the existence of a superfamily of cell surface proteins that share a high degree of structural similarity and probably also functional similarity (36–38). The members of this superfamily of cell surface proteins, collectively known as the integrins can be grouped on the basis of the identity of their β subunit. The first group includes the very late activation antigen (VLA) proteins (50), fibronectin receptor (41), collagen receptor (51), and laminin receptor (52). The second group includes LFA-1, MAC-1, and gp 150/95 (53), and the third group includes the vitronectin receptor (42), and gp IIb/IIIa of platelets (44; see Table 3.1).

The thymus is composed of a heterogenous group of cells. The cell surface molecules CD4 and CD8 which recognize antigens in association with class II and class I MHC, respectively, have been used to define the

TABLE 3.1. Integrin Receptor Superfamily

	Protein	MW	Known Ligands
$\alpha_x\beta_1$	VLA-1	200/120	?
	Collagen receptor (VLA-2)	150/120	Collagen Type I
	VLA-3 (CSAT)	150/120	LN, FN, (Collagen Type IV)
	VLA-4	140/120	?
	Fibronectin receptor (VLA-5)	160/120	FN
	Laminin receptor	150/120	LN
$\alpha_x\beta_2$	LFA	180/95	ICAM
	MAC-1	170/95	C3bi
	150/95	150/95	?
$\alpha_x\beta_3$	Vitronectin receptor	150/90	VN
	GP IIb/IIIa	140/90	FB, vWF, FN, VN

Abbreviations = MW, molecular weight; LN, laminin; FN, fibronectin; ICAM, intercellular adhesion molecule; VN, vitronectin; vWF, von Willebrand factor.

thymocyte subsets. Among peripheral T-lymphocytes, CD4 and CD8 are expressed in a mutually exclusive manner. In contrast, on thymic lymphocytes, the same two surface markers occur in all four possible combinations. the earliest ("immature") thymocytes lack both markers, and can recolonize the thymus of an irradiated host upon adoptive transfer. These CD4-8- cells are a minor subpopulation which make up 2 to 5% of normal thymocytes. Included in the immature subpopulation are large CD4+8+ blast cells which constitute approximately 10 to 15% of the total thymocyte pool. The thymus also contains minor subpopulations of cells that share the CD-8+ and CD4+8- phenotypes with peripheral T cells, and among these cells are found the thymocytes that exhibit "mature" T cell function. The majority of thymocytes, however, can be classified as "nonmature." These small cells express the surface phenotype CD4+8+, and have not been conclusively shown to have either precursor activity or mature T cell function. The role of both CD4+8+ subsets in T cell differentiation is not known; two main hypotheses suggest that they are either intermediates in differentiation or alternatively, dead-end cells that all die in the thymus leaving no progeny (54–58).

The interaction of developing T-lymphocytes with extracellular matrix molecules has been the focus of several studies in our laboratory. We have shown that a population of normal thymocytes carry fibronectin receptors on their surface. Using antibodies directed against CD4 and CD8, we determined the cell surface phenotype of the fibronectin adherent thymocytes. However, the physiological function of the fibronectin receptor on thymocytes is currently unknown. We speculate that the presence of these receptors on lymphocytes may be important for their differentiation. For example, several types of myeloid cells have been shown to be capable of interacting with fibronectin (59–63) or to carry fibronectin receptors (64). It has been suggested that attachment of myeloid cells to fibronectin-coated surfaces triggers the cell to phagocytize C3b and C3bi coated particles (65,66) as a result of rapid and reversible activation of complement receptors mediated by the RGD-binding site of fibronectin (67). Accordingly, fibronectin receptors may play an important role in the differentiation of monocytes into a more active phagocytic cell.

Fibronectin has been shown to promote cell migration, particularly during embryonic development (68). Either RGD-containing peptides or antibodies to the integrin complex inhibit embryonal cell migration (7,69). Bone marrow derived precursor T cells enter the thymus, migrate towards the subcapsular region and then back towards the medullary region where they exit as mature T-lymphocyte (70). An important unanswered question is whether fibronectin or other extracellular matrix molecules are involved in this migration. Alternatively, the interaction of erythroid, myeloid, or lymphoid progenitor cells with extracellular matrix molecules may represent a mechanism whereby these cells are anchored in the bone marrow until they receive the proper signals for differentiation and migration (64, 71). Indeed, reticulocytes (72) or uninduced murine erythroleukemia cells attach to fibronectin and this adhesion is lost upon differentiation (73). A similar mechanism may be functioning in the thymus.

Homotypic lymphocyte aggregation is stimulated by treatment with phorbol esters and this aggregation can be blocked by monoclonal antibody to LFA-1 (74). It is therefore conceivable that, like LFA-1, fibronectin or vitronectin receptor activity may be regulated by cell activation. Chemotactic factors, antigen activation or hormone-type stimuli could induce the expression of fibronectin or vitronectin receptors on circulating lymphocytes, enabling them to traverse connective tissue and enter lymphoid organs or sites of inflammation. Indeed, deposits of fibronectin have been found in lymph nodes (75) and buccal mucosa associated with T-lymphocytes (76), and it has been proposed that the connective tissue of the thymic interlobular septae could be the major pathway for the traffic of lymphocytes both into and out of the thymus (77). We and others (14) have found by immunofluorescence abundant fibronectin in this region. We have studied the interaction of thymocytes with extracellular matrix glycoproteins with the goal of increasing our understanding of T-cell differentiation and immune function as well as possibly some of the cellular interactions involved in the inflammatory process.

Fibronectin-adherent Thymocytes

The observation that a number of lymphoma cell lines adhere to fibronectin lead us to investigate the interaction of normal lymphocytes with extracellular matrix molecules (78). We have identified a subpopulation of thymocytes that binds specifically to fibronectin. Isolated thymocytes were incubated on substrates coated with various concentrations of fibronectin, vitronectin, laminin, collagen type I, or bovine serum albumin. Increasing concentrations of fibronectin on the substrate led to a concomitant increase in the number of thymocytes attached per well, reaching a plateau at approximately 10 µg of fibronectin/ml of coating buffer. In contrast, substrates carrying the other proteins did not support the adhesion of thymic lymphocytes. Attached cells were quantitated and represented approximately 10% of the total number of thymocytes. Subsequent assessment of the non-adherent thymocytes showed that all of the fibronectin-adherent cells had been removed in the first panning.

To further assess the specificity of the adhesion of this population of thymocytes to fibronectin, thymocytes were incubated on fibronectin-coated substrates in the presence of a peptide having the sequence of glycine-arginine-glycine-aspartic acid-serine-proline, the demonstrated cell-attachment site of fibronectin (33,34). This peptide significantly inhibited the binding of thymocytes to fibronectin. The peptide glycine-arginine-glycine-glutamic acid-serine-proline, which does not have cell attachment-promoting activity, had no effect on the ability of thymocytes to bind fibronectin. These results strongly suggest that this population of thymocytes binds specifically to fibronectin and the interaction appears to be mediated by the same RGD sequence which is recognized by the fibronectin receptor on fibroblasts, indicating that the receptor on thymocytes is the same as that on fibroblasts or is a member of the same family of RGD-directed receptors.

Immunoprecipitation of Fibronectin Receptors

Having made this observation, we next wanted to identify whether indeed the thymocytes expressed a receptor for fibronectin. Using a rabbit antiserum raised against the fibronectin receptor present on human fibroblasts, a receptor for fibronectin was immunoprecipitated from a fibronectin-adherent, murine thymic lymphoma cell line (79) and from fibronectin adherent thymocytes. In both cases, SDS-PAGE of the precipitate under nonreducing conditions revealed two radiolabeled components with approximate molecular weights of 175 and 150 kD. When reduced, these two bands migrated as a closely spaced doublet with molecular weights around 160 kD (80).

This receptor appears slightly larger when compared with the fibronectin receptor described on human osteogenic sarcoma cells (41) or compared with the receptor on normal rat kidney fibroblasts or NIH 3T3 cells (140 kD reduced; 160 and 125 kD nonreduced). When immunoprecipitates from equal numbers of surface-iodinated, fibronectin-adherent and non-adherent thymocytes were compared, only trace amounts of labeled receptor could be obtained from the preparation of non-adherent cells.

Trypsin Sensitivity of Lymphocyte Fibronectin Receptor

To begin to characterize the fibronectin receptor on thymocytes, we asked whether the ability of fibronectin-adherent thymocytes or the T-lymphoma cell line, WR16.1, to attach to fibronectin substrate could be affected by pretreatment of the cells with trypsin. Figure 3.1 demonstrates that both mouse thymocytes and the T-lymphoma cells lose their capacity to recognize fibronectin following trypsinization, whereas mouse 3T3 fibroblasts do not. These data suggested that differences may exist between fibroblast and thymocyte fibronectin receptor.

Immunoprecipitation of fibronectin receptor from trypsinized and non-trypsinized WR16.1 cells showed that the major effect of trypsin was on the high molecular weight component, changing its apparent molecular weight from 175 kD to about 163 kD. However, a small change of approximately 5 kD was observed in the lower subunit as well. In agreement with the functional resistance of the fibroblast receptor to trypsin, the

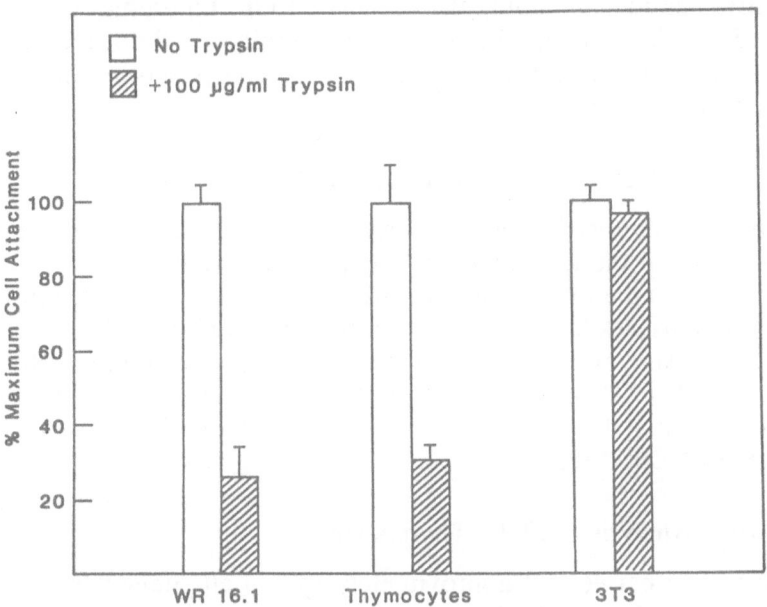

FIGURE 3.1. Effect of trypsin treatment on the adhesion of cells to fibronectin-coated substrates. Cells were pre-incubated in the presence or absence of 100 μg/ml trypsin for 30 minutes at 37 °C. Following three washes with 500 μg/ml soybean trypsin inhibitor, T-lymphoma cells (WR16.1) (5 $\times$ 10⁵ cells/well), thymocytes (2 $\times$ 10⁶ cells/well), or mouse 3T3 fibroblasts (5 $\times$ 10⁴ cells/well) were allowed to attach to microwells which had been coated with fibronectin (10 μg/ml). Maximal attachment of each cell type to fibronectin was set at 100%, and the adhesion of trypsinized cells is expressed relative to those values. Mean $\pm$ SEM are expressed (n = 8).

electrophoretic mobility of the 3T3 fibroblast fibronectin receptor was unaffected by the pretreatment of the cells with trypsin.

Isolation of Functional Fibronectin Receptor

Using affinity chromatography of cell extracts over a matrix of insolubilized fibronectin, a cell surface receptor can be isolated from fibroblasts (41) or from tissue (81). Because of the difficulty in obtaining large numbers of fibronectin-adherent mouse thymocytes and since we had shown the thymocyte and T-lymphoma receptor to have similar properties, we applied the affinity chromatography procedure to detergent extracts of the WR16.1 cells that had been surface iodinated. The material eluted from the fibronectin matrix with an RGD-containing synthetic peptide appeared on SDS-PAGE under nonreducing conditions as two components with apparent molecular weights of 175 and 150 kD. When reduced, a closely spaced doublet of polypeptide components having molecular weights around 160 kD could be visualized. No material was eluted with a control peptide containing the sequence RGE. Liposomes containing the 175/150 kD protein showed specific affinity for fibronectin-coated dishes. In contrast, such receptor-loaded liposomes showed no affinity for vitronectin (80).

Cell Surface Phenotype of Fibronectin Adherent Thymocytes

Because the fibronectin adherent thymocytes (FNR+) represent approximately 10% of the total thymocyte population, we determined the cell surface phenotype of this subpopulation. We found that adhesion to fibronectin enriches for immature "double negative" (CD4-8-) cells. The FNR+ cells were also slightly enriched for a subpopulation of cells that express both CD4 and CD8 ("double-positive" cells) and depleted of thymocytes that have the more mature "single-positive" phenotypes, CD4+8- or CD4-8+ (82).

Cell Cycle Analysis of FNR+ Thymocytes

It has been reported that approximately 15% of all young adult thymocytes are proliferating with an average cell cycle time of 9 h (83,84). Thymic lymphoblasts fall mainly into two subpopulations. CD4-8- and CD4+8+, and fibronectin adherence selects for these thymocyte subsets. Because this suggests an association between FNR expression and proliferation, we carried out cell cycle analysis using acridine orange staining on the total and FNR+ cells. Among unselected thymocytes, 83% were in G_0. In contrast, FNR+ cells were much more frequently in cycle. We found a threefold increase in the number of cells in G2/M in the FNR+ cells when compared to unselected thymocytes. In agreement with the cell cycle data, FNR+ cells incorporated significantly more [3H] thymidine. Thus, there is a strong bias towards proliferation in the FNR⁺ thymocyte population.

Cell Surface Phenotype of FNR⁺ Blast Cells

To determine whether FNR expression is associated with all thymic blast cells or only with subpopulations of blast cells, it was necessary to examine blast cells in isolation. Large thymocytes were separated from small cells by velocity sedimentation using a BSA density gradient. Approximately 17% of the thymocytes were large blast cells. Of the total isolated large cells, about 50% attached to fibronectin; this reflects extensive enrichment for FNR⁺ cells by this procedure. The large cells were greatly enriched for cells in cycle, as judged by [³H]TdR incorporation; and both FNR⁺ and FNR⁻ blast cells incorporated thymidine equally. In contrast, the isolated small cells did not incorporate [³H]TdR, nor did they attach well to fibronectin. This suggests that while not all blast cells are FNR⁺, most FNR⁺ cells are blast cells.

Two-color FACS analysis of the isolated blast cells showed that the isolated FNR⁺ blast cells were mostly CD4⁻8⁻ and CD4⁺8⁺ while CD4⁺8⁻ cells were strikingly absent. The selection of blast cells on the basis of FNR expression had only a subtle effect on CD4⁻8⁺ cells.

Attachment of CD4⁻8⁻ Thymocytes to Fibronectin

Since FNR⁺ thymocytes are enriched for cells with the CD4⁻8⁻ phenotype, purified double-negative cells were made. These double-negative cells were then assessed for their ability to bind to fibronectin-coated dishes. An estimated 20 ± 5% of CD4⁻8⁻ cells bound specifically to fibronectin-coated dishes (compared to 10% of total thymocytes) (78). After detachment of the FNR⁺, double-negative cells with the RGD-containing peptide, the cells were stained with antibodies directed against interleukin-2 receptor (IL-2R) and J11d, which define functionally distinct subsets of CD4⁻8⁻ cells (85,86). We found that 74% of these FNR⁺ CD4⁻8⁻ cells were positive for J11d and 36% carried IL-2R on their surface, whereas the FNR⁻ subset was 45% positive for J11d and 22% positive for IL-2R. Thus, the fibronectin panning enriched for cells that express J11d and IL-2R and, therefore, cells which have precursor potential.

In conclusion, we have found FNR expressed preferentially on early (J11d⁺CD4⁻8⁻) precursor cells and on CD4⁺8⁺ blast cells, and selectively lacking on functionally mature (CD4⁻8⁺ and CD4⁺8⁻) cells and cells committed to intrathymic death (small CD4⁺8⁺ cells). On this basis, we propose that FNRs may play a role in early differentiation events before lineage commitment occurs. Then, after some critical branch point(s) in the differentiation pathway, the cells down-regulate the FNR, and those which are destined to exit as functional T-cells also upregulate the T cell receptor (Fig. 3.2). Although the exact surface phenotype of the selectable

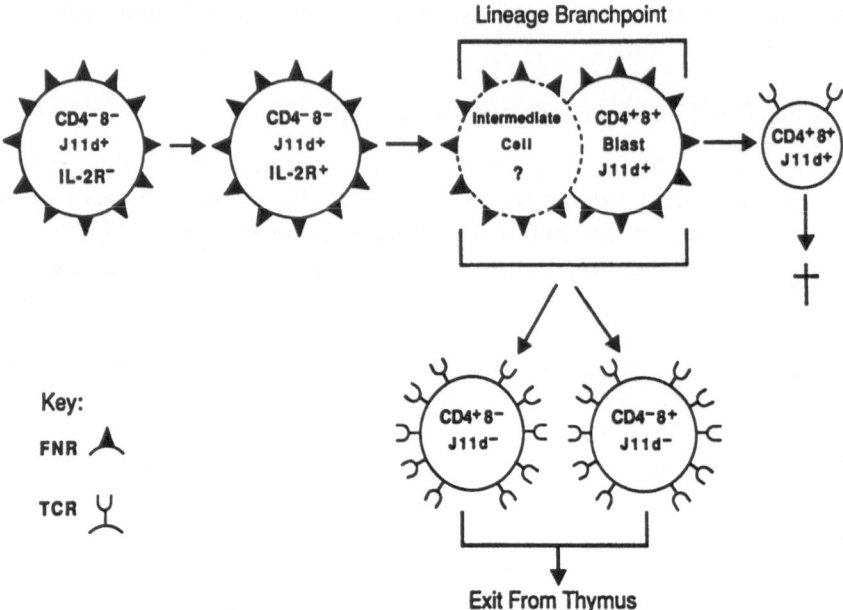

FIGURE 3.2. Model of FNR expression during thymocyte differentiation. The model proposes that FNRs (▲) are expressed on early CD4⁻8⁻ cells that are J11d⁺. After transiently expressing the IL-2R, the cells reach a branch point at which their developmental fate is determined. The cell labeled *Intermediate cell?* indicates that it is still not clear whether the functionally mature cells and the small CD4⁺8⁺ cells are both derived from CD4⁺8⁺ blasts or from some common precursor that is not CD4⁺8⁺. However, recent data suggests that indeed the intermediate cell expresses both CD4 and CD8 (87,88). We propose that the cells of this branch point are FNR⁺ and lack high-level expression of T-cell antigen receptor (Ψ). These cells can become small CD4⁺8⁺ cells that are believed to be committed to cell death (†). Alternatively, they can give rise to CD4⁺8⁻ or CD4⁻8⁺ cells which lack FNR and express high levels of T-cell antigen receptor. We propose that the loss of both FNR and J11d is associated with maturation of functional T cells that are then competent to leave the thymus. (From *J Cell Biol* 106:2183–2190, 1988, permission of the Rockefeller University Press).

cells remains unknown, it is probable, based on the data presented here, that they express FNRs. Such receptors would provide a means of obtaining purified populations of these selectable cells. In addition, an understanding of the adhesive interactions that take place in the thymus may be vital in our understanding of the events that dictate cell commitment.

References

1. Stutman O: Intrathymic and extrathymic T cell maturation. *Immunol Rev* 42:138, 1978.
2. Haynes BF: The human thymic microenvironment. *Adv Immunol* 36:87, 1984.
3. Kyewski BA, Rouse RV, Kaplan HS: Thymocyte rosettes: Multicellular complexes of lymphocytes and bone marrow-derived stromal cells in the mouse thymus. *Proc Natl Acad Sci USA* 79:5646, 1982.
4. Turley EA: The control of adrenocortical cytodifferentiation by extracellular matrix. *Differentiation* 17:93, 1980.
5. Chiquet M, Eppenberger HM, Turner DC: Muscle morphogenesis: Evidence for an organizing function of exogenous fibronectin. *Develop Biol* 88:220, 1981.
6. Hochman J, Levy E, Mador N, Gottesman MM, Shearer GM, Okon E: Cell adhesiveness is related to tumorigenicity in malignant lymphoid cells. *J Cell Biol* 99:1282, 1984.
7. Boucaut JC, Darribere T, Poole TJ, Aoyama H, Yamada KM, Thiery JP: Biologically active synthetic peptides as probes of embryonic development: A competitive peptide inhibitor of fibronectin function inhibits gastrulation in amphibian embryos and neural crest cell migration in avian embryos. *J Cell Biol* 99:1822, 1984.
8. Gospodarowicz D, Greenburg G, Birdwell CR: Determination of cellular shape by the extracellular matrix and its correlation with the control of cellular growth. *Cancer Res* 38:4155, 1978.
9. Chen LB: Alterations in cell surface LETS protein during myogenesis. *Cell* 10:393, 1977.
10. Hay E: Collagen and embryonic development, in Hay E (ed.): *Cell Biology of the Extracellular Matrix*. New York, Plenum Press, pp 379–409, 1981.
11. Evan CW, Davies MDJ: The influence of cell adhesiveness on the migratory behavior of murine thymocytes. *Cell Immunol* 33:211, 1977.
12. Otteskog P, Friskoff J, Sundqvist KG: Morphology and microfilament organization in human blood lymphocytes. *Esp Cell Res* 137:111, 1982.
13. Sundqvist KG, Wanger L: Anchorage and lymphocyte function. II. Contact with non-cellular surface, cell density and T-cell activation. *Immunology* 43:573, 1981.
14. Berrih S, Savino W, Cohen S: Extracellular matrix of the human thymus: Immunofluorescence studies on frozen sections and cultured epithelial cells. *J Histochem Cycochem* 33:655, 1985.
15. Sundqvist KG, Otteskog P: Anchorage and lymphocyte function. Collagen and maintenance of motile shape in T cells. *Immunology* 58:365, 1986.
16. Savagner P, Imhof BA, Yamada KM, Thiery J-P: Homing of hemopoietic precursor cells to the embryonic thymus: Characterization of an invasive mechanism induced by chemotactic peptides. *J Cell Biol* 103:2715, 1986.
17. Grinnell F, Hays DG, Minter D: Cell adhesion and spreading factor. Partial purification and properties. *Exp Cell Res* 110:175, 1977.
18. Ruoslahti E, Hayman EG: Two active sites with different characteristics in fibronectin. *FEBS Lett* 97:221, 1979.
19. Pearlstein E: Plasma membrane glycoprotein which mediates adhesion of fibroblasts to collagen. *Nature* 262:497, 1976.
20. Klebe RJ: Isolation of a collagen-dependent cell attachment factor. *Nature* 250:248, 1974.

21. Hynes RO, Yamada KM: Fibronectins: Multifunctional modular glycoproteins. *J Cell Biol* 95:369, 1982.
22. Holmes R: Preparation from human serum of an α-1 protein which induces the immediate growth of unadapted cells in vitro. *J Cell Biol* 32:297, 1967.
23. Barnes D, Wolfe R, Serrero G, McClure D, Sako G: Effects of a serum spreading factor on growth and morphology of cells in serum-free medium. *J Supramol Struct* 14:47, 1980.
24. Hayman EG, Engvall E, A'Hearn E, Barnes D, Pierschbacher M, Ruoslahti E: Cell attachment on replicas of SDS polyacrylamide gels reveals two adhesive plasma proteins. *J Cell Biol* 95:20, 1982.
25. Hayman EG, Pierschbacher MD, Ohgren Y, Ruoslahti E: Serum spreading factor (vitronectin) is present at the cell surface and in tissues. *Proc Natl Acad Sci USA* 80:4003, 1983.
26. Miller EJ: Chemistry of the collagens and their distribution, in Piez KA, Reddi AH, (eds): *Connective Tissue Biochemistry*. Amsterdam, Elsevier, pp 41–78, 1983.
27. Bornstein P, Sage H: Structurally distinct collagen types. *Annu Rev Biochem* 49:957, 1980.
28. Rubin K, Johansson S, Pettersson I, Ocklind C, Obrink B, Hook M: Attachment of rat hepatocytes to collagen and fibronectin: A study using antibodies directed against cell surface components. *Biochem Biophys Res Commun* 91:86, 1979.
29. Terranova VP, Rohrbach DH, Martin GR: Role of laminin in the attachment of PAM 212 (epithelial) cells to basement membrane collagen. *Cell* 22:719, 1980.
30. Couchman JR, Hook M, Rees DA, Timpl R: Adhesion, growth, and matrix production by fibroblasts on laminin substrates. *J Cell Biol* 96:177, 1983.
31. Carlsson RNK, Engvall E, Freeman A, Ruoslahti E: Laminin and fibronectin in cell adhesion: Enhanced adhesion of cells from regenerating liver to laminin. *Proc Natl Acad Sci USA* 78:2403, 1981.
32. Pierschbacher MD, Hayman EG, Ruoslahti E: A synthetic peptide with the cell attachment activity of fibronectin. *Proc Natl Acad Sci USA* 80:1224, 1983.
33. Pierschbacher MD, Ruoslahti E: The cell attachment activity of fibronectin can be duplicated by small synthetic fragments of the molecule. *Nature* 309:30, 1984a.
34. Pierschbacher MD, Ruoslahti E: Variants of the cell recognition site of fibronectin that retain attachment-promoting activity. *Proc Natl Acad Sci USA* 81:5985, 1984b.
35. Pierschbacher MD, Hayman EG, Ruoslahti E: The cell attachment determinant in fibronectin. *J Cell Biochem* 28:115, 1985.
36. Ruoslahti E, Pierschbacher MD: Arg-Gly-Asp: A highly versatile cell recognition signal. *Cell* 44:517, 1986.
37. Ruoslahti E, Pierschbacher MD: New perspectives in cell adhesion: RGD and integrins. *Science* 238:491, 1987.
38. Hynes RO: Integrins: A family of cell surface receptors. *Cell* 48:549, 1987.
39. Pierschbacher MD, Hayman EG, Ruoslahti E: Location of the cell-attachment site in fibronectin with monoclonal antibodies and proteolytic fragments of the molecule. *Cell* 26:259, 1981.

40. Yamada KM, Kennedy DW: Dualistic nature of adhesive protein function: Fibronectin and its biologically active peptide fragments can autoinhibit fibronectin function. *J Cell Biol* 99:29, 1984.
41. Pytela R, Pierschbacher MD, Ruoslahti E: Identification and isolation of a 140 kilodalton cell surface glycoprotein with properties expected of a fibronectin receptor. *Cell* 40:191, 1985a.
42. Pytela R, Pierschbacher MD, Ruoslahti E: A 125/115 KD cell surface receptor specific for vitronectin interacts with the Arg-Gly-Asp adhesion sequence derived from fibronectin. *Proc Natl Acad Sci USA* 82:5766, 1985b.
43. Dedhar S, Ruoslahti E, Pierschbacher MD: A cell surface receptor complex for collagen type I recognizes the Arg-Gly-Asp sequence. *J Cell Biol* 104:585, 1987.
44. Pytela R, Pierschbacher MD, Ginsberg MH, Plow EF, Ruoslahti E: Platelet membrane glycoprotein IIb/IIIa is a member of a family of Arg-Gly-Asp-specific adhesion receptors. *Science* 231:1559, 1986.
45. Pierschbacher MD, Ruoslahti E: Influence of stereochemistry of the sequence Arg-Gly-Asp-Xaa on binding specificity in cell adhesion. *J Biol Chem* 262:17294, 1987.
46. Argraves WS, Pytela R, Suzuki S, Millan JL, Pierschbacher MD, Ruoslahti E: cDNA sequences from the α subunit of the fibronectin receptor predict a transmembrane domain and a short cytoplasmic peptide. *J Biol Chem* 261:12922, 1986.
47. Argraves WS, Suzuki S, Arai H, Thompson K, Pierschbacher MD, Ruoslahti E: Amino acid sequence of the human fibronectin receptor. *J Cell Biol* 105:1183, 1987.
48. Horwitz A, Duggan K, Buck C, Beckerle MC, Burridge K: Interaction of plasma membrane fibronectin receptor with talin—a transmembrane linkage. *Nature* 320:531, 1986.
49. Hirst R, Horwitz A, Buck C, Rohrschneider L: Phosphorylation of the fibronectin receptor complex in cells transformed by oncogenes that encode tyrosine kinases. *Proc Natl Acad Sci USA* 83:6470, 1986.
50. Hemler ME: Adhesive protein receptors on hematopoietic cells. *Immunol Today* 9:109, 1988.
51. Kunicki TJ, Nugent DJ, Staats SJ, Orchekowski RP, Wayner EA, Carter WG: The human fibroblast class II extracellular matrix receptor mediates platelet adhesion to collagen and is identical to platelet glycoprotein Ia-IIa complex. *J Biol Chem* 263:4516, 1988.
52. Gelsen KR, Dillner L, Engvall E, Ruoslahti E: The human laminin receptor is a member of the integrin family of cell adhesion receptors. *Science* 241:1228, 1988.
53. Springer TA, Dustin ML, Kishimoto TK, Marlin SD: The lymphocyte function-associated LFA-1, CD2 and LFA-3 molecules: cell adhesion receptors of the immune system. *Annu Rev Immunol* 5:223, 1987.
54. Ceredig R, MacDonald HR: Intrathymic differentiation: some unanswered questions. *Surv Immunol Res* 4:87, 1985.
55. Fowlkes BJ, Mathieson BJ: Intrathymic differentiation: thymocyte heterogeneity and the characterization of early T-cell precursors. *Surv Immunol Res* 4:96, 1985.
56. von Boehmer H: The selection of the alpha, beta heterodimeric T-cell receptor for antigen. *Immunol Today* 7:333, 1986.

57. Adkins B, Mueller C, Okada CY, Reichert RA, Weissman IL, Spangrude GJ. Early events in T-cell maturation. *Annu Rev Immunol* 5:325, 1987.
58. Sprent J, Webb SR: Function and specificity of T cell subsets in the mouse. *Adv Immunol* 41:39, 1987.
59. Bevilacqua MP, Amrani D, Mosesson MW, Bianco C: Receptors for cold-insoluble globulin (plasma fibronectin) on human monocytes. *J Exp Med* 153:42, 1981.
60. Hosein B, Bianco C: Monocyte receptors for fibronectin characterized by a monoclonal antibody that interferes with receptor activity. *J Exp Med* 162:157, 1985.
61. Gudewicz PW, Molnar J, Lai MZ, Beezhold DW, Siebring Jr GE, Credo RB, Lorand L: Fibronectin-mediated uptake of gelatin-coated latex particles by peritoneal macrophages. *J Cell Biol* 87:427, 1980.
62. Van De Water L, Schroeder S, Crenshaw III EG, Hynes R: Phagocytosis of gelatin-latex particles by a murine macrophage line is dependent on fibronectin and heparin. *J Cell Biol* 90:32, 1981.
63. Cardarelli PM, Blumenstock FA, Saba TM, Rourke FJ: Fibronectin enhanced attachment of gelatin-coated erythrocytes to isolated hepatic Kupffer cells *J Leukocyte Biol* 36:477, 1984.
64. Giancotti FG, Comoglio PM, Tarone G: Fibronectin-plasma membrane interaction in the adhesion of hemopoietic cells. *J Cell Biol* 103:429, 1986.
65. Wright SD, Craigmyle LS, Silverstein S: Fibronectin and serum amyloid P component stimulate C3b- and C3bi-mediated phagocytosis in cultured human monocytes. *J Exp Med* 158:1338, 1983.
66. Pommier CG, Inada S, Fries LF, Takahashi T, Frank MM, Brown EG: Plasma fibronectin enhances phagocytosis of opsonized particles by human peripheral blood monocytes. *J Exp Med* 157:1844, 1983.
67. Wright SD, Meyer BC: Fibronectin receptor on human macrophages recognizes the sequence Arg-Gly-Asp-Ser. *J Exp Med* 162:762, 1985.
68. Thiery J-P, Duband JL, Tucker GC: Cell migration in the vertebrate embryo: role of cell adhesion and tissue environment in pattern formation. *Annu Rev Cell Biol* 1:91, 1985.
69. Bronner-Fraser M: Alteration in neural crest migration by a monoclonal antibody that affects cell adhesion. *J Cell Biol* 101:610, 1985.
70. Scollay R, Shortman K: Cell traffic in the adult thymus, in Watson J, Marbrook J (eds): *Recognition and Regulation in Cell-Mediated Immunity*. New York, Marcel Decker, 1985.
71. Patel VP, Lodish HF: The fibronectin receptor on mammalian erythroid precursor cells: Characterization and developmental regulation. *J Cell Biol* 102:449, 1986.
72. Patel VP, Ciechanover A, Platt O, Lodish HF: Mammalian reticulocytes lose adhesion to fibronectin during maturation to erythrocytes. *Proc Natl Acad Sci USA* 82:440, 1985.
73. Patel VP, Lodish HF: Loss of adhesion of murine erythroleukemia cells to fibronectin during erythroid differentiation. *Science* 224:996, 1984.
74. Rothlein R, Springer TA: The requirement for lymphocyte function-associated antigen-1 in homotypic leukocyte adhesion stimulated by phorbol ester. *J Exp Med* 163:1132, 1986.

75. D'Ardenne AJ, Burns J, Sykes BC, Kirkpatrick P: Comparative distribution of fibronectin and type III collagen in normal human tissues *J Pathol* 141:55, 1983.
76. Matthews JB, Potts AJC, Tjejdosiewicz LK: Relationship between fibronectin and lymphoid cells in buccal mucosa, labial salivary glands and palatine tonsil. *J Oral Pathol* 15:103, 1986.
77. Kendall MD: The cells of the thymus, in Kendall MD (ed): *The Thymus Gland.* New York, Academic Press, pp 63–83, 1981.
78. Cardarelli PM, Pierschbacher MD: T-Lymphocyte differentiation and the extracellular matrix: Identification of a thymocyte subset that attaches specifically to fibronectin. *Proc Natl Acad Sci USA* 83:2647, 1986.
79. Raschke WC: Transformation by Abelson murine leukemia virus: Properties of the transformed cells. Cold Spring Harbor Symposia on Quantitative Biology. Vol. XLIV, pp 2287, 1980.
80. Cardarelli PM, Pierschbacher MD: Identification of fibronectin receptors on T-lymphocytes. *J Cell Biol* 105:499, 1987.
81. Pytela R, Pierschbacher MD, Argraves S, Suzuki S, Ruoslahti E: Arg-Gly-Asp adhesion receptors. *Meth Enzymol* 144:475, 1987.
82. Cardarelli PM, Crispe IN, Pierschbacher MD: Preferential expression of fibronectin receptors on immature thymocytes. *J Cell Biol* 106:2183, 1988.
83. Metcalf D: Structure of the thymus, in Metcalf D (ed): *The Thymus, Recent Results in Cancer Research.* Berlin, Springer-Verlag New York pp 1–17, 1966.
84. Rothenberg E, Lugo JP: Differentiation and cell division in the mammalian thymus. *Dev Biol* 112:1, 1985.
85. Crispe IN, Moore M, Husmann LA, Smith LF, Bevan MJ, Shimonkevitz RP: Differentiation potential of subsets of CD4⁻8⁻ thymocytes. *Nature (London)* 329:336, 1987.
86. Shimonkevitz RP, Husmann LA, Bevan MJ, Crispe IN: Expression of the interleukin-2 receptor precedes the differentiation of immature thymocytes. *Nature (London)* 329:157, 1987.
87. Fowlkes BJ, Schwartz RH, Pardoll DM: Deletion of self-reactive thymocytes occurs at a CD4⁺8⁺ precursor stage. *Nature* 334:620, 1988.
88. MacDonald HR, Hengartner H, Pedrazzini T: Intrathymic deletion of self-reactive cells prevented by neonatal anti-CD4 antibody treatment. *Nature* 335:174, 1988.

4

Mechanism Regulating Recruitment of CD11b/CD18 to the Cell Surface is Distinct From That Which Induces Adhesion in Homotypic Neutrophil Aggregation

JILL P. BUYON, MARK R. PHILIPS,
STEVEN B. ABRAMSON, SETH G. SLADE,
GERALD WEISSMANN, AND ROBERT WINCHESTER

Introduction

In response to activation by certain stimuli, neutrophils undergo changes in surface properties that result in their formation of specific cellular aggregates (1). This phenomenon is being intensively studied as a model of induced homotypic cell-to-cell interaction. Evidence that heterodimeric glycoproteins of the Leu-CAM family are involved in mediating neutrophil aggregation rests on several findings; (i) most monoclonal antibodies (MAb) to CD18, the common 95 kD beta subunit ($\beta2$) abolish neutrophil aggregation (2); (ii) several of the MAb to CD11b, the 165 kD alpha subunit (also referred to as αM, CR3, or MAC-1 inhibit this response (2,3); and (iii) neutrophils from individuals genetically deficient in these molecules cannot be stimulated to aggregate (4).

An important property of the Leu-CAM family is that its molecules, notably gp 165/95 (CD11b/CD18) and gp150/95 (CD11c/CD18), in addition to their presence on the surface membrane, exist in presynthesized reserves situated both in the specific (5) and gelatinase (6) associated intracytoplasmic granules. Upon activation by certain ligands such as formyl-methionyl-leucyl-phenylalanine (fMLP) the contents of the granules are added to the low level of CD11b/CD18 or CD11c/CD18 molecules constitutively expressed on the cell membrane in a process termed *recruitment, upregulation,* or *"prevalence modulation"*. Numerous laboratories have confirmed increases of CD11b/CD18 of 2- to 10-fold over resting values (5,7–10).

Accordingly, the simplest original explanation of activated cell-to-cell adhesion was that it resulted from the recruitment of new CD11b/CD18 molecules from intracellular sites to the cell surface. Were this to be true, aggregation would be a quantitative phenomenon, dependent on the attainment of a threshold concentration of adhesive surface structures. The implications of this classic view are threefold: Firstly, adhesiveness would be an intrinsic property of CD11b/CD18 molecules. Secondly, all molecules of CD11b/CD18 would be functionally identical. Thirdly, regulation of the recruitment of new CD11b/CD18 molecules to the cell surface would control the aggregation response.

An alternative hypothesis may be formulated in which aggregation results from an induced modification of the CD11b/CD18 molecules present on the cell surface. The implications of this hypothesis are: Firstly, that at least two distinct functional forms of CD11b/CD18 exist. Secondly, only one of these forms is rendered competent to mediate aggregation by an activation event that results in an altered functional state of CD11b/CD18. Thirdly, the metabolic pathway that regulates this event is distinct from the pathway that controls recruitment of CD11b/CD18 molecules to the surface.

We reasoned that the most simple test of these hypotheses involves determining whether the molecules of CD11b/CD18 that were newly recruited to the surface exhibited the same property of mediating homotypic aggregation that characterized the CD11b/CD18 molecules constitutively expressed on the membrane. Secondly, we reasoned that exposure of cells to different pharmacologic agents may dissect the mechanisms underlying the process of recruitment of new molecules to the cell surface from those concerned with induction of the aggregation response.

We provide several lines of evidence that the quantitative changes in CD11b/CD18 that accompany neutrophil activation are insufficient to account for the homotypic aggregation response, a finding incompatible with the classic hypothesis and consistent with the alternative hypothesis of an "activated" and "inactivated" species. Since phosphorylation of either the alpha or beta chains of the Leu-CAM family is an obvious potential mechanism to render these molecules capable of promoting cell adhesion functions, we have initiated studies of this posttranslational modification.

Results and Discussion

Reagent Antibodies

Initially, studies were done to characterize the monoclonal antibodies used in subsequent experiments (11). By indirect immunofluorescence, MAb MN41 and VIM12 which both recognize the CD11b antigen (12,13), reacted similarly with surface CD11b/CD18 on cells maintained at 4 °C.

TABLE 4.1. MAbs MN41 and VIM12 Do Not Complete for the Same Epitope on CD11b/CD18

MAb*	Inhibition of MN41-FITC Binding
MN41	>80% (N=3)
VIM12	<5% (N=3)
MOPC21	<5% (N=3)

*PMN were incubated with each MAb (1:400) at 22 °C for 15 minutes followed by washing off excess MAb and staining with MN41 directly conjugated to FITC.

The epitopes recognized on CD11b/CD18 by MAb MN41 and VIM12 were both equally expressed during recruitment from intracellular pools following warming to 37 °C for 5 minutes (206% of constitutive level using MN41, 195% of constitutive level using VIM12), after stimulation with fMLP (10^{-7} M) for 5 minutes (327 and 314%) and after stimulation with fMLP for 30 minutes (534 and 561%). Crossblocking experiments demonstrated that MAb MN41 and VIM12 which both immunoprecipitate a heterodimer of 165/95kD recognize different epitopes on the CD11b molecule (Table 4.1). As seen in the representative aggregation tracings in Figure 4.1, the presence of MAb MN41 inhibited stimulated neutrophil aggregation by 78% (N = 6) of that observed for cells in the presence of control MAb MOPC21 with an irrelevant specificity. In contrast, there was no inhibition of the aggregation response when cells were incubated with MAb VIM12 (N = 4).

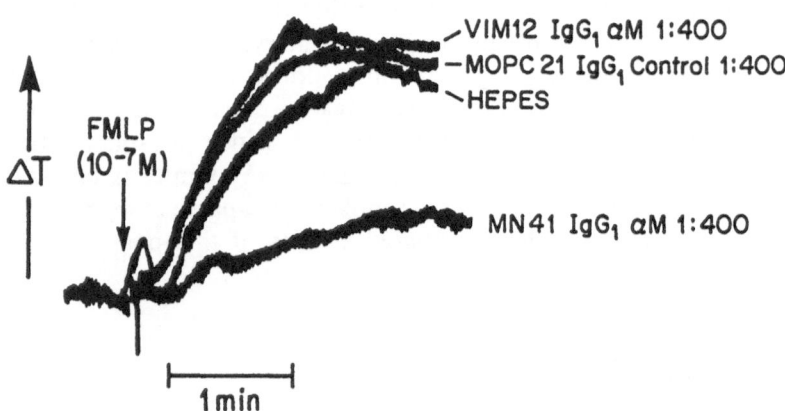

FIGURE 4.1. The effect of MAb MN41 on fMLP induced neutrophil aggregation. 1.25×10^6 peripheral blood neutrophils were incubated with buffer in the presence or absence of various MAbs (1:400) for 30 seconds at 22 °C prior to the addition of cells to the aggregometer. Cells were warmed to 37 °C and fMLP 10^{-7} M was added. Aggregation was measured as the change in the transmittance of light through the stirred (900 rpm) suspensions of cells in a dual chamber aggregometer.

Different Kinetics for Recruitment and Aggregation

Using MAb MN41 directly conjugated to fluorescein isothiocyanate (MAb MN41-FITC) to measure total cell surface CD11b/CD18, a divergence between the kinetics of recruitment and neutrophil aggregation was observed in certain circumstances (Table 4.2). When neutrophils were maintained at 22 °C for 15 minutes the quantity of expressed CD11b on the neutrophil surface increased by 60%, yet under equivalent conditions the neutrophils did not aggregate. Similarly, in cells maintained at 37 °C for 5 minutes there was a doubling of surface CD11b in the absence of a detectable aggregation response. After exposure to fMLP for 30 seconds at 37 °C the cells reached 30% of the maximal aggregation response, at which time immunofluorescence studies revealed that the total surface CD11b was not different from the doubling induced by warming alone. By 1 minute of exposure to fMLP, the total surface expression of CD11b increased to 210% of constitutive levels; however, the aggregation response was now maximal. By 5 minutes of exposure to fMLP there was a threefold increase in surface CD11b at a time when the cells had begun to disaggregate. Expression of CD11b continued to increase, reaching fivefold by 30 minutes of exposure to fMLP, long after disaggregation was found.

Selective Inhibition of Aggregation or Recruitment

To analyze further the apparent discordance in the kinetics of these two processes, sodium salicylate previously demonstrated to inhibit the aggregation response (14), was evaluated for its effect on recruitment of CD11b/CD18 molecules (15). When neutrophils were pretreated with 3 mM sodium salicylate, surface expression of CD11b reached the previ-

TABLE 4.2. Contrasting Kinetics of CD11b/CD18 Expression and Aggregation

Incubation Conditions*			Constitutive CD11b/CD18 (%)	Aggregation Response (%)
Temp. (°C)	Time	Stimulus		
22	15 min.	0	160	0
37	5 min.	0	200	0
37	30 sec.	fMLP	200	30
37	1 min.	fMLP	210	100
37	5 min.	fMLP	300	Disaggregation
37	30 min.	fMLP	600	Disaggregation

*Peripheral blood neutrophils were isolated at 4 °C. Cells were maintained under the various conditions for time periods as stated. Formyl-methionyl-leucyl-phenylamine (fMLP) was used at 10⁻⁷ M. Total surface CD11b/CD18 was measured by the binding of MAb MN41-FITC. The constitutive level of CD11b/CD18 is assigned 100%, and is defined as the expression of CD11b/CD18 on cells maintained at 4 °C.

TABLE 4.3. Dissociation Between Recruitment of CD11b/CD18 and Stimulated Aggregation

Conditions	Recruitment of CD11b/CD18	Neutrophil Aggregation
Sodium salicylate 3 mM*	Present	Inhibited
Tetracaine 1 mM**	Inhibited	Present
Neutroplasts***	Not detectable	Present

*Neutrophils were incubated with sodium salicylate and then exposed to fMLP (10^{-7} M) for 1 minute and then subjected to indirect immunofluorescence staining with MAb MN41 and cytofluorographic analysis. Parallel experiments were done to evaluate the presence of sodium salicylate on the aggregation response elicited by fMLP (10^{-7} M).
**Equivalent conditions in the presence of 1 mM tetracaine.
***Neutroplasts were prepared as previously described (17). CD11b/CD18 expression on neutroplasts exposed to fMLP (10^{-7} M) or buffer alone for 1 to 30 minutes was measured by indirect immunofluorescence staining with MAb MN41 and cytofluorographic analysis. Parallel experiments were done to evaluate the aggregation response of neutroplasts elicited by fMLP (10^{-7} M).

ously demonstrated doubling expected after 1 minute of exposure to fMLP. Surprisingly, surface expression of CD11b in the presence of sodium salicylate was enhanced after 30 minutes of exposure to fMLP. However, the aggregation response of neutrophils exposed to this drug was significantly inhibited (55% of control $p < 0.01$) (Table 4.3). Two other nonsteroidal anti-inflammatory agents (NSAIDS)—piroxicam and indomethacin—had parallel effects, significantly inhibiting the aggregation response. Both were without effect on the surface recruitment of CD11b.

Additional evidence that increased surface expression of CD11b is not required to promote neutrophil aggregation was obtained by incubation of neutrophils with 1 mM tetracaine (15,16) (Table 4.3, Fig. 4.2). Recruitment of CD11b was inhibited by 68%, yet under identical conditions the aggregation response was enhanced reaching 122% of control.

Two Functionally Distinct Forms of CD11b/CD18

The dissociation between the functional and immunoreactive assays for CD11b (aggregation vs. recruitment) observed upon warming and exposure to NSAIDS were consistent with the view that recruitment of CD11b is not sufficient to promote homotypic adhesion. The possibility that there were two functionally distinct pools of CD11b/CD18, one constitutively expressed on the surface membrane and the other intracellular but translocatable to the cell surface was next investigated. The approach involved first the blocking of constitutively expressed surface CD11b molecules with saturating concentrations of unlabelled MAb MN41 under conditions (20 °C) where detectable aggregation did not occur. This was followed by extensive washing of the cells to remove any unbound an-

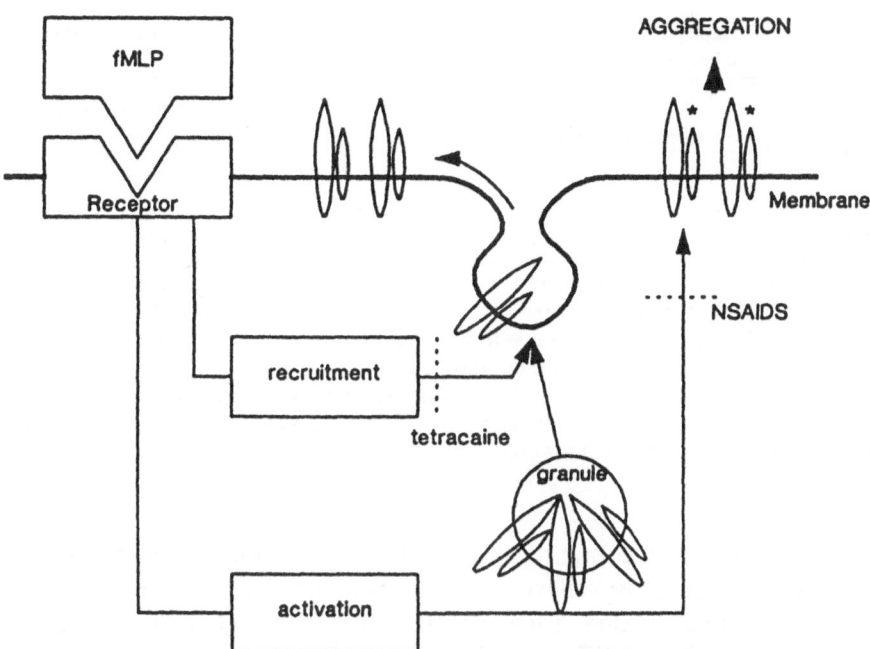

FIGURE 4.2. Proposed model of distinct regulatory mechanisms for recruitment and activation of CD11b/CD18 molecules. The activation process induced by the fMLP ligand may be limited to the plane of the membrane.

tibody from the supernatant. The cells were then stimulated with fMLP. Both aggregation and the quantity of newly expressed CD11b/CD18 were measured. The latter was determined by binding of directly conjugated MAb MN41-FITC. The results (11) demonstrated that despite removal of unbound MAb MN41, profound inhibition of fMLP stimulated aggregation was observed. This response was compared with the normal aggregation response obtained by cells incubated and washed in the same experiment with either MAb MOPC21 directed to an irrelevant structure or MAb VIM12 reactive with an epitope on CD11b which causes no inhibition of aggregation (Table 4.4). The efficacy of the washing steps to remove unbound MAb MN41 was critical and established in controlled experiments by staining of freshly isolated neutrophils with supernatant from the successive washes in indirect immunofluorescence, using rabbit anti-mouse FITC. The mean relative fluorescence for cells incubated with the second wash was at the background level obtained for cells incubated with the negative control MAb MOPC21. Of central importance, the blockade of pre-existing CD11b molecules with unconjugated MAb MN41 did not alter the kinetics of the expression of new surface CD11b as measured by MAb MN41-FITC (11).

Two additional important conclusions derived from this experiment, taken in the context of the preceding observations. Firstly, that the ac-

TABLE 4.4. Aggregation Response of Neutrophils (Measured in CM²) in the
Presence or Absence of MAbs in the Supernatant

MAb	Continuous Presence of MAbs*	Cells Preincubated with MAbs, then Washed**
MN41	2.8[a]	3.7
VIM12	12.5	12.6
MOPC21	13.3	11.0

*1.25 × 10⁶ neutrophils isolated at 4 °C were incubated with saturating concentrations
of MAbs (1:400) as stated for 30 seconds prior to the addition of fMLP (10⁻⁷ M).
**1.25 × 10⁶ PMN were incubated with MAb for 15 minutes at 22 °C and washed twice
to remove excess MAb prior to the addition of fMLP.
[a]For each aggregation tracing the integral of the aggregation tracings up to 2 minutes
after the addition of fMLP was measured in cm² by planimetry. The results represent the
mean value (N = 8) obtained for each condition.

tivation process is compartmentalized and presumably membrane associated in that only cell surface membrane associated CD11b/CD18
molecules are activated. Secondly, the inductive signal that results in
activating the CD11b/CD18 molecules is either short-lived or requires a
time dependent integration of CD11b/CD18 molecules with a specific
membrane environment because molecules of CD11b/CD18 newly recruited to the surface are incapable of mediating a full aggregation
response.

To emphasize that the mechanism of recruiting CD11b/CD18 molecules from intracellular reserves is entirely dissociable from the mechanism of activating CD11b/CD18 molecules, experiments were performed
using neutroplasts. These are in vitro constructs of the neutrophil consisting of plasma membrane enclosing cytoplasm devoid of granules and
the nucleus. There is also disruption of part of the normal cytoskeletal
elements (17). Although neutroplasts express CD11b/CD18 on their surface, they are incapable of any increase in surface expression after exposure to fMLP for 1 to 30 minutes. However, as in the studies described
above using tetracaine, the neutroplasts were fully capable of aggregation
upon exposure to fMLP (Table 4.3). Although granule membrane fuses
with the plasmalemma during the in vitro construction of the neutroplasts, any resultant increase of surface CD11b did not confer the ability
of these neutroplasts to aggregate in the absence of an additional activating stimulus.

Model of Distinct Regulatory Mechanisms

Taken together these results suggest a model with two distinct mechanisms regulating the expression and function of CD11b/CD18 in homotypic aggregation (Fig. 4.2). One mechanism, that of recruitment from

intracellular reserves, does not occur at 4 °C, proceeds upon isolation where cells are maintained at or above room temperature, and is inhibitable by tetracaine. The second mechanism, the activation mechanism, does not proceed at 4 °C, room temperature or upon warming to 37 °C, requires the addition of exogenous stimuli, and is inhibitable by sodium salicylate and at least some of its congeners. It resides in the neutroplast and is likely to be membrane associated.

Studies on the Molecular Basis of Adhesiotope Induction

A variety of molecular phenomena can be postulated to account for the alteration in the state of CD11b/CD18 that results in appearance of the property of adhesion. These include, for example, the association of CD11b/CD18 with another molecular species in the membrane or an exposure of a cryptic adhesiotope mediated by cleavage of a fragment from either subunit of the heterodimer by an extracellular protease, or a conformational change in the extracellular domains by a chemical modification of the cytoplasmic portion, such as phosphorylation, among many possible processes.

Based on the above observations which suggested that neutrophil aggregation results from an activation of CD11b/CD18 and not simply its quantitative change, we pursued the hypothesis that a posttranslational phosphorylation event renders the molecule capable of promoting this aggregation response. Phosphorylation of either alpha and/or beta chains was considered likely based on the published sequences of the molecules (18,19).

As seen in Figure 4.3, in unstimulated neutrophils radiolabelled with [^{32}P]-orthophosphate and immunoprecipitated with MAb 60.3 (directed to CD18 structures), two species of α chains, a predominant CD11b 165kD subunit and a minor 150kD CD11c subunit were found to be constitutively phosphorylated. In contrast, there is no constitutive phosphorylation of the common CD18 beta subunit. However, in the presence of phorbol myristate acetate (PMA), a direct and sustained activator of protein kinase C, phosphorylation of the beta chain is evident after 1 minute of incubation (20). This becomes maximal by 10 minutes. The beta chain remains phosphorylated up to 30 minutes after the addition of PMA. Of interest, this time course is similiar to that of PMA-induced neutrophil aggregation. Specifically, in terms of the aggregation response which differs from that induced by the ligand fMLP there is a lag phase of 1 to 1.5 minutes followed by a rapid increase in the number of neutrophil aggregates which is maximal at 3 minutes and persists without disaggregation. In contrast, phosphorylation of the beta subunit was not detected after the addition of fMLP for 30 seconds, 90 seconds, or 10 minutes. Of interest, the kinetics of the fMLP-induced neutrophil aggre-

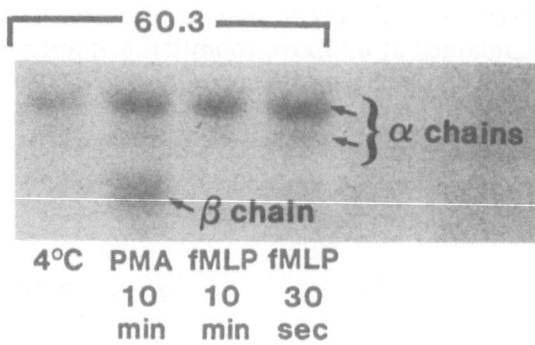

FIGURE 4.3. Phosphorylation of alpha and beta chains of the Leu-CAM family on neutrophils in the presence or absence of PMA or fMLP. 150×10^6 neutrophils were incubated with 1 mCi ^{32}P at 37 °C for 1.5 hours. PMA at 1 μg/ml or fMLP 10^{-7} M was added to appropriate cell aliquots for various time intervals as stated. Cells were then lysed in buffer containing: 50 mM NaCl, 10 mM HEPES, 300 mM sucrose, 2.5 mM MgCl$_2$, 1% NP-40, 1 mM PMSF, 5 mM NaF, 2.5 mM sodium pyrophosphate, 0.5 mM sodium metavanadate, 1 μM leupeptin, 1 μM pepstatin, 0.5 mM ATP, 50 μg/ml DNase I.

gation response differ from that induced by PMA. In the former, there is no lag phase, the response peaks earlier at 1.5 minutes and disaggregation occurs. The lack of phosphorylation in response to fMLP may be due to the simultaneous activation of a phosphatase specific for the β chain. In this regard, it is relevant to know which kinase is operative as there are differences between protein tyrosine phosphatase subtypes and serine/threonine phosphatase subtypes (21). One implication of these results is that PMA either inhibits the activation of a phosphatase which is not inhibited by fMLP or that PMA does not activate such a phosphatase. This difference would account for the persistence of aggregation induced by PMA and the parallel persistence of phosphorylation.

Conclusion

The central conclusion of these studies is that a mechanism exists for the activation of CD11b/CD18 to a state that mediates homotypic neutrophil aggregation and that this mechanism is distinct from that leading to a quantitative increase of CD11b/CD18 molecules on the cell surface. Several lines of evidence support the finding that the increased surface expression of CD11b/CD18 is neither necessary nor sufficient to account for the phenomenon of neutrophil aggregation. Firstly, the kinetics of recruitment of CD11b/CD18 to the cell surface do not parallel those of aggregation. Secondly, it is clear that inhibition of the recruitment mechanism does not prevent the aggregation response as supported by the experiments using tetracaine. Similarly, neutroplasts which were incapable of increased CD11b/CD18 surface expression were able to aggregate in response to fMLP. Thirdly, as demonstrated by the studies using sodium salicylate and other NSAIDS the converse is true, that is recruitment can proceed normally in the absence of an aggregation response. Fourthly, resting neutrophils treated with unlabelled MAb MN41 and then thoroughly washed are incapable of aggregation, again despite the increase of unlabelled CD11b/CD18 newly recruited to the surface after warming or in response to fMLP. Thus, it follows that all molecules of CD11b/CD18 present on the cell surface are not equivalently competent to mediate the active event of aggregation. Moreover, the mechanism of fMLP aggregation appears to only allow the use of CD11b/CD18 that are both immediately present in the membrane and "primed" at the time of stimulation.

Therefore, we investigated the possibility that a posttranslational modification such as phosphorylation of either the alpha or beta chains of the CD11b/CD18 molecule confers adhesive function. The CD11b and CD11c α chains were found to be constitutively phosphorylated. Stimulation of the neutrophils with PMA induced readily detectable phosphorylation of the β chain which was not phosphorylated in unstimulated cells. The time course of phosphorylation of the common β chain was similar to the slower kinetics of persistent homotypic aggregation induced by PMA. In contrast, the much faster and reversible kinetics of ligand-receptor mediated aggregation induced by fMLP were not associated with this pattern of accumulative and sustained β-chain phosphorylation. Additionally, phosphorylation of the β chain was not observed in isolated neutrophils maintained at 37 °C, a condition which results in prevalence modulation but is insufficient to promote aggregation. Currently, we are investigating other possibilites to account for the activation.

In summary, these studies demonstrate the existence of two functionally distinct forms of CD11b/CD18, only one of which is rendered specifically competent to mediate adhesion by an activation event. Available evidence suggests that the process that results in this activation following binding of fMLP to its receptor is compartmentalized, probably membrane associated, inhibitable by NSAIDS, and constrained by time. The process itself is heterogeneous, as revealed by qualitative and quantitative differences in responses to PMA or fMLP, emphasizing the distinctive behavior of CD11b/CD18 molecules located in different anatomic sites. Induced phosphorylation of the common Leu-CAM β chain might be relevant to at least one variety of aggregation reflecting the operation of a system of kinases and phosphatases involved in promoting activation.

References

1. Hoffstein ST, Friedman RS, Weissmann G: Degranulation, membrane addition, and shape change during chemotactic factor-induced aggregation of human neutrophils. *J Cell Biol* 95:234, 1982.
2. Anderson DC, Miller LJ, Schmalstieg FC, Rothlein R, Springer TA: Contributions of the Mac-1 glycoprotein family to adherence-dependent granulocyte functions: structure-function assessments employing subunit-specific monoclonal antibodies. *J Immunol* 137:15, 1986
3. Buyon JP, Hopkins P, Abramson SB, Winchester R: Differential participation of epitopes on alpha and beta chains of the LFA family in neutrophil aggregation as defined by workshop monoclonal antibodies (MoAbs), in McMichael AJ et al. (eds); *Leukocyte Typing III. White Cell Differentiation Antigens.* Oxford, Oxford University Press, pp 844–848, 1987.
4. Anderson DC, Schmalstieg FC, Finegold MJ, Hughes BJ, Rothlein R, Miller LJ, Kohl S, Tosi MF, Jacobs RL, Waldrop TC, Goldman AS, Shearer WT, Springer TA: The severe and moderate phenotypes of heritable Mac-1, LFA-1 deficiency: Their quantitive definition and relation to leukocyte dysfunction and clinical features. *J Infect Dis* 152:668, 1985.
5. Todd RF, III, Arnaout MA, Rosin RE, Crowley CA, Peters WA, Babior BM: Subcellular localization of the large subunit of Mol (Molα; formerly gp110), a surface glycoprotein associated with neutrophil adhesion. *J Clin Invest* 74:1280, 1984.
6. Jones D: Characterization of a new mobilizable Mac-1 (CD11/CD18) pool that co-localizes with gelatinase in human neutrophils, in Springer TA et al. (eds.), *Leukocyte Adhesion Molecules.* New York, Springer Verlag, ch 8, 1989.
7. Berger M, O'Shea J, Cross AS, Folks TM, Chused TM, Brown EJ, Frank MM: Human neutrophils increase expression of C3bi as well as C3b receptors upon activation. *J Clin Invest* 74:1566–1571, 1984.
8. Arnaout MA, Spits H, Terhorst C, Pitt J, Todd, RF III: Deficiency of a leukocyte surface glycoprotein (LFA-1) in two patients with Mol deficiency: Effects of cell activation on Mol/LFA-1 surface expression in normal and deficient leukocytes. *J Clin Invest* 74:1291–1300, 1984.
9. Miller LJ, Bainton DF, Borregaard N, Springer TA: Stimulated mobilization of monocyte Mac-1 and p150,95 adhesion proteins from an intracellular vesicular compartment to the cell surface. *J Clin Invest* 80:535–544, 1987.
10. O'Shea JJ, Brown EJ, Seligmann BE, Metcalf JA, Frank MM, Gallin JI: Evidence for distinct intracellular pools of receptors for C3b and C3bi in human neutrophil. *J Immunol* 134:2580–2587, 1984.
11. Buyon JP, Abramson SB, Philips MR, Slade SG, Ross GD, Weissmann G, Winchester RJ: Dissociation between increased surface expression of Gp165/95 and homotypic neutrophil aggregation. *J Immunol* 140:3156–3160, 1988.
12. Eddy A, Newman SL, Cosio F, LeBien T, Michael A: The distribution of the CR3 receptor on human cells and tissue as revealed by a monoclonal antibody. *Clin Immunol Immunopathol* 31:371, 1984.
13. Bernstein ID, Self S: Joint report of the myeloid section of the Second International Workshop on Human Leukocyte Differentiation Antigens, in Reinherz EL, Haynes BF, Nadler LM, Bernstein ID (eds): *Leukocyte Typing II* Vol. 3. New York, Springer-Verlag pp 1–25, 1986.

14. Kaplan H, Edelson H, Korchak H, Given W, Weissmann G. Effects of non-steroidal anti-inflammatory agents on human neutrophil functions in vitro and in vivo. *Biochem Pharmacol* 33:371–378, 1984.

15. Philips MR, Buyon JP, Winchester RJ, Weissmann G, Abramson SB: Up-regulation of the iC3b Receptor (CR3) is neither necessary nor sufficient to promote neutrophil aggregation. *J Clin Invest* 82:495–501, 1988.

16. Goldstein IM, Lind S, Hoffstein S, Weissmann G: Influence of local anesthetics upon human polymorphonuclear leukocyte function in vitro: reduction of liposomal enzyme release and superoxide anion production. *J Exp Med* 146:483–494, 1977.

17. Korchak HM, Roos D, Giedd KN, Wynkoop EM, Vienne K, Rutherford LE, Buyon JP, Rich AM, Weissmann G: Granulocytes without degranulation: Neutrophil function in granule-depleted cytoplasts. *Proc Natl Acad Sci USA* 80:4968–4972, 1983.

18. Kishimoto TK, O'Connor K, Lee A, Roberts TM, Springer TA: Cloning of the β subunit of the leukocyte adhesion proteins: Homology to an extracellular matrix receptor defines a novel supergene family. *Cell* 48:681–690, 1987.

19. Corbi AL, Kishimoto TK, Miller LJ, Springer TA: The human leukocyte adhesion glycoprotein Mac-1 (complement receptor type 3, CD11b) α subunit, cloning, primary structure, and relation to the integrins, Von Willebrand factor and factor B. *J Biol Chem* 263:12403–12411, 1988.

20. Buyon JP, Abramson SB, Phillips M, Slade SG, Weissmann G, Winchester RJ: Phosphorylation of the alpha and beta chains of the CD11/CD18 integrin family: constitutive and activated states. *Clin Res* 37:425A, 1989.

21. Tonks NK, Diltz CD, Fischer EH: Characterization of the major protein-tyrosine-phosphatases of human placenta. *J Biol Chem* 263:6722, 1988

Part 2
Adhesion in Disease and Therapy

5

Mechanisms of Neutrophil Emigration*

C.M. Doerschuk, R.K. Winn, and J.M. Harlan

Introduction

Neutrophils adhere to endothelium by neutrophil-dependent or endothelial-dependent mechanisms (1). The neutrophil-dependent mechanism operates when the neutrophil is directly stimulated by activating agents such as bacterial chemotactic peptides, complement fragments, or tumor necrosis factor/cachectin (TNF). Neutrophil-dependent adherence is mediated by the interaction of neutrophil CD11b/CD18 (Mac-1, Mo1) and endothelial intercellular adhesion molecule-1 (ICAM-1) (2).

Endothelial-dependent mechanisms involve induced endothelial cell surface changes that promote neutrophil binding. Thrombin and leukotrienes C4 and D4 induce a rapid, protein synthesis-independent proadhesive alteration in the endothelial cell surface (3, 4). Thrombin-induced neutrophil adherence is not dependent on CD11/CD18 (5). Lipopolysaccharide (LPS) (6), interleukin-1 (IL-1) (7), and TNF (8) provoke synthesis of endothelial cell surface proteins that promote leukocyte binding. These proadhesive proteins have been designated endothelial-leukocyte adhesion molecules (E-LAMs) (9). Neutrophil adherence to E-LAMs is only partially inhibited by anti-CD18 monoclonal antibodies (MAb) (10).

From these in vitro studies, it appears that there may be a significant CD11/CD18-independent component of neutrophil adherence to endothelium when endothelial-dependent mechanisms are involved. However, there is at present no evidence that CD11/CD18-independent mechanisms of adherence are sufficient to promote neutrophil emigration in vivo. Infusion of anti-CD18 MAb prevents neutrophil emigration induced by IL-1, a protein synthesis-dependent response that is presumably

---

*This work was supported by MRC of Canada 4219 and USPHS HL 30542. Dr. Doerschuk is the recipient of an American Lung Association Fellowship and an Edward Livingston Trudeau Scholar Award. Dr. Harlan is an Established Investigator of the American Heart Association.

mediated by E-LAMs (11). Moreover, in leukocyte adhesion deficiency (LAD), the syndrome resulting from congenital deficiency of CD11/CD18 (12, 13), neutrophils fail to emigrate to sites of infection, despite the fact that these individuals likely generate stimuli that induce endothelial-dependent mechanisms of adherence (e.g., leukotrienes, thrombin, IL-1, and TNF).

The discrepancy between the in vitro studies that demonstrate CD11/CD18-independent neutrophil binding to endothelium and the in vivo studies that show only CD11b/CD18-dependent neutrophil emigration prompted us to search for conditions in which CD11/CD18-independent adherence leads to emigration in vivo. We now report that pneumonia induced in rabbits by *Streptococcus pneumoniae* induces neutrophil emigration into alveoli that is not inhibited by anti-CD18 MAb.

Methods

Reagents

Monoclonal antibody 60.3 is an IgG2a murine antibody directed to CD18 (14). It was purified by protein A-sepharose chromatography and suspended in sterile phosphate-buffered saline (PBS) at 4 mg/ml. Endotoxin levels in the preparation were <1 ng/ml by limulus amoebocyte lysate assay. *Streptococcus pneumoniae* from a clinical isolate were grown on blood agar plates. Organisms ($1-3 \times 10^9$/ml) were suspended in PBS and mixed with $1/10$ volume of colloidal carbon (India Ink, Pelican, Hanover, Germany) to identify their distribution in the inflammatory lesions. *Escherichia coli* from a clinical isolate were grown on blood agar plates and were suspended in PBS (10^9/ml) prior to use. Endotoxin (*E. coli* 055:B5, Sigma, St. Louis, MO) was prepared by diluting a stock solution (1 mg/ml) to 10 μg/ml with saline and mixed with $1/10$ volume of monasteral blue (Sigma, St. Louis, MO).

Induction of Inflammatory Foci

New Zealand white rabbits (1.7–2.2 kg) were anesthetized using intramuscular injection of ketamine hydrochloride (25–40 mg/kg) and acepromazine maleate (2–3 mg/kg) and a catheter was placed in the marginal ear vein for sampling of peripheral blood. Monoclonal antibody 60.3 (2 mg/kg) or saline was given intravenously prior to induction of inflammatory foci in the lung or the abdominal wall. After 30 minutes, the trachea was isolated and a narrow flexible tube was inserted through a small incision.

Protocol A

Under fluoroscopic guidance, *E. coli* endotoxin (0.1 ml/kg of 10 μg/ml solution containing monasteral blue) was instilled in 1 to 2 segments of the right lower lobe. *Streptococcus pneumoniae* (0.1 ml/kg of a solution containing $1-3 \times 10^9$ organisms/ml and colloidal carbon) were instilled through a second tube into 1 to 2 segments of the left lower lobe. The tube was removed and the neck incision was sutured.

Inflammatory foci were also induced in abdominal wall of the same animals. The skin of the ventral abdominal wall was anesthetized with subcutaneous injection of xylocaine. Polyvinyl sponges (1.5 × 1.5 × 0.3 cm) were infiltrated with *Strep. pneumoniae* organisms or *E. coli* endotoxin. The sponges were then placed within the subcutaneous fascial plane of the ventral abdominal wall through a midline incision. The incision was sutured and the animals were placed ventrally.

Blood samples for complete blood cell counts were taken before placement of the stimuli and at 0.5, 1, 2, 3, and 4 hours after induction of inflammatory foci.

Protocol B

In these experiments, 2 ml of a suspension containing 1×10^9 organisms/ml of *Strep. pneumoniae* or *E. coli* was instilled in the trachea above the carina to induce a bilateral pneumonia with a single organism. MAb 60.3 (2 mg/kg) or saline were injected intravenously 10 minutes prior to intratracheal instillation of organisms.

Quantitation of Neutrophil Infiltrates by Bronchoalveolar Lavage

The animals were studied 4 hours after induction of inflammatory foci. In rabbits receiving *Strep. pneumoniae* in left lung and *E. coli* endotoxin in right lung (Protocol A), the alveolar neutrophil infiltrate was evaluated by bronchoalveolar lavage (BAL) of the segment(s) containing the inflammatory stimuli. These animals were given heparin (200 U/kg intravenously prior to sacrifice by pentobarbital overdose. The inferior vena cava was cut and the thoracic organs were removed. Each lung was separated from the mediastinal structures and the blood was allowed to drain. The pneumonic region identified by the black colloidal carbon or monasteral blue staining was lavaged by inserting a catheter into the bronchus and instilling PBS containing 5% acid citrate dextrose until the region was fully distended. The lavage fluid was aspirated and the procedure repeated twice. The lavage fluids were pooled and the volumes were measured. Cell counts and differentials were determined using a hemocytometer and cytospins.

In animals receiving intratracheal *Strep. pneumoniae* or *E. coli* organisms (Protocol B) the alveolar neutrophil infiltrate was evaluated by BAL of the entire left lung. The animals were sacrificed by intravenous pentobarbital overdose. The thoracic cavity was opened and the right hilum was clamped. The left lung was then lavaged four times with 10 ml of PBS. The lavage fluid was analyzed for cell counts and differential.

Morphometric Quantitation of Neutrophil Infiltrates

In Protocol A, the neutrophil infiltrates in the lung and the abdominal wall were also evaluated using morphometric techniques. In these animals, the heart was tied off and the lungs fixed in situ by intratracheal instillation of 5% glutaraldehyde in phosphate buffer (pH 7.2). In all animals, the entire ventral abdominal wall was removed and plated in 10% formalin.

The pneumonic regions and the inflamed areas of the abdominal wall were identified by colloidal carbon or monasteral blue staining and tissue blocks were taken from these foci. Paraffin-embedded 8 μm sections were prepared and stained with hematoxylin and eosin. Randomly selected fields were examined by an observer without knowledge of treatment.

In the pneumonias, the number of neutrophils and red blood cells was counted in randomly selected alveoli that contained colloidal carbon or monasteral blue microscopically. The number of neutrophils per 100 alveoli and the ratio of neutrophils to red blood cells was calculated and the mean $\pm$ SE was determined for each stimulus in each group. The groups were compared using a nonparametric analysis of variance.

In the stained tissues of the abdominal wall that surrounded the sponges, the number of neutrophils per field at 500 $\times$ magnification (diameter = 370 μm, area = 0.108 mm²) were counted in five randomly selected fields of connective tissue that were located within 800 μm of the stained surface where the sponge had been placed. The number of neutrophils per 0.108 mm² tissue was calculated for each type of stimulus and the three experimental groups were compared using nonparametric analysis of variance.

Results

Protocol A

In Protocol A, inflammatory foci were induced simultaneously in the lung by intrabronchial instillation of *Strep. pneumoniae* organisms in the left lung and *E. coli* endotoxin in the right lung and in the abdominal wall by implantation of sponges soaked with *Strep. pneumoniae* organisms and sponges soaked with *E. coli* endotoxin. Animals pretreated

TABLE 5.1. Effect of MAb 60.3 on White Blood Cell Counts in Infected Animals

	Cell Count ($\times 10^6$/ml)	
Time (hrs)	Saline	MAb 60.3
-0.5	5.1 ± 0.4	6.4 ± 1.4
0	4.8 ± 0.5	5.5 ± 1.2
0.5	4.5 ± 0.9	6.7 ± 0.4
1	4.2 ± 0.5	5.8 ± 0.7
2	4.0 ± 0.4	9.0 ± 1.5
3	5.0 ± 1.0	13.7 ± 5.5
4	4.7 ± 0.9	11.9 ± 3.2

Inflammatory foci were induced simultaneously in the lung and the abdominal wall (Protocol A) of three rabbits pretreated with MAb 60.3 (2 mg/kg) or six rabbits pretreated with saline. Total white blood cell counts were determined one-half hour prior to (−0.5) and for 4 hours after induction of the inflammatory foci.

TABLE 5.2. Effect of MAb 60.3 on Neutrophil Emigration into Alveoli Induced by *Strep. Pneumoniae* vs *E. Coli* Endotoxin

	Neutrophils in BAL ($\times 10^6$)	
	E. coli Endotoxin	*Strep. pneumoniae*
Saline	38.6	34.6
MAb 60.3	0.8	29.5

Inflammatory foci were induced simultaneously in the lung and abdominal wall (Protocol A) of a rabbit pretreated with MAb 60.3 (2 mg/kg) or saline. After 4 hours, cell counts and differential were determined on bronchoalveolar lavage (BAL).

with MAb 60.3 (2 mg/kg) developed a leukocytosis after 2 hours, whereas the circulating white blood cell count remained at baseline in the saline-treated animals (Table 5.1). Pretreatment with MAb 60.3 markedly reduced neutrophil emigration into alveoli induced by the intrabronchial instillation of *E. coli* endotoxin in the right lung, but had no effect on neutrophil accumulation in alveoli following intrabronchial instillation of *Strep. pneumoniae* in the left lung (Table 5.2 and Table 5.3). Monoclonal antibody 60.3 prevented neutrophil emigration into both the *E. coli* endotoxin- and the *Strep. pneumoniae* -soaked sponges in the abdominal wall (Table 5.3).

Protocol B

In Protocol B, live *E. coli* or *Strep. pneumoniae* organisms were instilled in the trachea. Pretreatment with MAb 60.3 inhibited neutrophil emigration into alveoli induced by *E. coli* by 78%, but had no significant effect on neutrophil emigration induced by *Strep. pneumoniae* (Table 5.4).

TABLE 5.3. Effect of MAb 60.3 on Neutrophil Accumulation in the Lung and the Abdominal Wall

Pretreatment	E. coli Endotoxin	Strep. pneumoniae
A. Lung (Neutrophils/100 alveoli)		
Saline	547 ± 129	584 ± 11
MAb 60.3	127 ± 45	592 ± 45
B. Abdominal wall (Neutrophils/HPF*)		
Saline	172 ± 10	204 ± 10
MAb 60.3	3 ± 1	4 ± 1

Inflammatory foci were induced simultaneously in the lungs and the abdominal wall (Protocol A) of two rabbits pretreated with MAb 60.3 (2 mg/kg) and two animals pretreated with saline. After 4 hours, neutrophil accumulation was determined by morphometry as described in Methods. Values are means ± SD of two animals.
*High-power field.

TABLE 5.4. Effect of MAb 60.3 on Neutrophil Emigration into Alveoli Induced by Strep. pneumoniae vs. E. coli

	Neutrophils in BAL ($\times$ 10^6)	
	E. coli	Strep. pneumoniae
Saline	42.0 ± 20.0	8.7 ± 3.7
MAb 60.3	9.3 ± 6.2*	5.9 ± 2.3**

Escherichia coli or Strep. pneumoniae were instilled in the trachea (Protocol B) of rabbits pretreated with MAb 60.3 (2 mg/kg) or saline. After 4 hours cell counts and differential were determined on bronochoalveolar lavage (BAL). Values represent means + SD of three rabbits in each group.
*$p = 0.05$ vs. saline-treated animals.
**not significant vs. saline-treated animals.

Conclusion

Pretreatment with the anti-CD18, MAb 60.3 markedly reduced neutrophil emigration into alveoli in response to E. coli endotoxin and E. coli organisms and into the abdominal wall in response to Strep. pneumoniae or E. coli endotoxin-soaked sponges. The potent inhibitory effect of MAb 60.3 on neutrophil emigration induced by these stimuli is similar to that reported previously for chemotactic peptides and leukotriene B4 (15), interleukin-1 (11), and endotoxin (16). The leukocytosis observed in the animals treated with MAb 60.3 was also noted by Price et al. (16), and mimics the response of LAD patients to infection (12, 13).

Surprisingly, MAb 60.3 did not significantly inhibit neutrophil emigration into alveoli induced by Strep. pneumoniae organisms. The fact that MAb 60.3 significantly reduced emigration induced by E. coli organisms argues against a unique effect of live organisms vs. peptide or lipid mediators as a stimulus to emigration. The failure of MAb 60.3 to prevent emigration into alveoli induced by *Strep. pneumoniae* is not likely due to incomplete inhibition of CD11/CD18 for several reasons. First, the dose of MAb 60.3 was sufficient to inhibit emigration in response to *E. coli* endotoxin or *E. coli* organisms, stimuli, which in the saline-treated animals, produced neutrophil accumulation in alveoli quantitatively similar to (endotoxin) or greater than (*E. coli*) that observed with *Strep. pneumoniae*. Second, we previously found that repeated doses of MAb 60.3 failed to inhibit emigration into alveoli induced by *Strep. pneumoniae* in the lung, even though saturating plasma levels of MAb were maintained throughout the experiment. In those experiments, plasma levels of MAb were proven to be saturating on analysis by flow cytometry by demonstrating that binding of MAb to blood neutrophils did not increase when additional MAb was infused. Finally, neutrophils removed by bronchoalveolar lavage of *Strep. pneumoniae* lungs were also analyzed by flow cytometry. Lavage neutrophils that had emigrated into alveoli exhibited binding (saturating) of MAb 60.3 identical to that of blood neutrophils isolated simultaneously.

We conclude from these studies that under some circumstances neutrophils can emigrate to extravascular tissue by a CD11/CD18-independent mechanism and that this response is stimulus-specific. The observation that the anti-CD18 MAb prevented *Strep. pneumoniae*-induced emigration in the abdominal wall, but not in the lung, might indicate an intrinsic difference between the mechanisms of neutrophil binding to pulmonary vs. systemic endothelium. Alternatively, accessory cells present in the lung (e.g., alveolar macrophages) might initiate a CD11/CD18-independent mechanism of adherence and emigration when stimulated by *Strep. pneumoniae*. In either case, the pulmonary response to *Strep. pneumoniae* appears to be qualitatively different from the response to *E. coli* organisms or *E. coli* endotoxin. Additional studies are required to determine whether stimuli other than *Strep. pneumoniae* can provoke CD11/CD18-independent neutrophil emigration in the lung or in other vascular beds.

References

1. Schwartz BR, Harlan JM: Mechanisms of leukocyte adherence to endothelium, in Ryan US (ed): *Endothelial Cells, Vol. II.* Boca Raton, FL, CRC Press, p. 213–226, 1988.
2. Smith CW, Rothlein R, Hughes B, Mariscalco M, Schmalstieg F, Anderson DC: Identification of an endothelial determinant for CD18-dependent neutrophil adherence. *FASEB J* 2:A1237, 1988.
3. Zimmerman GA, McIntyre TM, Prescott SM: Thrombin stimulates the adherence of neutrophils to human endothelial cells in vitro. *J Clin Invest* 76:2235, 1985.

4. McIntyre TM, Zimmerman GA, Prescott SM: Leukotrienes C4 and D4 stimulate human endothelial cells to synthesize platelet-activating factor and bind neutrophils. *Proc Natl Acad Sci USA* 83:2204, 1986.

5. Zimmerman GA, McIntyre TM: Neutrophil adherence to human endothelium in vitro occurs by CDw18 (Mol, Mac-1/LFA-1/GP150,95) glycoprotein-dependent and -independent mechanisms. *J Clin Invest* 81:531, 1988.

6. Schleimer, RP, Rutledge BK: Cultured human vascular endothelial cells acquire adhesiveness for neutrophils after stimulation with interleukin-1, endotoxin, and tumor-producing phorbol diesters. *J Immunol* 136:649, 1986.

7. Bevilacqua MP, Pober JS, Wheeler ME, Mendrick D, Cotran RS, Gimbrone MA, Jr: Interleukin-1 acts on cultured human vascular endothelial cells to increase the adhesion of polymorphonuclear leukocytes, monocytes, and related leukocyte cell lines. *J Clin Invest* 76:2003, 1985.

8. Gamble JR, Harlan JM, Klebanoff SJ, Lopez AF, Vadas MA: Stimulation of the adherence of neutrophils to umbilical vein endothelium by human recombinant tumor necrosis factor. *Proc Natl Acad Sci USA* 82:8667, 1985.

9. Bevilacqua MP, Pober JS, Mendrick DL, Cotran RS, Gimbrone MA, Jr: Identification of an inducible endothelial-leukocyte adhesion molecule. *Proc Natl Acad Sci USA* 84:9238, 1987.

10. Pohlman TH, Stanness KA, Beatty PG, Ochs HD, Harlan JM: An endothelial cell surface factor(s) induced in vitro by lipopolysaccharide, interleukin-1, and tumor necrosis factor-alpha increases neutrophil adherence by CDw18-dependent mechanism. *J Immunol* 136:4548, 1986.

11. Rampart M, Williams TJ: Evidence that neutrophil accumulation induced by interleukin-1 requires both local protein biosynthesis and neutrophil CD18 antigen expression in vivo. *Br J Pharmacol* 94:1143, 1988.

12. Anderson DC, Springer TA: Leukocyte adhesion deficiency: An inherited defect in the Mac-1, LFA-1, and p150,95 glycoproteins. *Ann Rev Med* 38:175, 1987.

13. Todd, RF, Freyer DR: The CD11/CD18 leukocyte glycoprotein deficiency. *Hematol Oncol Clin N Am* 2:13, 1988.

14. Beatty PG, Ledbetter JA, Martin PG, Price TH, Hansen JA: Definition of a common leukocyte cell-surface antigen (Lp95-150) associated with diverse cell-mediated immune functions. *J Immunol* 131:2913, 1983.

15. Arfors K-E, Lundberg C, Lindbom L, Lundberg K., Beatty PG, Harlan JM: A monoclonal antibody to the membrane glycoprotein complex CD18 inhibits polymorphonuclear leukocyte accumulation and plasma protein leakage in vivo. *Blood* 69:338, 1987.

16. Price TH, Beatty PG, Corpuz SR: In vivo inhibition of neutrophil function in the rabbit using monoclonal antibody to CD18. *J Immunol* 12:4174, 1987.

6

Leukocyte Adhesion Deficiency as a Model for the Study of the Functional Role of LFA-1

A. Fischer, B. Lisowska-Grospierre,
F. Mazerolles, F. Le Deist, N. Perez,
M.T. Dimanche-Boitrel, and C. Griscelli

Introduction

In the past years, a new inherited immunodeficiency has been identified and been named the *leukocyte adhesion deficiency* (LAD) (1). Leukocyte adhesion deficiency is characterized by a defective expression of three related leukocyte adhesion molecules, i.e., LFA-1, Mac-1, and p150,95 which are heterodimers, the α chain being specific and the β chain common to the three molecules.

The basic defect has been shown to reside in the β subunit synthesis leading to either no β chain synthesis or the synthesis of an abnormal subunit (2–5). The main functional consequences of LAD are related to a defective in vivo adhesion of phagocytic cells to endothelial cells and to the extracellular matrix leading to uncontrolled bacterial infections (6).

In this chapter, we review some new data dealing with the heterogeneity of the structural defect, with the dysfunctions of LFA-1⁻ T lymphocytes and with the therapy of LAD.

Heterogeneity of the Structural Defect Underlying Leukocyte Adhesion Deficiency

Anderson et al. (1) and Kishimoto et al. (7) have shown that there are two types of the LAD, one known as the *severe phenotype* characterized by a complete lack of α-β dimer expression and the second known as *the moderate phenotype* characterized by low but detectable α-β expression.

Further heterogeneity of LAD has been found by biosynthetic studies of LFA-1 and α and β chains and β chain RNA expression in PHA-induced blasts and in EBV-transformed B cells from LAD patients (8). We found in the patients we have studied three different phenotypes:

1. No β chain mRNA, no β chain precursor.
2. Low amounts of β chain mRNA with no detectable β chain precursor.
3. Normal β subunit mRNA level of normal size associated with detectable β precursor synthesis and extremely low α-β complex membrane expression (8).

The β chain precursor detected in the cell lysates from patient type 3 was not glycosylated and was no longer detected after chase indicating rapid degradation. Interestingly, β chain-specific mRNA transcription could be enhanced in this case by reagents such as PMA or γ interferon, but this was not associated with increase in membrane α-β complex expression.

These data point to the degree of heterogeneity of the disease. Further studies of sequencing of abnormal β chain gene may give new insights into the functional sites of the β chain.

T Cell Dysfunction in Leukocyte Adhesion Deficiency

A variety of T cell anomalies have been reported to occur in association with the severe phenotype of LAD (1, 9,–11). Defective T cell cytotoxicity is the most frequently encountered anomaly. In some cases, a T helper cell defect for antibody production has also been reported (12).

We have further studied the activation, helper function and adhesion of LFA-1⁻ T cells (13). In coculture experiments of LFA-1⁻ T cells with HLA identical LFA-1⁺ monocytes, it appears that antigen-specific T cell activation does occur, an observation that has also been made in the presence of LFA-1⁻ monocytes. In contrast, the mixture of either LFA-1⁻ T lymphocytes or monocytes with HLA identical LFA-1⁺ B lymphocytes gives rise to suboptimal in vitro specific antibody production to influenza virus. LFA-1⁻ B lymphocytes appear nonfunctional but such a result could be the consequence of defective in vivo priming as well.

Similar poor functions of monocytes and T lymphocytes in antibody production assay could be reproduced by preincubating purified LFA-1⁺ cells with saturating concentrations of anti-LFA-1 antibodies (25/3 or TS1/22) (12, 13). B cell function is not directly inhibited by these antibodies. In a mirror image, anti-ICAM-1 antibody 87H10 is able to block in vitro antibody production to influenza virus at the monocyte and B cell levels.

The adhesion of LFA-1⁻ T lymphocytes to B lymphocytes is impaired as judged by measurements of cell conjugate formation (11). This impairment is not due to a delay in attachment as shown by kinetic studies of conjugate formation. In contrast, there is no difference in binding of LFA-1⁻ or LFA-1⁺ B lymphocytes to T lymphocytes.

This is in correlation with the apparent unidirectional functional interaction between LFA-1 on T lymphocytes and ICAM-1 (+ x) on B lymphocytes. Since resting T lymphocytes poorly express ICAM-1, such results may indicate that this interaction is required at the onset of T-B cell interaction while LFA-1 molecules on B cells do not encounter ligands on helper T cells.

The residual binding capacity of LFA-1⁻ T lymphocytes can be abrogated by anti-CD2 antibodies, an observation that confirms the existence of two adhesive pathways for T cells and indicates that CD2-mediated adhesion is independent of the expression of LFA-1 molecules.

Interestingly, activated LFA-1⁻ T lymphocytes (6 day-cultured T cells in the presence of ionomycine, PMA and IL2) can normally bind B cells in contrast to resting T cells. Activated LFA-1⁻ T cells can also efficiently bind LFA-1⁻ B cells. Such adhesion capacity is only partially blocked by anti-CD2 antibodies (60% inhibition) while anti-LFA-1 and anti-ICAM-1 antibodies exert no inhibitory activity. Activated LFA-1⁻ T cells express comparable amounts of CD2 to activated LFA-1⁺ T cells while LFA-1 is still not expressed (13).

These results led to several conclusions:

1. Defective adhesion of LFA-1⁻ T cells to B cells and other target cells that is observed for CD4+ and CD8+ T cells as well may account for defective cytotoxic and cognate helper functions.
2. The conditions of T cell binding to target cells are more stringent than those of T cell binding to antigen presenting cells.
3. The existence of a second T cell adhesion pathway and the LFA-1 independent adhesion ability of activated T cells can underline the limited in vivo consequences of poor LFA-1 expression by T cells.

In addition,

4. Since activated LFA-1⁻ T cells coated with anti-CD2 antibody can still bind B cells, such cells might use a third adhesion pathway.
5. The ICAM-1 molecule does not appear to be a ligand for other molecules on the T cell surface than LFA-1 (14).

Therapy of Leukocyte Adhesion Deficiency

About 80% of the 24 known patients affected with the severe type of LAD have died before the age of 2 years because of infections (9).

Allogeneic bone-marrow transplantation (BMT) has thus been attempted to cure the disease. We have performed HLA identical BMT in three patients. Engraftment occurred following aggressive conditioning regimen. In all three, normalization of leukocyte functions was observed. One died 1 year posttransplant accidentally and the last is doing well now 7 years posttransplant with a stable mixed chimerism (15). We have also attempted to perform HLA partially nonidentical BMT in three other patients (16). They received myeloablative and immunosuppressive chemotherapy prior to infusion of a T cell depleted bone marrow. In the three cases, engraftment occurred with complete (in two) or partial (in one) correction of both granulocytic and lymphocytic function. A full chimerism has been achieved in one, mixed in two. The three patients are alive and well 4, 3, and 2½ years, respectively, posttransplant.

We were struck by the successful and stable engraftment of T cell depleted, HLA incompatible marrow in these patients since this has been only observed previously in patients with severe combined immunodeficiency.

Because of these results and because of the known in vivo (17) and in vitro inhibitory effects of anti LFA-1 antibodies (18) we decided to utilize a monoclonal anti LFA-1 antibody to try to prevent graft failure in patients receiving HLA partially incompatible BMT (19).

The antibody (25/3, mouse anti α-subunit [20] was first infused at a dose of 0.1 mg/kg every other day (five doses). Such a schedule gave rise to low serum concentrations of antibody (<0.4 μg/ml) possibly not high enough to exert full inhibitory activity. Therefore, the antibody is now infused at a dose of 0.2 mg/kg on 10 consecutive days between day -3 to $+6$ of transplantation. Serum trough levels reach a mean of 2 μg/ml (21). Antibody infusions are well tolerated and did not induce immunization.

Fifty-one children have received an HLA nonidentical BMT with the use of the anti-LFA-1 antibody. The bone marrow was T-cell depleted in all occasions. Underlying diseases were immunodeficiencies (ID) (SCID excluded) in 20, metabolic diseases in 12, Fanconi anemia in 4 and leukemia in 15. The overall rate of engraftment is 73% and the actuarial survival with a functional graft 51% (mean follow up 16 months). Results are similar in the different group of patients except for those with osteopetrosis where the rate engraftment is poorer and for those with Fanconi anemia who are at risk for graft vs. host disease.

Although a number of failures occur, the use of the anti-LFA-1 antibody has significantly reduced the rate of graft failure (from 90–25% in the group of patients with ID). Simultaneously, Ferrara et al. have shown that the in vivo infusion of an anti-LFA-1 antibody in the mouse prevent, in most of the cases (87%), the rejection of fully H-2 incompatible bone marrow (22).

The in vivo mechanism of action of the antibody is not known. It does not kill leukocytes as no alteration in leukocyte cell counts is observed under therapy (21). In contrast to in vitro observations, there is a partial modulation of LFA-1 expression on lymphocyte surfaces (21). We speculate that in addition to modulation, the antibody acts through functional blocking of residual host T cells.

According to in vitro data (see paragraph 2), one would not expect an in vivo therapy with an anti-LFA-1 antibody to suppress an ongoing immune reaction since pre-activated T cells bind target cells independently of LFA-1 mediated adhesion. This indicates that anti-LFA-1 antibody in vivo can only be used for prevention of unwanted immune responses.

References

1. Anderson DC, Springer TA: Leukocyte adhesion deficiency: an inherited defect in the Mac-1, LFA-1 and p150,95 glycoproteins. *Ann Rev Immunol* 38:176–194, 1987.
2. Springer TA, Thompson WS, Miller LJ, Schmalstieg FC, Anderson DC: Inherited deficiency of the Mac-1, LFA-1, p150,95 glycoprotein family and its molecular basis. *J Exp Med* 160:1901–1908, 1984.
3. Dana N, Clayton LK, Tennen DG, Pierce MW, Lachmann PJ, Law SA, Arnaout MA: Leukocytes from four patients with complete or partial leu-Cam deficiency contain the common beta subunit precursor and beta subunit messenger RNA. *J Clin Invest* 79:1010–1015, 1987.
4. Lisowska-Grospierre B, Bohler MC, Fischer A, Mawas C, Springer TA, Griscelli C: Defective membrane expression of the LFA-1 complex may be secondary to the absence of the β-chain in a child with recurrent bacterial infection. *Eur J Immunol* 16:205–208, 1986.
5. Dimanche MT, Le Deist F, Fischer A, Arnaout MA, Griscelli C, Lisowska-Grospierre B: LFA-1 β chain synthesis and degradation in patients with leukocyte-adhesive proteins deficiency. *Eur J Immunol* 17:417–419, 1987.
6. Anderson DC, Schmalstieg FC, Fineglod MJ, Hughes BJ, Rothlein R, Miller LJ, Kohl S, Tsoi MF, Jacobs RL, Waldrop TC, Goldman AS, Shearer WT, Springer TA: The severe and moderate phenotypes of heritable Mac-1, LFA-1, deficiency: Their quantitative definition and relation to leukocyte dysfunction and clinical features. *J Inf Dis* 152:668–689, 1985.
7. Kishimoto TK, Hollander N, Roberts TM, Anderson DC, Springer TA: Heterogeneous mutations in the β subunit common to the LFA-1, Mac-1 and p150,95 glycoproteins cause leukocyte adhesion deficiency. *Cell* 50:193–202, 1987.
8. Dimanche-Boitrel MT, Guyot A, de Saint Basile G, Fischer A, Griscelli C, Lisowska-Grospierre B: Heterogeneity in the molecular defect leading to the leukocyte adhesion deficiency. *Eur J Immunol* 18:1575–1579, 1988.
9. Fischer A, Lisowska-Grospierre B, Anderson DC, Springer TA: Leukocyte adhesion deficiency: Molecular basis and functional consequences. *Immunodef Rev.* 1:39–54, 1988.
10. Kohl S, Loo LS, Schmalstieg FC, Anderson DC: The genetic deficiency of leukocyte surface glycoprotein Mac-1, LFA-1, p150,95 in humans is associated with defective antibody-dependent cellular cytotoxicity in vitro and defective protection against herpes simplex virus infection in vivo. *J Immunol* 197:1688–1694, 1986.
11. Mentzer SJ, Bierer BE, Anderson DC, Springer TA, Burakoff SJ: Abnormal cytolytic activity of LFA-1 deficient human cytolytic T lymphocyte clones. *J Clin Invest* 78:1387–1391, 1986.

12. Fischer A, Durandy A, Sterkers G, Griscelli C: Role of LFA-1 in antigen-specific helper T lymphocyte-B lymphocyte interaction. *J Immunol* 136:3198–3201, 1986.
13. Mazerolles F, Lumbroso C, Lecomte O, Le Deist F, Fischer A: The role of LFA-1 in the adherence of T lymphocytes to B lymphocytes. *Eur J Immunol* 18:1229–1234, 1988.
14. Marlin SD, Springer TA: Purified intercellular adhesion molecule-1 (ICAM-1) is a ligand for LFA-1. *Cell* 51:813–820, 1987.
15. Fischer A, Pham HT, Descamps B, Lisowska-Grospierre B, Durandy A, Virelizier JL, Gerota I, Perez N, Sceinmetzler C, Griscelli C: Bone marrow transplantation for inborn error of phagocytic cell associated with defective adherence, chemotaxis and oxidative response during opsonized particle phagocytosis. *Lancet* 2:473–473, 1983.
16. Le Deist F, Blanche S, Keable H, Gaud C, Pham H, Descamps B, Wahn V, Griscelli C, Fischer A: Successful HLA non identical bone marrow transplantation in three patients with LAD. *Blood*, In press.
17. Heagy W, Waltenbaugh C, Martz E: Potent ability of anti LFA-1 monoclonal antibody to prolong allograft survival. *Transplantation* 37:520–523, 1984.
18. Hildreth DEK, Gotch FM, Hildreth PDK, McMichael AJ: A human lymphocyte function associated antigen involved in cell-mediated lympholysis. *Eur J Immunol* 13:202–206, 1983.
19. Fischer A, Griscelli C, Blanche S, Veber F, Le Deist F, Lopez M, Delaage M, Olive D, Mawas C, Janossy G: Prevention of graft failure by an anti LFA-1 monoclonal antibody in HLA mismatched bone marrow transplantation. *Lancet* 2:1058–1061, 1986.
20. Olive D, Charmot D, Dubreuil P, Mawas C: Human lymphocyte functional antigens, in Feldmann M (ed): *Human T Cell Clones. A New Approach to Immune Regulation.* Clifton, NJ Humana Press, pp 173–187, 1986.
21. Perez N, Le Deist F, Fischer A: In vivo infusion of anti LFA-1 antibody. Pharmakinetic studies, *Bone Marrow Transplantation*, In Press.
22. Ferrara J, Down J, Van Dijken P, Mauch P, Burakoff S: Anti LFA-1 in vivo improves survival after T-cell depleted BMT. *J Cell Biochem* suppl.12C:99 (Abstract 207), 1988.

7

Role of Cellular Adhesion in Inflammatory Cutaneous Disorders

GUNHILD LANGE VEJLSGAARD, NILS LAUGE HANSEN, ELISABETH RALFKIAER, AND ROBERT ROTHLEIN

Introduction

The efforts to understand the physiology of cell adhesion has revealed new information in the inflammatory response. One finding which is important in the understanding of cell adhesion was the identification of the ligand for LFA-1, intercellular adhesion molecule-1 (ICAM-1) (1,2).

It has been shown that ICAM-1 expression on different cell lines is influenced by different cytokines (3–6). Interleukin-1 (IL-1) or gamma-interferon (IFN-gamma) increased ICAM-1 expression on dermal fibro-blasts four to fivefold. Cultured human endothelial cells also have increased expression of ICAM-1 after incubation with IL-1, IFN-gamma, tumor necrosis factor (TNF) or lymphotoxin (LT). ICAM-1 induction is the only known antigen-expression system on endothelial cells in which IL-1 and INF-gamma have been shown to have the same influence.

In normal skin the keratinocytes do not express ICAM-1, but cultured human epidermal keratinocytes show up to 50,000 ICAM-1 monoclonal antibody binding sites per cell (6). Incubation with IFN-gamma or TNF increased the ICAM-1 expression while IL-1 and IFN-beta did not. The induced ICAM-1 expression requires both mRNA and protein synthesis. An inhibitor of protein synthesis or an inhibitor of mRNA synthesis, cycloheximide and actinomycid D respectively, abolish the effect of IL-1 and IFN-gamma on ICAM-1 expression on fibroblasts (3).

The recognition of ICAM-1 expression on cells involved in immune and inflammatory reactions and the influence of the different inflammatory mediators mentioned, strongly suggest that ICAM-1 is involved in inflammatory and immunological responses (7,8). Cryostat sections of various skin lesions, all with inflammation as a major component, were studied for the expression of ICAM-1 (9,10).

Results and Conclusions

The difference in ICAM-1 expression in allergic and toxic related inflammation is interesting (Table 7.1). Hapten-containing patches were applied on individuals known to be sensitive towards the particular antigen and in addition toxic patches with an irritant chemical were applied to normal volunteers. The patches were left for 48 hours and then removed. At 48 hours, the clinical manifestations were at a maximum for both types of reactions, and histologically all had a heavy mononuclear cell infiltrate in the dermis. ICAM-1 expression showed great differences. All patients with the allergic patches showed ICAM-1 expression on the keratinocytes, the endothelial cells and on a proportion of the mononuclear cells in the dermal infiltrate. The first ICAM-1 expression on keratinocytes was found as early as 4 hours after the introduction of the patch. After 24 hours, seven out of eight patients had ICAM-1 positive keratinocytes and at 48 hours eight out of eight patients were positive. In contrast, the first expression of HLA-DR on the keratinocytes were seen at 48 hours in three out of eight patients. Thus there was an earlier and more pronounced expression of ICAM-1 compared to the expression of HLA-DR on the keratinocytes.

In the toxic patch test only 1 out of 14 biopsies showed ICAM-1 positive keratinocytes at 48 hours and the expression of HLA-DR was totally absent. These finding suggest that ICAM-1 plays a role in the specific immune response in the allergic reactions while the role of ICAM-1 in the non specific inflammatory response in the irritant contact reactions is absent. Normal skin with non reactive control patches showed no ICAM-1 expression on the keratinocytes, while scattered expression on the endothelial cells were demonstrated, the same was observed in biopsies from normal skin (9).

Biopsies from allergic contact eczema, lichen planus, pemphigoid and pemphigus, all benign cutaneous disorders characterized by more chronic

TABLE 7.1. Expression of ICAM-1 on Keratinocytes in Allergic and Toxic Patch Test

Hours after Administration	Number of Biopsies with Keratinocytes Expressing ICAM-1 (no. of patients)	
	Allergic Patch Test	Toxic Patch Test
4	3(6)	0(4)
8	3(9)	1(3)
24	7(8)	1(3)
48[a]	8(8)	1(14)
72	7(8)	1(3)

[a]Patch was removed at 48 hours.

inflammation in the dermis, showed without exception, ICAM-1 expression on the keratinocytes. Cells in the mononuclear cell infiltrate also expressed ICAM-1, and at sites with heavy mononuclear cell infiltration in lichen planus the keratinocytes of the overlying epidermis showed pronounced ICAM-1 expression. A high percentage of the lichen planus biopsies coexpressed HLA-DR (9).

In psoriasis, another cutaneous skin disease with inflammation as a major component, the ICAM-1 expression on the keratinocytes were studied under photochemical treatment. Keratinocytes from biopsies of normal skin of the patients stayed ICAM-1 negative during treatment, indicating that the treatment by itself did not induce ICAM-1 expression. All biopsies from psoriatic lesions before treatment had ICAM-1 positive keratinocytes. In addition scattered ICAM-1 positive cells were seen in the dense mononuclear cell infiltrate of the dermis. During treatment a strong correlation between the severity of the lesion and ICAM-1 expression was seen. Patients going into remission showed no ICAM-1 expression on the keratinocytes at the site of the former psoriatic plaque and in the case of relapse, ICAM-1 expression was again present (Table 7.2). The HLA-DR expression on the keratinocytes was variable, but no HLA-DR positive biopsies were ICAM-1 negative (10).

In the previously mentioned benign cutaneous inflammatory disorders, there was a strong correlation between the severity of the clinical manifestations and the ICAM-1 expression on the keratinocytes. Furthermore there was a strong correlation between the density of the mononuclear cell infiltrate in the dermis and the ICAM-1 expression on the keratinocytes. Expression of HLA-DR on keratinocytes was only seen in connection with severe lesions (9,10).

These consistent findings on benign disorders are not comparable to the result on cutaneous lymphomas. Cutaneous T cell lymphomas are also characterized by a heavy mononuclear cell infiltrate. However the ICAM-1 expression on keratinocytes was less frequent while the expression of HLA-DR on the keratinocytes was more frequent than in benign disorders. In addition there was a tendency of keratinocytes expressing

TABLE 7.2. Expression of ICAM-1 on Keratinocytes in Psoriatic Plaques During PUVA Treatment, Representative Example in One Patient.

Time after Treatment	Psoriatic Skin	Normal Skin
start	+ +	−
1 week	+	−
2 weeks	+	−
6 weeks	−	−
7 weeks	+ +[a]	−
10 weeks	−	−

[a]Clinical relapse.

ICAM-1 in the early stages of cutaneous T cell lymphomas to lose this expression in the more advanced stages of this disease (9).

In conclusion the expression of ICAM-1 on the keratinocytes of the epidermis and the expression of ICAM-1 on the mononuclear cells in the dermal infiltrate is highly correlated to the severity of the clinical manifestation and also to the histological changes when dealing with allergic reactions and benign cutaneous disorders. However in irritant reactions on the skin, there is no ICAM-1 expression on the keratinocytes whereas scattered mononuclear cells in dermal infiltrate do express ICAM-1. In cutaneous T cell lymphomas there is a tendency that the keratinocytes lose their ICAM-1 expression when the disease progresses to more advanced stages. Thus ICAM-1 expression on keratinocytes can be used as a diagnostic tool between the allergic and toxic skin reaction. Furthermore, the role of ICAM-1 in the inflammatory skin disorders does suggest that ICAM-1 antagonists could play a major role in the future treatment of a broad spectrum of cutaneous disorders. Studies are being done to map out the immunological role of ICAM-1 in the epidermal-dermal interaction with regard to future treatment modalities.

References

1. Rothlein R, Dustin ML, Marlin SD, Springer TA: A human intercellular adhesion molecule (ICAM-1) distinct from LFA-1. *J Immunol* 137:1270, 1986.
2. Marlin SD, Springer TA: Purified intercellular adhesion molecule-1 (ICAM-1) is a ligand for lymphocyte function-associated antigen 1 (LFA-1). *Cell* 51:813, 1988.
3. Dustin ML, Rothlein R, Bhan AK, Dinarello CA, Springer TA: Induction by IL-1 and interferon-gamma: Tissue distribution, biochemistry, and function of a natural adherence molecule (ICAM-1). *J Immunol* 137:245, 1986.
4. Pober JS, Gimbrone MA, Lapierre LA, Mendrick DL, Fiers W, Rothlein R, Springer TA: Overlapping patterns of activation of human endothelial cells by interleukin 1, tumor necrosis factor, and immune interferon. *J Immunol* 137:1893, 1986.
5. Pober JS, Lapierre LA, Stolpen AH, Brock TA, Springer TA, Fiers W, Bevilacqua MP, Mendrick DL, Gimbrone MA. Activation of cultured human endothelial cells by recombinant lymphotoxin: comparison with tumor necrosis factor and interleukin 1 species. *J Immunol* 138:3319, 1987.
6. Dustin ML, Singer KH, Tuck DT, Springer TA: Adhesion of T lymphoblasts to epidermal keratinocytes is regulated by interferon gamma and is mediated by intercellular adhesion molecule 1 (ICAM-1). *J Exp Med* 167:1323, 1988.
7. Dougherty GJ, Murdoch S, Hogg N: The function of human intercellular adhesion molecule-1 (ICAM-1) in the generation of an immune response. *Eur J Immunol* 18(1):35, 1988.
8. Sanders ME, Makgoba MW, Sharrow SO, Stephany D, Springer TA, Young HA, Shaw S: Human memory T lymphocytes express increase levels of three cell adhesion molecules (LFA-3, CD2, and LFA-1) and three other molecules (UCHL1, CDw29, and Pgp-1) and have enhanced IFN-gamma production. *J Immunol* 140:1401, 1988.

9. Lange Vejlsgaard G, Ralfkiaer E, Avnstorp C, Czatkowski M, Marlin SD, Rothlein R: Kinetics and characterization of intercellular adhesion molecule-1 (ICAM-1) expression on keratinocytes in various inflammatory skin lesions and malignant cutaneous lymphomas. *J Am Acad Dermatol* :782, 1989.
10. Lisby S, Ralfkiaer E, Rothlein R, Lange Vejlsgaard G: Intercellular adhesion molecule-1 (ICAM-1) expression correlated to inflammation. *Br J Dermatol* 120:479, 1989.

8

Characterization of a New Mobilizable Mac-1 (CD11b/CD18) Pool That Co-Localizes with Gelatinase in Human Neutrophils*

DOUGLAS H. JONES, FRANK C. SCHMALSTIEG,
HAL K. HAWKINS, BEAN L. BURR,
HELEN E. RUDLOFF, SHARON KRATER,
C. WAYNE SMITH, AND DONALD C. ANDERSON

Introduction

The CD11/CD18 complex consists of three glycoprotein heterodimers [Mac-1, LFA-1, and p150,95 (1–3)] that share identical β subunits non-covalently associated with immunologically distinct α subunits. Mac-1 (CD11b/CD18) is the receptor for iC3b opsonized particles (4–6) and plays a major role in adhesion and adhesion-dependent functions of human myeloid cells (4,7–9). In normal unstimulated neutrophils and monocytes, Mac-1 is present in intracellular compartments as well as on the cell surface (3,10–13). Exposure to chemotactic stimuli (C5a, f-Met-Leu-Phe [fMLP], or Leukotriene B$_4$) or secretagogues (phorbol myristate acetate [PMA] or calcium ionophore A23187) elicits a 5- to 10-fold increase in surface binding of monoclonal antibodies (MAbs) directed at Mac-1 (3,4,14,15) associated with a two to threefold increase in neutrophil adhesiveness (7,16). This surface protein "upregulation" is maximal within 10 minutes at 37 °C and is not impeded by protein or mRNA synthesis inhibitors, suggesting that Mac-1 is translocated from latent intracellular pools to the cell surface in response to inflammatory stimuli.

Increasing evidence indicates that granules or vesicles are intracellular compartments for Mac-1 and p150,95 in neutrophils or monocytes (8,10–

*Supported by National Institutes of Health Grants No. AI19031, AI23521, DE07875, and HL41408, the National Cystic Fibrosis Foundation, and by the US Department of Agriculture-Agricultural Research Service/Children's Nutrition and Research Center, Baylor College of Medicine and Texas Children's Hospital. Dr. Anderson was the recipient of a Research Career Development Award (KO4-AI/AM 00105) during the performance of these investigations.

13,16–19). Fluorescein-tagged MAb directed at the α subunit of Mac-1 demonstrate a diffuse granular fluorescent pattern in permeabilized neutrophils (12). Subcellular fractionation studies show Mac-1 to be present in fractions enriched for secondary or "tertiary" granules as well as in membrane fractions (8,11–13). Petrequin et al. (17) demonstrated correlation between extracellular release of gelatinase and stimulated surface expression of Mac-1, suggesting the possible existence of intracellular Mac-1 pools associated with gelatinase containing organelles (20). More recently, Lacal and co-workers (13) demonstrated co-sedimentation of gelatinase and Mac-1 (Mo-1) in subcellular fractions of neutrophils resolved on sucrose gradients.

Our studies provide evidence for a distinct and major intracellular pool of Mac-1, which co-sediments with gelatinase-rich electron dense granules and which is preferentially mobilized to the plasma membrane by chemotactic stimuli. As defined by density, this pool is distinct from a vitamin B_{12} binding protein and lactoferrin-rich granular pool from which minimal translocation of Mac-1 occurs in response to chemotactic factors. Immunofluorescence and ultrastructural immunocytochemistry confirm the co-localization of Mac-1 and gelatinase in a subset of granules in intact neutrophils. In additional studies, utilizing the neutrophil from the human neonate as a model for impaired Mac-1 expression following stimulation with inflammatory mediators (16,21), we examined the cellular content and subcellular distribution of Mac-1 in neonatal cells prior to or following their exposure to fMLP. Our findings suggest that diminished surface expression of Mac-1 on stimulated neonatal cells reflects diminished translocation of this glycoprotein from this novel intracellular pool associated with gelatinase-rich granules. This abnormality illustrates the physiologic importance of this intracellular pool for surface Mac-1 expression.

Results

Effects of fMLP Concentration and Time on Extracellular Release of Granule-Associated Constituents

Initially, the effects of fMLP on Mac-1 binding by OKM-1 MAb and extracellular release of granule constituents were compared in dose or time-response experiments. The relative effects of fMLP concentration are shown in Figure 8.1a Purified neutrophils maintained at 4 °C prior to stimulation were exposed to a range of concentrations of fMLP (10^{-11}–10^{-8} M) for 15 min at 37 °C. Expression of Mac-1 and release of vit B_{12} B.P., lactoferrin, and gelatinase were minimally effected by cell incubation at 37 °C alone or in response to 10^{-11} to 10^{-9} M fMLP for 15 minutes.

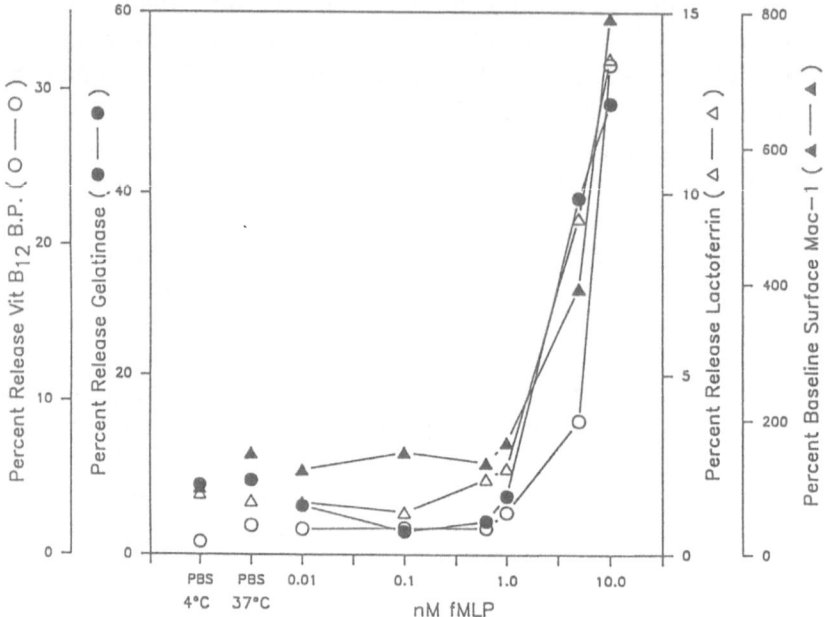

FIGURE 8.1a The effect of fMLP concentration on Mac-1α expression, and the extracellular release of vit B_{12} B.P., lactoferrin, and gelatinase. Results shown represent the mean of two experiments. Neutrophils were stimulated with a range of fMLP concentrations at 37 °C, for 15 minutes. Controls were run at 4 and 37 °C in PBS without fMLP. The surface expression of Mac-1α was determined by cytofluorometric assessments of OKM-1 binding and is expressed as the percentage of baseline condition (4 °C) mean fluorescence channel. Enzyme determinations (38) are expressed as the percentage of total cell content released. As shown, a "threshold" concentration of fMLP (5×10^{-9} M) elicited significant increases in OKM-1 binding and the extracellular release of vit B_{12} B.P., lactoferrin, and gelatinase. Under all experimental conditions shown, <2% extracellular release of MPO was detected (not shown).

Exposure to $\geq 5 \times 10^{-9}$ M fMLP elicited an increase above baseline for OKM-1 MAb binding as well as release of vit. B_{12} B.P., lactoferrin, and gelatinase. However, following stimulation with 10^{-8} M fMLP, significantly greater release of gelatinase ($44 \pm 4\%$ of total, n = 7) was observed as compared to release of vit B_{12} B.P. ($20 \pm 7\%$ of total, n = 9; 0.005 < p < 0.01) and release of lactoferrin ($22 \pm 11\%$ of total, n = 3; 0.05 < p < 0.1). To compare the kinetics of Mac-1 surface expression and granule release, cell suspensions were exposed to 5×10^{-9} M fMLP at 37 °C for varying time intervals (Fig. 8.1b). In these experiments, the increased expression of Mac-1α (OKM-1 binding) over time paralleled the extracellular release of vit B_{12} B.P. lactoferrin, and gelatinase. Half-maximal responses for each were achieved by 6 to 8 minutes and maximal

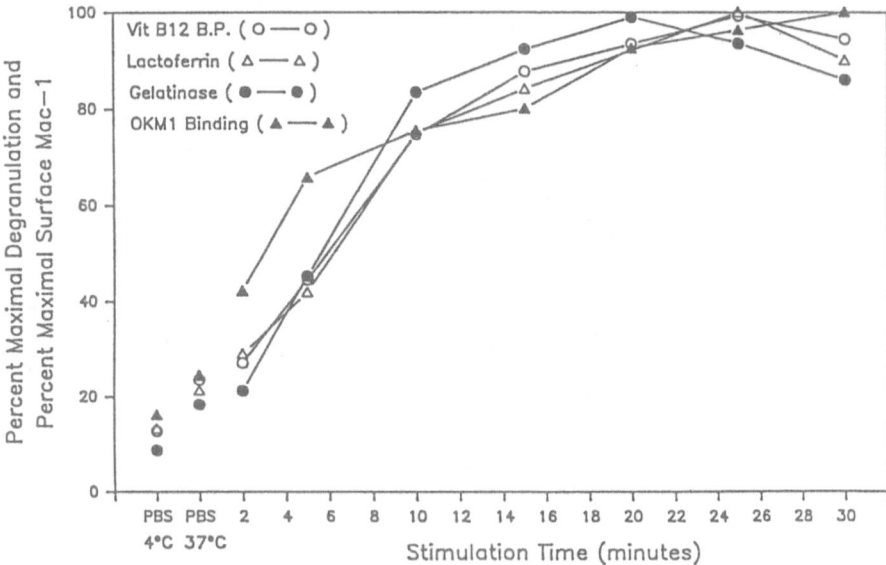

FIGURE 8.1b. Time course of Mac-1α expression and release of vit B_{12} B.P., lactoferrin, or gelatinase. Results shown represent the mean of two experiments. Neutrophils were exposed to fMLP (5×10^{-9} M, 37 °C), for intervals of 2, 5, 10, 15, 20, 25, or 30 minutes. Controls were run at 4 and 37 °C × 30 minutes, in PBS without fMLP. The surface expression of Mac-1α was determined by cytofluorometric assessments of OKM-1 binding and results are expressed as the percentage of maximum response (mean fluorescence channel). Enzyme determinations (38) are also expressed as percent of maximal release values. As shown, increased binding of OKM-1 paralleled the extracellular release of vit B_{12} B.P., lactoferrin, and gelatinase over the time intervals tested. Under these experimental conditions, <2% release of MPO was detected (not shown).

responses were evident following 14 to 20 minutes incubation. Under all experimental conditions, release of the primary granule marker MPO was <2% above baseline values.

Distribution of Mac-1α in Subcellular Fractions Resolved on Discontinuous Sucrose Gradients

Resolution of Granule, Membrane, and Cytosolic Markers

The location of Mac-1α in unstimulated neutrophils was investigated by comparing the distribution of Mac-1α (quantitated by immunoblot analysis as described previously (22,23) with the distribution of granule-associated, membrane, or cytosolic markers in subcellular fractions resolved on discontinuous sucrose gradients. As shown in Figure 8.2a, four

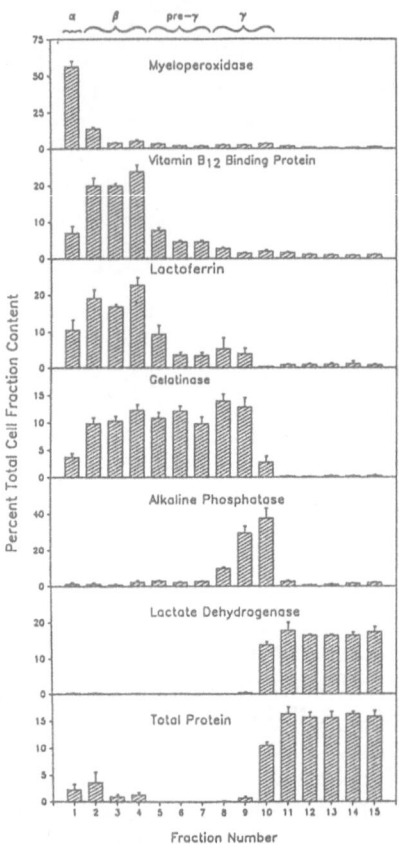

FIGURE 8.2a. Enzyme and protein distribution among 15 subcellular fractions. Nitrogen cavitates (250 psi $\times$ 20 min) of 2×10^8 unstimulated neutrophils in relaxation buffer (34) were subjected to discontinuous sucrose gradient centrifugation (sucrose layers: 30% wt/wt, 35% wt/wt, 40% wt/wt, 45% wt/wt) and the absolute quantities of enzymes and total protein were determined in each of 15 fractions (0.763 ml) (38). Results are expressed as the percentage of total cell fraction content (sum of 15 fractions) and represent the mean $\pm$ S.E. of eight experiments (except lactoferrin, n = 2). The locations of α, β, pre-γ and γ regions are indicated by the overlying brackets. For these experiments, the mean $\pm$ S.E. percent of each total cell protein or enzyme marker recovered in cavitation supernatants was as follows: MPO; 93 $\pm$ 13, vit B_{12} B.P.; 75 $\pm$ 5, lactoferrin ; 70, gelatinase; 45 $\pm$ 5, alkaline phosphatase; 88 $\pm$ 9, LDH; 85 $\pm$ 9, total protein; 81 $\pm$ 5. The recovery of each market in subcellular fractions (total of all fractions) expressed as mean $\pm$ S.E. percentage of the content of cavitation supernatants was as follows: MPO; 63 $\pm$ 10, vit B^{12} B.P.; 56 $\pm$ 5, lactoferrin; 52, gelatinase; 95 $\pm$ 17, alkaline phosphatase; 81 $\pm$ 7, LDH; 100 $\pm$ 18, and total protein; 103 $\pm$ 9. Mean $\pm$ S.E. total cell values/10^6 neutrophils were as follows: MPO; 2.5 $\pm$ 0.5 Δ 450 nm min^{-1}, vit B^{12} B.P.; 358 $\pm$ 40 pg, lactoferrin; 3.4 μg, gelatinase; 0.11 $\pm$ 0.01 μg gelatin min^{-1}, alkaline phosphatase; 0.08 $\pm$ 0.01 Δ 420 nm min^{-1}, LDH; 0.9 $\pm$ 0.2 Δ 340 nm min^{-1}, and total protein 32 $\pm$ 5 μg.

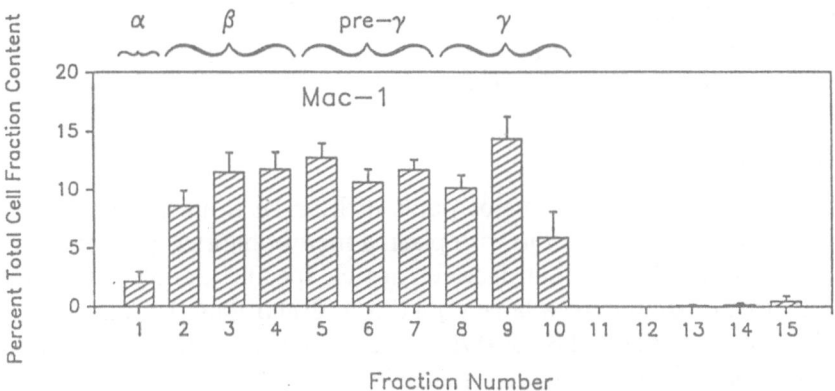

FIGURE 8.2b. Distribution of Mac-1α in subcellular fractions. Cavitates of 2 × 10^8 neutrophils were subjected to discontinuous sucrose gradient centrifugation and the quantity of Mac-1α in each of 15 fractions (0.763 ml) was determined by a quantitative immunoblot assay as previously described in detail (22,23). Briefly, Mac-1 was immunoprecipitated with monoclonal antibodies (MAbs) directed at the Mac-1α (OKM-1) (39) or the β subunit (TS1/18) (39). The immunoprecipitated proteins were subjected to discontinuous sodium dodecyl sulfate polyacrylamide gel electrophoresis (SDS-PAGE) and subsequently electroblotted to nitrocellulose paper. Glycoproteins bound to nitrocellulose were incubated with Concanavalin A followed by subsequent incubations with rabbit anti-concanavalin A and horseradish peroxidase-conjugated goat anti-rabbit IgG. Antibody complexes were detected by reaction with 4-chloro-1-napthol plus H_2O_2. Quantitation of electroblots was accomplished by computer image analysis. Results shown represent the mean ± S.E. of the same eight separate experiments utilized for the enzyme and total protein resolution in Figure 8.2a. The locations of α, β, pre-γ, and γ regions are indicated by the overlying brackets.

distinct regions (α, β, pre-γ, and γ) were identified. The location of MPO identified the primary granule-rich α region (fraction #1; density 1.200 g/ml) which corresponded to a visible α band. The β region (fractions #2–4; density 1.198–1.176 g/ml) contained the majority of the specific granule constituents vit B_{12} B.P. and lactoferrin and contained a distinctly visible β band. The granule marker gelatinase was also identified in the vit B_{12} B.P.-rich and lactoferrin-rich β region, but in addition was contained in lower density fractions. A unique pre-γ region which was rich in gelatinase but poor in vit B_{12} B.P. and lactoferrin was identified in fractions #5 to 7 (density 1.166–1.151 g/ml). The location of the plasma membrane marker alkaline phosphatase defined the γ region (fractions #8–10; density 1.151–1.078 g/ml), also identified by one or two visible thin bands. Only small amounts of granular constituents were identified in the LDH-rich cytosolic fractions #11 to 15 (density <1.078), indicating that granules remained largely intact during cell disruption. Electron micrographs of material pelleted from the α, or β and pre-γ regions demonstrated granular structures morphologically consistent with azurophilic (1°) or specific (°) granules, respectively.

Resolution of Mac-1 Pools

As shown in Figure 8.2b, major pools of Mac-1α in unstimulated cells were identified in association with plasma membrane-rich γ fractions (30 $\pm$ 3%) and vit B$_{12}$ B.P. and lactoferrin-rich specific granule β fractions (32 $\pm$ 4%). Additionally, a large pool (35 $\pm$ 2%) of Mac-1α was detected in the gelatinase-rich, vit B$_{12}$ B.P./lactoferrin-poor pre-γ fractions. Similar distributions of gelatinase were identified in β (33 $\pm$ 2%), pre-γ (33 $\pm$ 1%) and γ (30 $\pm$ 2%) fractions. The distributions of Mac-1 were identical when employing either anti-Mac-1α (OKM-1) or β (TS1/18) MAbs in the immunoblot procedure. The mean $\pm$ S.E. percent of total cell Mac-1α recovered in the subcellular fractions was 105 $\pm$ 37. The absolute total cell content of Mac-1α protein was 0.95 $\pm$ 0.34 μg/10^6 cells (mean $\pm$ S.E.).

Effects of fMLP on Redistribution of Mac-1α and Secretion of Granule Markers

To investigate the mechanisms by which chemotactic stimuli increase Mac-1 expression on neutrophil surfaces (1,3,4,11), the quantitative distributions of Mac-1α and graunle-associated markers were analyzed in subcellular fractions of neutrophil suspensions prior to and following their exposure to fMLP at 10^{-8} M for 15 minutes at 37 °C (Fig. 8.3). To adjust for variation in the total Mac-1 recovery between unstimulated and stimulated cells in each experiment, Mac-1α was expressed as the mean percentage of total Mac-1 contained in all fractions. Following fMLP exposure, resulting in 5.8 $\pm$ 0.9-fold increase OKM-1 binding to surface Mac-1, a significant reduction of both Mac-1α ($-14.4 \pm 2.6\%$, $p < 0.005$) and gelatinase ($-18.7 \pm 2.6\%$, p 0.005) in the pre-γ fractions was observed coincident with an increase of Mac-1α in γ fractions ($+11.1 + 4.2\%$, $0.05 < p < 0.1$) and the extracellular secretion of 41 $\pm$ 7% ($p < 0.005$) of total cell gelatinase. The secretion of gelatinase from the pre-γ region was significantly ($p > 0.05$) greater than from the β region, indicating relatively greater mobilization of gelatinase-rich, vit B$_{12}$ B.P. poor granules in response to this stimulus. Changes in intracellular Mac-1α following chemotactic stimulation did not parallel those of vit B$_{12}$ B.P. or lactoferrin. In fact, while the β fractions #2 to 4 represent the major source from which vit B$_{12}$ B.P. and lactoferrin were secreted extracellularly in response to fMLP, there was little change in Mac-1α in these fractions (net decrease of 2.3 $\pm$ 5.2%, $p > 0.1$). Incubation of neutrophils at 37 °C, without fMLP, resulted in no significant degranu-

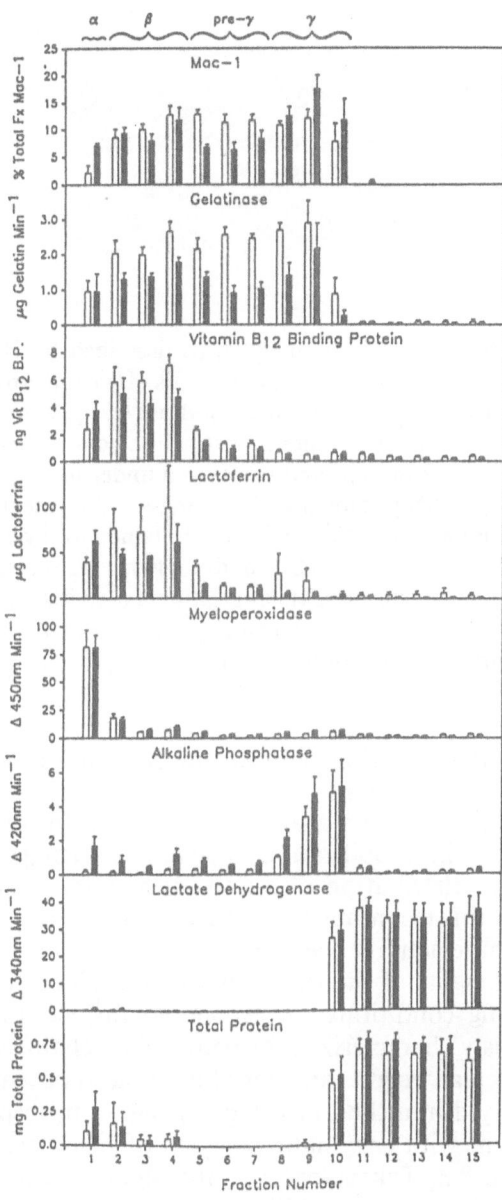

FIGURE 8.3. Subcellular distributions of Mac-1α, gelatinase, vit B_{12} B.P. lacto-ferrin, MPO, alkaline phosphatase, lactate dehydrogenase, and total protein in unstimulated or fMLP (10^{-8} M, 15 min, 37 °C) stimulated neutrophils are com-pared. Cavitates of 2×10^8 cells were resolved on discontinuous sucrose gradients and 15 fractions (0.763 ml) were collected (see Methods). Results shown represent the mean ± S.E. of five experiments (for lactoferrin mean with range, n = 2, LDH n = 4). The location of the α, β, pre-γ, and γ regions are indicated by the overlying brackets.

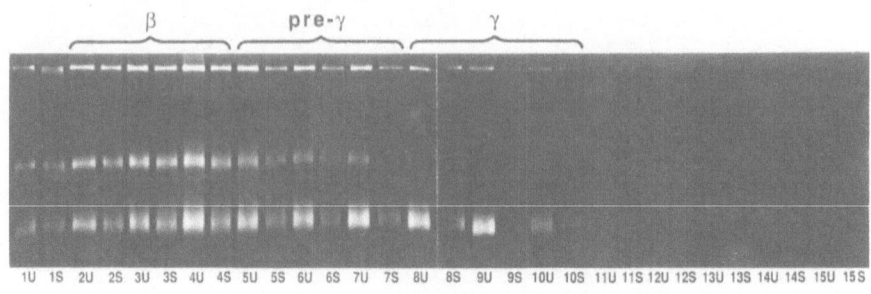

FIGURE 8.4. Gelatin substrate gel analysis, as described by Hibbs (24), of the subcellular fractions of unstimulated neutrophils (U) or neutrophils stimulated with 10^{-8} M fMLP at 37 °C × 15 minutes (S). Briefly, 25 µl aliquots of subcellular fractions resolved over sucrose gradients were subjected to SDS PAGE employing gels containing gelatin (2 mg/ml) and performed under nonreducing conditions. These gels were sequentially stained with Coomassie blue, incubated at 37 °C × 3 hours to elicit gelatinase activity, and then destained to reveal areas of gelatin hydrolysis. The locations of α, β, pre-γ, and γ regions are indicated by the overlying brackets. The molecular weights of gelatinase activities in subcellular fractions were consistent with previous characterizations of neutrophil gelatinase activity under nonreducing conditions (24).

lation from the pre-γ fractions and no translocation of Mac-1 from the pre-γ region (data not shown).

Gelatin substrate gels were also used to confirm the enzymatic observations of preferential mobilization of gelatinase-rich granules from the pre-γ region. The distribution of gelatinase activity in subcellular fractions detected by this technique was similar to that in enzymatic analyses; almost all gelatinase activity was identified in fractions #2 to 9 (Fig. 8.4). The molecular weights of the three gelatinase moieties identified under these nonreducing conditions correspond to the 92, 130 and 225 kD proteins previously characterized for neutrophil gelatinase (24). Activity of all three molecular weight moieties of unreduced neutrophil gelatinase was preferentially diminished in the pre-γ region in response to fMLP, findings consistent with enzymatic analyses of gelatinase activity as described in Figure 8.3. Taken together, the above results define a novel granular pool preferentially mobilized in response to a chemotactically relevant concentration of fMLP.

Immunofluorescence Localization of Mac-1 and Gelatinase

Indirect immunofluorescence microscopy of whole neutrophils using OKM-1 and polyclonal anti-gelatinase antibody revealed both distinct and overlapping patterns of staining. OKM-1 exhibited both membranous and granular staining while anti-gelatinase was limited to intracellular granular structures. While most fluorescent granules exhibited staining with both fluorochromes (Fig. 8.5), occasional granules stained only for anti-gelatinase (Fig. 8.5 c,d). In many instances, the pattern of fluorescent staining of an individual granule was distinct for each antibody; ring-shaped staining patterns for OKM-1 were commonly seen suggesting that Mac-1 is distributed in the perigranular membrane (Fig. 8.6), whereas gelatinase was identified as a solid density positioned within the interior of corresponding granules.

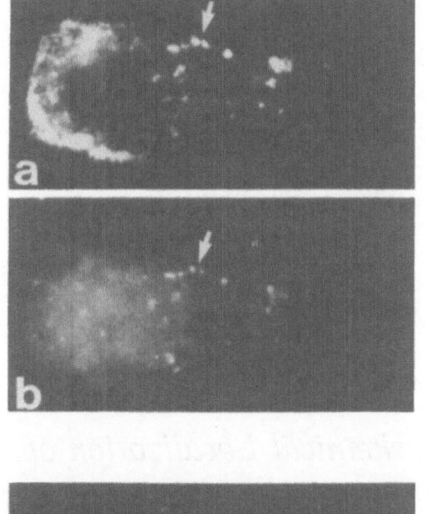

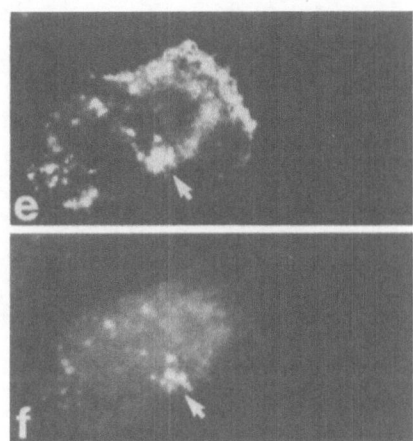

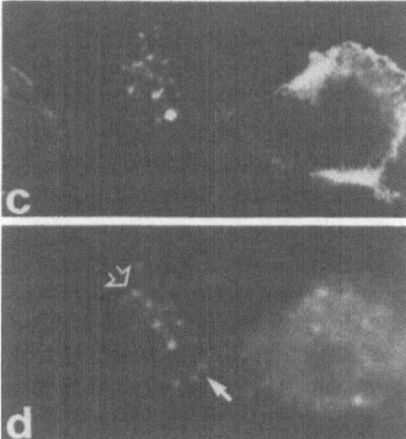

FIGURE 8.5. Immunofluorescent study of the Mac-1α and gelatinase distributions in human neutrophils spread on a glass surface using in the same cells FITC-labeled anti-rabbit immunoglobulin to detect anti-gelatinase and RITC-labeled anti-mouse IgG to detect bound OKM-1 (38). Photographs *a, c,* and *e,* Mac-1; *b, d,* and *f,* gelatinase. Arrows in *a, b,* and *e, f* indicate apparent co-localization; arrows in *d* indicate co-localization in a row of granules (*open arrow*) and gelatinase positive granules without Mac-1 (*solid arrow*).

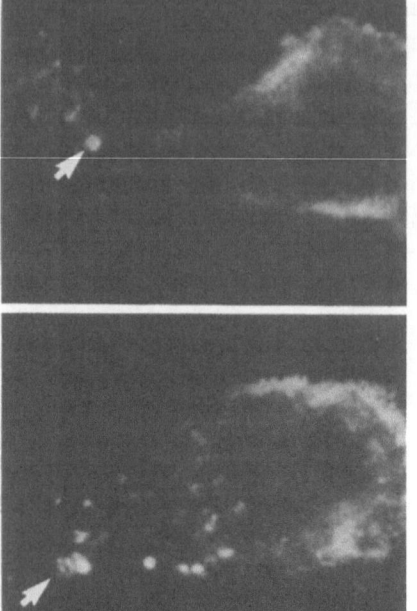

FIGURE 8.6. Immunofluorescent study of the Mac-1α distribution in human neutrophils spread on a glass surface using RITC-labeled anti-mouse IgG to detect bound OKM-1. Arrows indicate perigranular staining.

Electron Microscopic Immunochemical Localization of Mac-1, Lactoferrin, and Gelatinase

Antiserum to Mac-1 labeled granules which had moderate to low electron density and tended to be seen in clusters, and were somewhat larger than typical secondary granules. Antiserum to gelatinase labeled small, round to oval, moderately electron dense granules. In paired sections, antibodies directed against Mac-1 and gelatinase usually labeled different granules, but sometimes marked identical granules (Fig. 8.7).

Effects of fMLP on the Subcellular Distribution of Mac-1 in Neonatal Neutrophils: Relationship to Cell Surface Expression of Mac-1

Recent studies have demonstrated diminished Mac-1 expression on the surface of neutrophils from human neonates following exposure to chemotactic factors (16,21). Therefore, neutrophils from human neonates provided a model with which to study the relationship between Mac-1 surface expression and the intracellular translocation of Mac-1 following chemotactic stimulation. Unstimulated neutrophils from human neonates

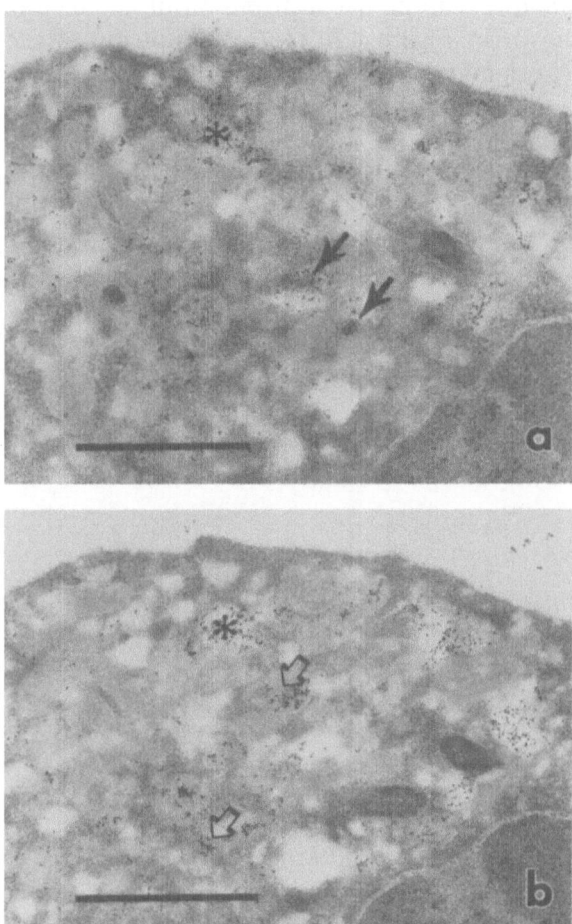

FIGURE 8.7 a,b. Electron micrographs of corresponding surfaces of two adjacent sections immunolabeled with colloidal gold using rabbit antiserum against Mac-1 (*a, upper*) and gelatinase (*b, lower*) followed by 5 nm goat-anti-rabbit-IgG-gold. Note that the granules labeled for Mac-1 occasionally show labeling for gelatinase as well (*asterisks*). Some granules label for Mac-1 but not for gelatinase (*arrows*) while others label only for gelatinase (*open arrows*). The bar indicates 0.5 μm. *a,* × 48,000; *b,* × 51,000.

(cord blood samples) contained comparable total Mac-1 quantities as compared to healthy adult donors (Total Mac-1: Neonate 5.96 + 0.53 μg/10⁷ cells, Adult 560 ± 0.69 μg/10⁷ cells; $p > 0.1$; n = 9), and similar distributions of Mac-1 following resolution over Percoll (25) or sucrose gradients (Neonate: 27, 38, or 35% of total Mac-1 contained in β, pre-γ, or γ fractions, respectively).

Demonstrations of the effect of fMLP stimulation on the surface expression of Mac-1 and its subcellular distribution in neonatal or adult neutrophils are shown in Figure 8.8. Up regulation by fMLP of OKM-1 binding sites was significantly less for neonatal as compared to adult cells (neonates: 3.5 ± 0.4, adult: 5.6 ± 0.5-fold increase; $p < 0.005$). In addition to this abnormality, the mean $\pm$ SEM percent of Mac-1 depleted from pre-γ fractions of neonatal as compared to adult cells was also significantly diminished (neonates: $-8 \pm 2\%$, adult $-16 \pm 3\%$; $p < 0.05$). However, among all neonatal or adult suspensions tested, pre-γ fractions represented the major if not only source from which Mac-1 was translocated in response to fMLP. The mean $\pm$ SEM percent of Mac-1 incorporated into plasma membrane fractions of neonatal cells was also less than that observed from adult cells (neonates: $+7 + 4\%$ adult: 13

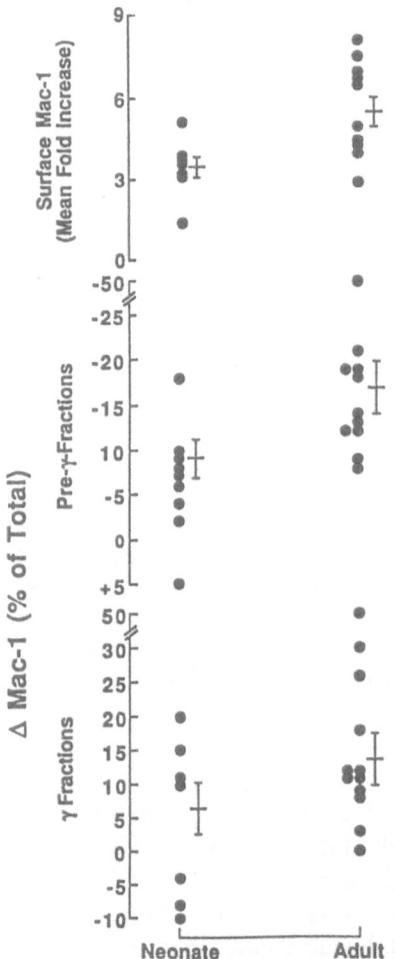

FIGURE 8.8. The effects of fMLP on surface expression and intracellular movement of Mac-1 in neonatal or healthy adult neutrophils. Results of individual experiments (•) and the mean $\pm$ SEM of group (bars) are indicated for studies of 8 neonates and 12 healthy adults.

± 4%; $p < 0.1$). While these mean group values are not statistically different, several neonatal suspensions demonstrated no incorporation of Mac-1 into plasma membrane fractions after exposure to fMLP.

Discussion

Our findings confirm the existence of major plasma membrane and specific granule (i.e., vit B_{12} B.P. and lactoferrin-rich) associated Mac-1 pools, as previously reported (8,11–13). In addition, we provide evidence for a novel intracellular Mac-1 pool, anatomically and functionally distinct from that associated with vit B_{12} B.P. and lactoferrin-rich specific granules. This pool sediments to a region between membrane-rich (γ region) and specific granule (β region) fractions. It contains organelles morphologically indistinguishable from electron-dense specific granules by transmission electron microscopy. The pre-γ fractions are gelatinase-rich and vit B_{12} B.P./lactoferrin-poor as compared to the β fractions which are rich in vit B_{12} B.P., lactoferrin, and gelatinase. Identification of neutrophil gelatinase in pre-γ, β, and γ subcellular fractions was confirmed by enzymatic as well as gelatin substrate gel analyses The association of gelatinase with granules was established by the finding that ≥90% of the gelatinase activity in granular fractions was removed by centrifugation. The anatomical co-localization of Mac-1 and gelatinase in electron-dense granules was established by indirect immunofluorescence and by ultrastructural immunocytochemistry using colloidal gold. Immunofluorescence studies showed a perigranular distribution of Mac-1 in association with a subset of gelatinase positive granules in permeabilized neutrophil preparations.

The distinct nature of this pre-γ fraction Mac-1 pool was further suggested by studies utilizing chemotactic stimulation. In response to fMLP, a significant reduction of Mac-1 contained in pre-γ fractions coincided with a significant increase in detectable Mac-1 in plasma membrane fractions, and a three to eightfold increase in surface Mac-1. In contrast, the quantity of Mac-1 contained in β fractions decreased minimally under these experimental conditions, even though most of the vit B_{12} B.P. and lactoferrin released extracellularly originated in these fractions. Furthermore, while chemotactic factors stimulated the release of gelatinase from β as well as pre-γ fractions, significantly more was released from pre-γ associated granules ($p < 0.05$). Substrate gel analysis of gelatinase in subcellular fractions was sufficiently sensitive to detect selective depletion of gelatinase from pre-γ fractions of cells following exposure to fMLP (Fig. 8.4). Also significantly greater overall release of gelatinase as compared to the specific granule marker vit B_{12} B.P. was consistently observed at or above threshold stimulus conditions. These findings are consistent with the existence of a gelatinase-rich, vit B_{12} B.P. and lactoferrin-poor granule subset preferentially mobilized in response to a chemotactic stimulus, as suggested previously by Petrequin et al. (17), Lacal (13) and others (26). Considered together, these results indicate that the pool of Mac-1 contained in pre-γ fractions is more responsive to chemotactic signaling than that contained in β fractions.

It appears that the gelatinase-rich granules identified in our studies are similar to the tertiary granules described previously (20). Further, the recent report of Lacal (13) described co-sedimentation of Mac-1 (Mo-1) and gelatinase in fractions of unstimulated neutrophils to a density greater than that of plasma membrane but less than that of lysozyme-rich granular fractions. A similar distribution for the T-200 glycoprotein complex of myeloid cells is suggested by the same report. In contrast, it appears that the gelatinase-rich pre-γ fractions described in our report are distinct from the alkaline phosphatase compartments recently defined by other investigators (27,28). Smith, Sharp, and Peters (27) originally described an alkaline phosphatase-rich intracellular granular compartment in human neutrophils, and Borregaard subsequently described an alkaline phosphatase containing compartment with a density slightly greater than that of plasma membranes, which was revealed in Triton-X 114 solubilized whole neutrophils or subcellular fractions (26). This compartment appeared to be mobilized to the plasma membrane much more readily than other granule subsets following stimulation with fMLP. Our studies have consistently documented a significant (29 $\pm$ 8% $p < 0.05$) but unexplained increase in alkaline phosphatase activity distributed over each subcellular fraction following stimulation with fMLP. However, employing our density gradient procedures and those identical to that of Borregaard (data not shown) (26), we have been unable to identify translocation of latent alkaline phosphatase from pre-γ fractions following chemotactic stimulation.

In our studies, surface Mac-1 binding sites increased 300 to 800% upon stimulation with fMLP, a finding associated with a net decrease of Mac-1 in pre-γ fractions of 14 $\pm$ 2.6% and a net increase of this protein in plasma membrane-rich (γ region) fractions of 11 $\pm$ 4.2%. Utilizing previous estimates of the amount of surface Mac-1 available to antibody (approximately 65,000 binding sites/cell or 18 ng/10^6 cells) (7), a translocation of 10 to 15% of total Mac-1 (500–1000 ng/10^6 cells) (equivalent to the amount depleted from pre-γ fractions by fMLP in our studies) could account for the incorporation in membrane of 50 to 150 ng of Mac-1/10^6 cells. Such amounts could in turn account for the three to eightfold elevation of surface Mac-1 detected by MAb binding. However, Mac-1 containing granules or vesicles that co-sediment with plasma membrane may also translocate to the plasma membrane and, thereby, increase surface expression of Mac-1 (undetectable in our studies). Furthermore, it has not been possible to determine the proportion of Mac-1 associated with plasma membrane (i.e., in γ fractions) that is expressed on the cell surface. A direct relationship between translocation of Mac-1 from the pre-γ fractions and surface Mac-1 expression is supported by studies on neutrophils obtained from human neonates. Upregulation by fMLP of OKM-1 binding sites and Mac-1 translocation from pre-γ fractions were both significantly diminished in neonatal as compared to adult neutrophils. Stimulated Mac-1 surface expression on neutrophils from human neonates was diminished 38% and Mac-1 translocation from pre-γ fractions was diminished 50%, when compared to neutrophils from healthy adult donors.

Studies in Mac-1 deficient patients have demonstrated a direct relationship between the quantity of surface Mac-1 expression and the levels of adhesive functions of chemotactically stimulated neutrophils (4). However, some recent reports describe disassociation between the degree of neutrophil adhesiveness and the degree of CD11b/CD18 surface expression under selected experimental conditions (29–31). Our investigation of neutrophil adherence to endothelial monolayers indicate enhancement of adherence by fMLP concentrations that are insufficient to increase binding of OKM-1 (32). Collectively, these findings suggest that conformational or biochemical alterations of the functional (adhesive) domain of Mac-1 may occur upon chemotactic stimulation. Further studies will be necessary to determine whether such alterations of surface Mac-1 contribute to chemotactic functions of neutrophils.

Within the limitations mentioned above, our studies are consistent with the concept that fusion of perigranular membrane with plasma membrane represents a physiologic mechanism by which inflammatory stimuli elicit enhanced surface expression of adherence proteins or other receptors on human neutrophils (33,34). Results of our immunofluorescence analyses and our observations utilizing immunogold techniques suggest that CD11b/CD18 subunits are distributed in perigranular membrane as well as plasma membrane. Based on available amino acid sequence data, the Mac-1β (CD18) subunit is known to contain a transmembrane domain (35). Further, Stevenson et al. (36) have shown that both and α and β subunits of Mac-1 (Mo-1) partition exclusively in a detergent-rich phase of specific granules or membranes of cavitated neutrophils. These findings are consistent with the concept that Mac-1 is an integral membrane protein as are other integrin proteins (37). Triton X-114 detergent partition studies performed in our laboratory confirmed that Mac-α and β (immunoprecipitated with OKM-1 or TS1/18) are in part, integral membrane proteins (F. Schmalstieg, unpublished data). Thus, it appears that chemotactic stimuli facilitate the mobilization of subsets of granules which undergo fusion with plasma membrane and, thereby, provide additional Mac-1 on the cell surface during degranulation. These intracellular granule-associated pools of Mac-1 may be required to ensure the continued availability of this molecule for multifaceted functions of neutrophils in the inflammatory response.

Acknowledgments. The authors wish to thank Margaret Hibbs, M.D. for her many helpful suggestions regarding the gelatinase assay and Candace Smith, M.D. for her work on the lactoferrin assay. We acknowledge Dr. David Dennison of the University of Texas Dental Branch, Houston, Texas for performing the fluorescence activated cell sorting and Linda Rehm of Texas Children's Hospital for electron microscopic immunostaining. The authors would also like to thank Dr. Timothy A. Springer, Dana-Farber Cancer Institute, Harvard School of Medicine for providing TS1/18 monoclonal antibody for use in our studies. We also acknowledge Debra Delmore for her expertise in gel photography, the secretarial assistance of Gaye Stokes and the technical assistance of Bonnie J. Hughes, Mary Dombrowski, and Kelley Dempsey.

References

1. Anderson DC, Springer TA: Leukocyte adhesion deficiency: An inherited defect in the Mac-1, LFA-1 and p150,95 glycoproteins. *Ann Rev Med* 38:175–194, 1987.
2. Sanchez-Madrid F, Simon P, Thompson S, Springer TA: Mapping of antigenic and functional epitopes on the alpha and beta subunits of two related glycoproteins involved in cell interactions, LFA-1 and Mac-1 *J Exp Med* 158:586–602, 1983.
3. Springer TA, Thompson WS, Miller LJ, Anderson DC: Inherited deficiency of the Mac-1, LFA-1, p150,95 glycoprotein family and its molecular basis. *J Exp Med* 160:1901–1918, 1984.
4. Anderson DC, Schmalstieg FC, Finegold MJ, Hughes BJ, Rothlein R, Miller LJ, Kohl S, Tosi MF, Jacobs RL, Waldrop TC, Goldman AS, Shearer WT, Springer TA: The severe and moderate phenotypes of heritable Mac-1, LFA-1, p150,95 deficiency: Their quantitative definition and relation to leukocyte dysfunction and clinical features. *J Infec Dis* 152:668–689, 1985.
5. Beller DI, Springer TA, Schreiber RD: Anti-Mac-1 selectively inhibits the mouse and human type three complement receptor. *J Exp Med* 156:1000–1009, 1982.
6. Wright SD, Rao PE, VanVoorhis WC, Craigmyle LS, Iida K, Talle MA, Westbery EF, Goldstein G, Silverstein SC: Identification of the C3bi receptor of human monocytes and macrophages with monoclonal antibodies. *Proc Natl Acad Sci USA* 80:5699–5703, 1983.
7. Anderson DC, Schmalstieg FC, Kohl S, Arnaout MA, Hughes BJ, Tosi MF, Buffone GJ, Brinkley BR, Dickey WD, Abramson JS, Springer TA, Boxer LA, Hollers JM, Smith CW: Abnormalities of polymorphonuclear leukocyte function associated with a heritable deficiency of high molecular weight surface glycoproteins (GP138): Common relationship to diminished cell adherence. *J Clin Invest* 74:536–551, 1984.
8. Arnaout MA, Spits H, Terhorst C, Pitt J, Todd RF: Deficiency of a leukocyte surface glycoprotein (LFA-1) in two patients with Mo1 deficiency. *J Clin Invest* 74:1291–1300, 1984.
9. Schmalstieg FC, Rudloff HE, Hillman GR, Anderson DC: Two dimensional and three dimensional movement of human polymorphonuclear leukocytes:

Two fundamentally different mechanisms of location. *J Leuk Biol* 40:677–708, 1986.

10. Miller LJ, Bainton DF, Borregaard N, Springer TA: Stimulated mobilization of monocyte Mac-1 and p150,95 adhesion proteins from an intracellular vesicular compartment to the cell surface. *J Clin Invest* 80:185–191, 1987.

11. Todd RF, Arnout MA, Rosin RE, Crowley CA, Peters WA, Babior BM: Subcellular localization of the large subunit of Mo1 (Mo1$_1$; formerly gp110), a surface glycoprotein associated with neutrophil adhesion. *J Clin Invest* 74:1280–1290, 1984.

12. O'Shea JJ, Brown EJ, Seligmann BE, Metcalf JA, Frank MM, Gallin JI: Evidence for distinct intracellular pools of receptors for C3b and C3bi in human neutrophils. *J Immunol* 134:2580–2587, 1985.

13. Lacal P, Pulido R, Sanchez-Madrid F, Mollinedo F: Intracellular location of T200 and Mo1 Glycoproteins in human neutrophils. *J Biol Chem* 263:9946–9951, 1988.

14. Tonnesen MG, Anderson DC, Springer TA, Knedler A, Avdi N, Henson PM: Adherence of neutrophils to cultured human microvascular endothelial cells. Stimulation by chemotactic peptides and lipid mediators and dependence upon the Mac-1, LFA-1, p150,95 glycoprotein family. *J Clin Invest* 83:637–646, 1988.

15. Berger M, O'Shea JJ, Cross AS, Folks TM, Chused TM, Brown EJ, Frank MM: Human neutrophil increase expression of C3bi as well as C3b receptors upon activation. *J Clin Invest* 74:1566–1571, 1984.

16. Anderson DC, Freeman KLB, Heerdt B, Hughes BJ, Jack RM, Smith CW: Abnormal stimulated adherence of neonatal granulocytes: Impaired induction of surface Mac-1 by chemotactic factors of secretagogues. *Blood* 70:740–750, 1987.

17. Petrequin PR, Todd RF, Devall LJ, Boxer LA, Curnutte JT: Association between gelatinase release and increased plasma membrane expression of the Mo1 glycoprotein. *Blood* 69:605–610, 1987.

18. Petrequin PR, Todd III RF, Smolen JE, Boxer LA: Expression of specific granule markers on the cell surface of neutrophil cytoplasts. *Blood* 67:1119–1125, 1986.

19. Bainton DF, Miller LJ, Kishimoto TK, Springer TA: Leukocyte adhesion receptors are stored in peroxidase-negative granules of human neutrophils. *J Exp Med* 166:1641–1653, 1987.

20. Dewald B, Bretz U, Baggiolini M: Release of gelatinase from a novel secretory compartment of human neutrophils. *J Clin Invest* 70:518, 1982.

21. Bruce MC, Bailey JE, Medvik K, Berger M: Impaired surface membrane expression of C3bi, but not C3b receptors in neonatal neutrophils. *Pediatr Res* 21:306–311, 1987.

22. Schmalstieg FC, Rudloff HE, Anderson DC: Binding of adhesive protein complex (LFA-1/Mac-1/p150,95) to Concanavalin A. *J Leuk Biol* 39:193–203, 1986.

23. Jones DH, Anderson DC, Burr BL, Rudloff HE, Smith CW, Schmalsteig FC: Quantitation of intracellular Mac-1 (CD11b/CD18) pools in human neutrophils. *J Leuk Biol* 44:535–544, 1988.

24. Hibbs MS, Hasty KA, Seyer JM, Kang AH, Mainardi CL: Biochemical and immunological characterization of the secreted forms of human neutrophil gelatinase. *J Biol Chem* 260:2493–2500, 1985.

25. Jones DH, Schmalstieg FC, Rudloff HE, Burr BL, Smith CW, Smith CL, Prieto C, Dennison DK, Anderson DC: Identification and quantitation of subcellular locations of Mac-1 (complement receptor-3) in neonatal neutrophils *Pediatr Res* 21:1410, 1987.

26. Borregaard N, Miller LJ, Springer TA: Chemoattractant-regulated mobilization of a novel intracellular compartment in human neutrophils. *Science* 237:1204–1207, 1987.

27. Smith GP, Sharp G, Peters TJ: Isolation and characterization of alkaline phosphatase-containing granules (phosphasomes) from human polymorphonuclear leukocytes. *J Cell Sci* 76:167–178, 1985.

28. Borgers M, Thone F, DeCree J, DeCock W: Alkaline phosphatase activity in human polymorphonuclear leukocytes. *Histochem J* 10:31–43, 1978.

29. Buyon JP, Abramson SB, Philips MR, Slade SG, Ross GD, Weissmann G, Winchester RJ: Dissociation between increased surface expression of Gp165,95 and homotypic neutrophil aggregation. *J Immunol* 140:3156–3160, 1988.

30. Vedder NB, Winn RK, Rice CL, Chi EY, Arfors KE, Harlan JM: A monoclonal antibody to adherence-promoting leukocyte glycoprotein, CD18, reduces organ injury and improves survival from hemorrhagic shock and resuscitation in rabbits. *J Clin Invest* 81:939–944, 1988.

31. Zimmerman GA, McIntyre TM: Neutrophil adherence to human endothelium in vitro occurs by CDw18 (Mo1, MAC-1/LFA-1/GP150,95) glycoprotein-dependent and independent mechanisms. *J Clin Invest* 81:531–537, 1988.

32. Smith CW, Rothlein R, Hughes BJ, Mariscalco MM, Rudloff HE, Schmalstieg FC, Anderson DC: Recognition of an endothelial determinant for CD18-dependent human neutrophil adherence and transendothelial migration. *J Clin Invest* 82:1746–1756, 1988.

33. Hoffstein ST, Friedman RS, Weissmann G: Degranulation, membrane addition, and shape change during chemotactic factor-induced aggregation of human neutrophils. *J Cell Biol* 95:234–241, 1988.

34. Borregaard N, Heiple JM, Simons R, Clark RA: Subcellular localization of the b-cytochrome component of the human neutrophil microbial oxidase: Translocation during activation *J Cell Biol* 97:54–61, 1983.

35. Kishimoto TK, O'Connor K, Lee A, Roberts TM, Springer TA: Cloning of the beta subunit of the leukocyte adhesion proteins: Homology to an extracellular matrix receptor defines a novel supergene family. *Cell* 48:681–690, 1987.

36. Stevenson KB, Nauseef WM, Clark RA: The neutrophil glycoprotein Mo1 is an integral membrane protein of plasma membranes and specific granules *J Immunol* 139:3759–3763, 1987.

37. Ruoslahti E, Pierschbacher MD: New perspectives in cell adhesion:RGD and integrins. *Science* 238:491–497, 1987.

38. Jones, DH, Schmalsteig FC, Hawkins HK, Burr BL, Rudloff HE, Krater SS, Smith CW, Anderson DC: A mobilizable Mac-1 (CD11/CD18) pool co-localizes with gelatinase in human neutrophils. *J Cell Physiol* submitted.

39. Anderson DC, Miller LJ, Schmalstieg FC, Rothlein R, Springer TA: Contributions of the Mac-1 glycoprotein family to adherence-dependent granulocyte functions: Structure-function assessments employing subunit-specific monoclonal antibodies. *J Immunol* 137:15–27, 1986.

9

Anti-Inflammatory Properties of Monoclonal Anti-Mo1 (CD11b/CD18) Antibodies In Vitro and In Vivo

Robert F. Todd III, Paul J. Simpson, and Benedict R. Lucchesi

Introduction

Mo1 is a heterodimeric glycoprotein (gp155,95) expressed on the plasma membrane of neutrophils (PMN), monocytes and certain macrophages, and a subset of large granular lymphoid cells (reviewed in 1). The identification and characterization of Mo1 was made possible by the generation of monoclonal antibodies specific for epitopes expressed on the higher molecular weight α-subunit or the lower molecular weight β-subunit. According to World Health Organization (WHO) nomenclature, antibodies recognizing the α-subunit of Mo1 are designated anti-CD11b while antibodies specific for the β-subunit are termed anti-CD18 (2). Mo1 (CD11b/CD18) is a member of a family of three structurally related glycoproteins that include LFA-1 (CD11a/CD18) and p150,95 (CD11c/CD18) (3). Each member of this family has a unique higher molecular weight α-subunit that is noncovalently linked with a structurally identical β-subunit (CD18) (Fig. 9.1). Considerable progress has been made in the biochemical characterization of Mo1, LFA-1, and p150,95 (3), and the genes encoding the CD11b and CD18 subunits of Mo1 have been cloned (4,5). Similarly, the functional significance of the CD11/CD18 glycoproteins has been deduced from the results of antibody blocking experiments in which selective functional defects are exhibited by normal cells pretreated with monoclonal antibodies specific for CD11 or CD18 epitopes. As shown in Table 9.1, murine monoclonal antibodies that recognize CD11b or CD18 epitopes inhibit a variety of adhesion-dependent functions of neutrophils and mononuclear phagocytes that include binding to complement C3bi-opsonized particles and certain unopsonized microorganisms (6–16) (preventing subsequent phagocytic, secretory, and oxidative burst responses that may accompany particle binding [6, 9–15, 17]), binding of monocytes to the coagulation factors fibrinogen and factor

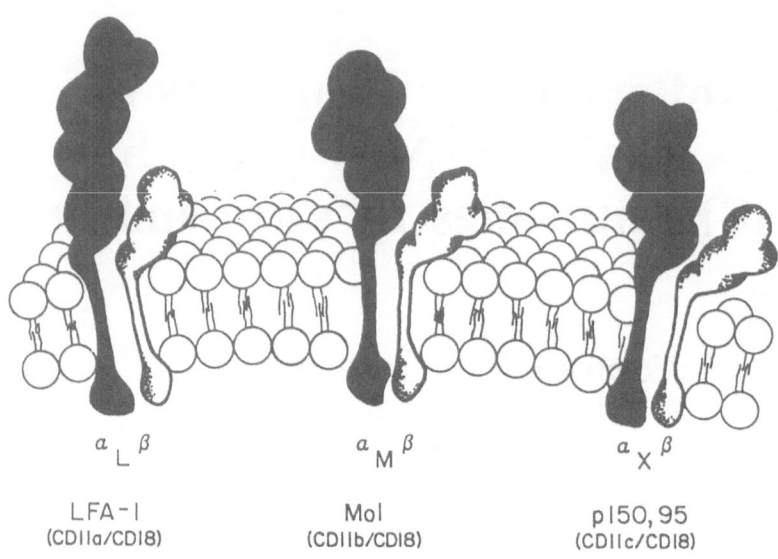

a_L β a_M β a_X β

LFA-1   Mol   p150,95
(CDIIa/CD18)   (CDIIb/CD18)   (CDIIc/CD18)

FIGURE 9.1. Schematic representation of the CD11/CD18 leukocyte glycoprotein family: LFA-1 (CD11a/CD18), Mol (CD11b/CD18), and p150,95 (CD11c/CD18). Each transmembrane glycoprotein consists of unique higher MW α subunit (αL, αM, αX) that is noncovalently associated with a structurally identical lower MW β subunit.

X (18,19), PMN aggregation (10, 17, 20) and chemotaxis (6,10), and the adherence and spreading PMNs or monocytes to nonphysiologic (coated or uncoated plastic and glass surface [10, 17, 21–23]) or physiologic substrates (monolayers of endothelial or epithelial cells [17, 22–25]). Adhesion-dependent, PMN-mediated cytotoxicity of endothelial or epithelial cells is also attenuated by pretreatment of PMNs with anti-CD11b or anti-CD18 specific reagents (24–26). These observations have led to the concept that the CD11/CD18 glycoproteins are plasma membrane receptor molecules that promote cellular adhesive function by the recognition of one or more ligands. In the case of the Mol glycoprotein, the existence of multiple distinct receptor sites may account for the Mol-dependent recognition of structurally unrelated ligands by PMNs (11, 13, 21). The fact that certain anti-CD11b antibodies block PMN-binding to C3bi-opsonized particles, but not substrate adhesion (or vice versa) supports the concept that Mol displays unique ligand-binding domains (21). Indeed, recent work by Wright et al. suggests the existence of at least two Mol receptor sites, one with an affinity for RGD-containing peptide sequences (27), and another which recognizes lipopolysaccharide (28). PMN binding to C3bi which contains an RGD sequence and is inhibitable by RGD-containing peptide fragments (27), is mediated by the RGD rec-

TABLE 9.1. Inhibitory Effects of Monoclonal Antibodies Specific for CD11b/ CD18 on Phagocyte-Mediated Function In Vitro

Phagocyte Function (In Vitro)	Function Blocked by:		References
	Anti-CD11b	Anti-CD18[b]	
PMN/MP[a]-particle binding:			
C3bi-opsonized particles	+	+	6–10
Yeast	+		11
Zymosan	+	+	11–13
Leishmania	+		14, 15
Legionella	+		16
PMN particle-stimulated secretion	+	+	6, 10, 12
PMN particle-stimulated resp. burst	+	+	10–13
PMN/MP-particle phagocytosis	+	+	6, 9, 10, 14, 15, 17
MP binding to coagulation factors:			
Fibrinogen	+	+	18
Factor X	+	+	19
PMN homotypic adhesion (Aggregation)	+	+	10, 17, 20
PMN FMLP-stimulated motility[c]	+	+	6, 10
PMN/MP-substrate adhesion			
Glass, plastic (coated/uncoated)	+	+	10, 17, 21–23
Endothelial cell monolayers	+	+	17, 22–24
Epithelial cell monolayers	+		25
PMN-mediated cellular cytotoxicity			
Endothelial cells		+	24
Epithelial cells	+		25
Perfused rat lungs (ex vivo)	+		26

[a]Abbreviations: PMN = neutrophil; MP = mononuclear phagocytes (monocytes or macrophages); FMLP = formyl-methionyl-leucyl-phenylalanine.
[b]Whereas blocking by anti-CD11b suggests a function attributable to the Mol glycoprotein, the inhibitory effect of anti-CD18 does not exclude the functional contribution of LFA-1 or p150,95 which also share the CD 18 β-subunit.
[c]In reference 10, anti-CD11b and CD18 antibodies failed to inhibit chemotaxis in the Boyden chamber assay, but did block motility in the "under agarose" assay.

ognition site of Mol while PMN-binding to certain unopsonized gram-negative microorganisms is mediated by the LPS recognition site (28). Observations from Wright's laboratory further suggest that PMN-binding to endothelial cell surfaces may also depend on an RGD recognition site of the Mol molecule (29).

The physiological significance of Mol expression by human PMNs has been furthered substantiated by the identification of individuals who are genetically deficient in their expression of the CD11/CD18 glycoproteins (30). CD11/CD18 deficient PMNs from these individuals exhibit the same defects in adhesion-dependent function as do normal PMNs pre-treated with antibody specific for CD11b or CD18 (30). Moreover, these cellular defects cripple the inflammatory response to certain microorganisms (generally bacteria) resulting in recurrent, sometimes fatal, infectious episodes. Whereas the Mo1-dependent, PMN-mediated inflammatory response may play a beneficial role as a component of the immune defense mechanism, the Mo1 glycoprotein may also promote other cellular inflammatory responses that are deleterious to the health of the host. As Malech and Gallin (31) recently reviewed, PMNs may inflict widespread tissue damage in a variety of noninfectious inflammatory diseases affecting multiple organs. The involvement of Mo1 in promoting these pathological effector responses of PMNs was suggested by the inhibitory effect of anti-CD11b and/or CD18 antibodies on PMN-mediated endothelial or epithelial cell cytotoxicity in vitro (17, 22–25). Anti-Mo1 (CD11b) antibody was also found to block pulmonary endothelial cell injury by activated human PMNs in an isolated perfused rat lung model of acute lung inflammation (26). These observations led to the notion that monoclonal antibodies specific for the CD11/CD18 glycoproteins might have therapeutic utility as anti-inflammatory agents capable of inhibiting the deleterious adhesion-dependent effector responses of PMNs. To test this concept, we determined whether the administration of anti-Mo1 (CD11b) monoclonal antibody could attenuate a destructive inflammatory process in vivo that is mediated (in part) by activated PMNs: myocardial ischemia-reperfusion injury.

Results

Ischemia-reperfusion injury is a destructive inflammatory process that occurs in an organ that is reperfused with blood after a period of sublethal ischemia (32,33). Whereas the mechanisms resulting in reperfusion injury are the subject of continued research and debate, secretory products of activated PMNs (attracted to ischemic tissue by chemotactic factors) such as reactive oxidative intermediates appear to contribute to organ endothelial damage (34). In support of the role of the PMN as an effector cell in ischemia-reperfusion injury are the results of Lucchesi et al. In this study, myocardial injury was attenuated when PMNs were eliminated from the blood perfusing the hearts of dogs subjected to cardiac ischemia (35). Moreover, agents that inhibit neutrophil activation and accumulation within the myocardium also resulted in reduced myocardial necrosis (36–40).

To assess the involvement of the Mo1 glycoprotein in promoting PMN-mediated ischemia-reperfusion injury, we employed the canine model of myocardial reperfusion injury as described by Lucchesi et al. (41). In this model, anesthetized mechanically ventilated dogs are subjected to thoractomy with exposure of the heart and the left circumflex coronary artery (LCA). Subtotal left ventricular ischemia is induced by ligature of the

LCA for a period of 90 minutes after which the ligature is loosened to allow reperfusion for an experimentally fixed duration (6–72 hours). To assess the magnitude of ischemia-reperfusion injury, the dogs are sacrificed and the LCA is perfused with the histochemical reagent, triphenyltetrazolium chloride (TPT), to delineate the viable myocardial tissue within the area of left ventricle at risk for infarction (i.e., the ischemic region perfused by the LCA). This is compared to nonviable myocardial tissue within the area at risk which does not react with TPT (35). The hearts are sectioned and the area of infarct relative to the area at risk is measured planimetrically (35). Without intervention, approximately 35 to 50% of the area of left ventricle perfused by the LCA (area at risk) is infarcted.

The effect of anti-Mo1 (CD11b) monoclonal antibody on this model of myocardial reperfusion injury was established in two protocols shown schematically in Figure 9.2. For these experiments, we employed the IgG1, k murine monoclonal antibody 904 (supplied by Coulter Immunology, Hialeah, FL) which is specific for a CD11b epitope expressed by an α-subunit of Mo1 found on human and dog PMNs (see Table 9.2). Preliminary studies had shown that pretreatment of human canine PMNs with 904 inhibited PMN aggregation (43) suggesting that the epitope recognized by 904 was included within a domain of the Mo1 glycoprotein that promotes cellular adhesion. In Protocol I (Fig. 9.2) (45), eight evaluable dogs received an intravenous infusion of 904 IgG as a single bolus infusion administered prior to reperfusion at a dose sufficient to produce antibody excess (1 mg/kg) during 6-hours of reperfusion; eight control

PROTOCOL I. EFFECT OF ANTI-Mo1 IgG vs. DILUENT CONTROL DURING 6 HR REPERFUSION.

PROTOCOL II. EFFECT OF ANTI-Mo1 vs. CONTROL F(ab')$_2$ ANTIBODY DURING 72 HR REPERFUSION.

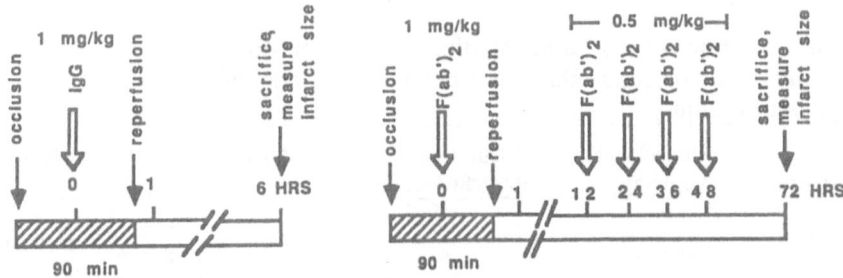

FIGURE 9.2. Protocols to test the effect of anti-Mo1 monoclonal antibody on canine myocardial reperfusion injury. See text for details of experimental protocols.

TABLE 9.2. Characteristics of Clone 904 Anti-Mo1 (CD11b) Monoclonal Antibody

- 904 is a murine IgG1, k monoclonal antibody specific for α_M (CD11b) epitope of Mo1 (21, 42).
- 904 cross-reacts with primate and canine Mo1 homolog (42, 43).
- In vitro blocking activity:
  Blocks human PMN adhesion and spreading to plastic (21).
  Blocks human PMN chemotaxis in response to FMLP (21).
  Blocks human PMN aggregation in response to FMLP (R.F. Todd III, unpublished).
  Blocks canine PMN aggregation in response to PMA (43).
  Blocks human PMN E. coli (LPS) binding site (44).
  Does not block human PMN C3bi (CR3) binding site (21, 44).
  Weakly blocks adhesion of human PMNs to human umbilical vein endothelial cell monolayers (29).
  Does not block adhesion of canine PMNs to canine endothelial cell monolayers (C.W. Smith, unpublished).

evaluable animals received diluent alone. As indicated by the results shown in Table 9.3, the administration of 904 IgG produced a significant reduction in myocardial infarct size as compared to diluent alone (25.8 ± 4.7% vs. 47.8 ± 5.7%, $p < 0.01$) (45). In the 904 treated group, there was also a modest reduction in the cellular inflammatory response in the myocardium adjacent to the infarcted tissue as manifested by reduced neutrophil infiltration (45). Between the 904 and diluent-treated groups, there were no significant differences in degree of initial myocardial ischemia (as assessed electrocardiographically), heart rate (HR), blood pressure (BP), left circumflex blood flow, or rate-pressure product (HR × BP; a measure of cardiac oxygen consumption) to account for the reduction in infarct size observed in the 904 treated group (45). Moreover, there was no significant reduction in the number of circulating PMNs (relative to baseline at the beginning of reperfusion) occurring during the 6 hours of reperfusion in the 904 treated group which might have resulted

TABLE 9.3. Intravenous Administration of 904 Anti-Mo1 Monoclonal Antibody Attentuates Myocardial Infarct Size in a Canine Model of Ischemia-Reperfusion Injury

Protocol[a]	Monoclonal Antibody	Duration of Reperfusion, hrs	Area of Infarct/Area at Risk, %	
I	—	6	47.6 ± 5.7[b]	$p < 0.01$
	904 IgG	6	25.8 ± 4.7	
II	IgG1 F(ab')$_2$	72	37.4 ± 5.8	$p < 0.025$
	904 F(ab')$_2$	72	21.6 ± 2.8	

[a]See protocol experiment design in Figure 9.2.
[b]N = 8 dogs in each of the 4 groups.

in diminished tissue injury. On the basis of these data, we concluded that intravenous administration of 904 anti-Mo1 monoclonal antibody prior to reperfusion could significantly attenuate the cellular inflammatory process leading to myocardial infarction when assessed at 6 hours of reperfusion (45).

Since the destructive inflammatory response to ischemia may extend beyond the 6 hours of reperfusion of Protocol I (47), we then initiated a second experimental protocol (Protocol II, Fig. 9.2) to determine if the administration of 904 anti-Mo1 monoclonal antibody could provide sustained protection against myocardial infarction when assessed after 72 hours of reperfusion in awake experimental subjects (46). To eliminate the possibility of any Fc receptor-mediated clearance of 904-coated PMN effector cells, we used the F(ab')2 fragment of 904 (prepared and supplied by Coulter Immunology) in Protocol II and randomized dogs (in a blinded fashion) between infusions of 904 F(ab')2 and a negative control IgG1 F(ab')2 (unreactive with human or dog PMNs). To maintain serum antibody excess during the first 48 hours of a 72-hour period of reperfusion, it was necessary to administer bolus infusions of F(ab')2 at a dose of 0.5 mg/kg at four 12-hour intervals after the initial 1 mg/kg infusion given just prior to reperfusion (46). The outcome of this second protocol is shown in Table 9.3 and again demonstrated a significant reduction in infarct size in the 904 treated group as compared to the control F(ab')2 treated individuals ($21.6 \pm 2.8\%$ vs. $37.4 \pm 5.8\%$; $p < 0.025$) (46). As observed in Protocol I, neither the 904 nor control F(ab')2 treated groups differed significantly in the degree of initial ischemia (as assessed by myocardial blood flow) or other hemodynamic parameters and the number of circulating neutrophils throughout the period of reperfusion. In Protocol II, unlike Protocol I, there was no significant difference in peri-infarction PMN infiltration (46), although the magnitude of the cellular inflammatory response was markedly reduced at 72 hours (Protocol II) as compared to 6 hours (Protocol I) in the control groups. From the results of Protocol II, we concluded that the administration of the F(ab')2 of 904 anti-Mo1 antibody could provide a sustained reduction in myocardial infarct size after ischemic insult when measured at 72 hours (46).

What is the mechanism by which 904 antibody protects ischemic, reperfused myocardium from infarction? As indicated above, there were no electrophysiological or hemodynamic differences between the 904 and control subjects to account for a reduction in relative infarct size in the 904-treated animals. Loss of 904-coated PMN effector cells by Fc-mediated clearance was not observed and was largely excluded by the use of F(ab')2 fragments. In Protocol I, but not in Protocol II, there was a quantitative reduction in the degree of PMN infiltration in peri-infarcted tissues suggesting that 904 administration may reduce the magnitude of the PMN-mediated destructive inflammatory response that occurs early during the course of reperfusion (but which subsides by 72 hours [48]).

Whereas recent observations from Wright et al. (29) and Smith et al. (unpublished data) indicate that the 904 antibody does not block human or dog PMN attachment to human or dog endothelial cell monolayers in vitro (as do certain other anti-CD11b or CD18-specific reagents [Table 9.1]), 904 does inhibit PMN spreading to artificial substrates (21) as well as the aggregation of human (RF Todd III, unpublished data) and dog PMNs (43) in vitro stimulated by FMLP or PMA, respectively (Table 9.2). On the basis of these in vitro and in vivo data, we conclude that 904 exerts its protective effect on ischemic reperfused myocardium by blocking a domain of the Mo1 glycoprotein involved in an adhesion-dependent, PMN-mediated inflammatory process that may include (but is not necessarily limited to) PMN aggregation. The extent to which 904 inhibits PMN adherence to and diapedesis through microvascular endothelium remains to be determined. In addition to our observations, the recent experience of several other laboratories has confirmed the anti-inflammatory properties of anti-CD11/CD18 reagents in vivo. As shown in Table 9.4, monoclonal antibodies specific for either CD18 or CD11b could inhibit the recruitment and infiltration of PMNs to sites of inflammation in rabbit skin (49, 50) and mouse peritoneum (51), respectively. The PMN-mediated reperfusion injury in ischemic cat intestine could be blocked by anti-CD18 antibody, 60.3 (52). The intravenous administration of this same antibody to rabbits resulted in a significant increase in the proportion of animals surviving an episode of hypovolemic shock with an attenuation of histologically detectable neutrophilic infiltration and endothelial cell damage in 60.3-treated animals as compared to untreated controls (53). These data support the concept that the PMN when activated by inflammatory factors can inflict significant tissue injury to a variety of organs and that a CD11/CD18-dependent adhesive interaction is a prerequisite for this injurious effector response.

TABLE 9.4. Inhibitory Effects of Monoclonal Antibodies Specific for CD11b/ CD18 on PMN-Mediated Noninfectious Inflammatory Responses In Vivo

Intravenous Administration of Monoclonal Antibody Caused:	Monoclonal Antibody (CD)	Reference
• Inhibition of recruitment of PMNs to sites of inflammatory stimuli in:		
Rabbit skin	60.3 (CD18)	49, 50
Mouse peritoneum	5C6 (CD11b)	51
• Attenuation of ischemia-reperfusion injury in:		
Cat intestine	60.3 (CD18)	52
Dog heart	904 (CD11b)	45, 46
• Enhanced survival of rabbits subjected to hypovolemic shock (with diminished vascular endothelial injury)	60.3 (CD18)	53

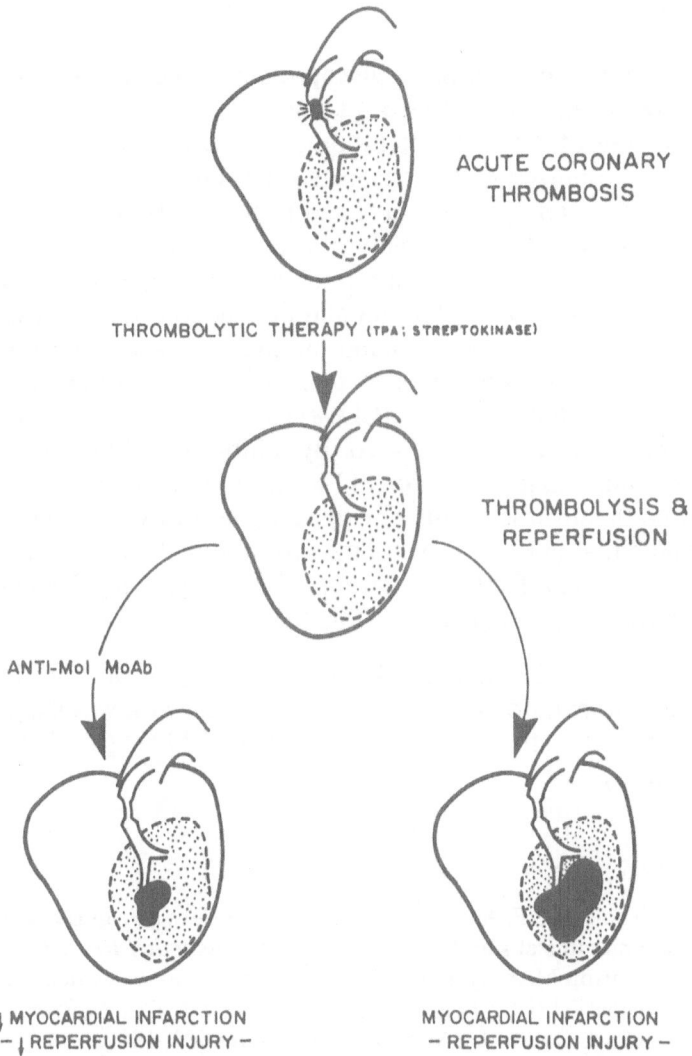

FIGURE 9.3. The administration of anti-Mo1 monoclonal antibody may have therapeutic utility in attenuating myocardial reperfusion injury. The stippled area represents the area of left ventricle at risk for myocardial infarction because it is perfused by coronary circulation in which a coronary thrombosis has occurred. The black area represents the area of left ventricle within the area at risk in which irreversible myocardial infarction has occurred. The size of the infarcted area may depend upon the magnitude and duration of the initial ischemic insult and upon the degree of subsequent reperfusion injury. (TPA = tissue plasminogen activator). In the setting of an acute coronary thrombosis in humans, the intravenous administration of anti-Mo1 monoclonal antibody may reduce infarct size due to reperfusion injury after thrombolytic therapy. A clinical trial to test this hypothesis is planned.

Conclusions

Mo1 is an adhesion-promoting glycoprotein expressed by PMNs and mononuclear phagocytes which, as a result of one or more receptor domains, facilitates the specific membrane interaction between these cells and various opsonized and unopsonized particulate ligands, soluble coagulation factors, and physiological substrata. By promoting cellular adhesive interactions, Mo1 can play a beneficial role in the host defense function of PMNs in response to microbial invasion, while also contributing to the deleterious recognition and destruction of host vascular endothelium that is mediated by inappropriately activated effector PMNs. We and others have demonstrated that murine monoclonal antibodies specific for the α-(CD11b) or β (CD18)-subunits of Mo1 can modulate PMN function in vitro and in vivo. By inhibiting pathologic adhesion-dependent inflammatory function of PMNs in vivo, these antibodies may prove to be therapeutic agents of considerable utility. A clinical trial to test the potential efficacy of 904 anti-Mo1 monoclonal antibody in reducing myocardial infarct size after reperfusion (resulting from emergent thrombolytic therapy) is planned (Fig. 9.3).

Acknowledgments. The authors thank Ms. Tracy Lockhart for her assistance in the preparation of the manuscript. This work was supported in part by National Institutes of Health Grants CA-39064 (R.F.T.) and HL-19782 (B.R.L.).

References

1. Dana N, Todd III RF, Arnaout MA: The Mo1 surface glycoprotein: Structure, function and clinical significance. *Pathol Immunopathol Res* 5:371, 1986.
2. Hogg N, Horton MA: Myeloid antigens: new and previously defined clusters, *In* McMichael AJ (ed.): *Leucocyte Typing III: White Cell Differentiation Antigens.* Oxford, Oxford University Press, p. 576–602, 1987.
3. Sanchez-Madrid F, Nagy JA, Robbins E, Simon P, Springer TA: A human leukocyte differentiation antigen family with distinct alpha subunits and a common beta subunit: the lymphocyte-function associated antigen (LFA-1), the C3bi complement receptor (OKM1/Mac-1), and the p150,95 molecule. *J Exp Med* 158:1785, 1983.
4. Kishimoto TK, O'Connor K, Lee A, Roberts TM, Springer TA: Cloning of the β subunit of the leukocyte adhesion proteins: Homology to an extracellular matrix receptor defines a novel supergene family. *Cell* 48:681, 1987.
5. Arnaout MA, Remold-O'Donnell E, Pierce MW, Harris P, Tenen DG: Molecular cloning of the α subunit of human and guinea pig leukocyte adhesion glycoprotein Mo1: Chromosomal localization and homology to the α subunits of integrins. *Proc Natl Acad Sci USA* 85:2776, 1988.
6. Arnaout MA, Todd III RF, Dana N, Melamed J, Schlossman SF, Colten HR: Inhibition of phagocytosis of complement C3- or immunoglobulin G-coated

particles and of C3bi binding by monoclonal antibodies to a monocyte-granulocyte membrane glycoprotein (Mo1). *J Clin Invest* 72:171, 1983.

7. Beller DI, Springer TA, Schreiber RD: Anti-Mac-1 selectively inhibits the mouse and human type three complement receptor. *J Exp Med* 156:1000, 1982.

8. Wright SD, Rao PE, Van Voorhis WC, Craigmyle LS, Iida K, Talle MA, Westberg EF, Goldstein G, Silverstein SC: Identification of the C3bi receptor of human monocytes and macrophages by using monoclonal antibodies. *Proc Natl Acad Sci USA* 80:5699, 1983.

9. Klebanoff SJ, Beatty PG, Schreiber RD, Ochs HD, Waltersdorph AM: Effect of antibodies directed against complement receptors on phagocytosis by polymorphonuclear leukocytes: Use of iodination as a convenient measure of phagocytosis. *J Immunol* 134:1153, 1985.

10. Anderson DC, Miller LJ, Schmalsteig FC, Rothlein R, Springer TA: Contributions of the Mac-1 glycloprotein family to adherence-dependent granulocyte functions: Structure-function assessments employing subunit-specific monoclonal antibodies. *J Immunol* 137:15, 1986.

11. Ross GD, Cain JA, Lachmann PJ: Membrane complement receptor type three (CR3) has lectin-like properties analogous to bovine conglutinin and functions as a receptor for zymosan and rabbit erythrocytes as well as a receptor for C3bi. *J Immunol* 134:3307, 1985.

12. Arnaout MA, Dana N, Pitt J, Todd III RF: Deficiency of two human leukocyte surface membrane glycoproteins (Mo1 and LFA-1). *Fed Proc* 44:2664, 1985.

13. Hickstein DD, Locksley RM, Beatty PG, Smith A, Stone DM, Root RK: Monoclonal antibodies binding to the human neutrophil C3bi receptor have disparate functional effects. *Blood* 67:1054, 1986.

14. Blackwell JRM, Ezekowitz RAB, Roberts MB, Channon JY, Sim RB, Gordon S: Macrophage complement and lectin-like receptors bind *Leishmania* in the absence of serum. *J Exp Med* 162:324, 1985.

15. Mosser DM, Edelson PJ: The mouse macrophage receptor for C3bi (CR3) is a major mechanism in the phagocytosis of *Leishmania* promastigotes. *J Immunol* 135:2785, 1985.

16. Payne, NR, Horwitz MA. Phagocytosis of *Legionella pneumophila* is mediated by human monocyte complement receptors. *J Exp Med* 166:1377, 1987.

17. Wallis WJ, Hickstein DD, Schwartz BR, June CH, Ochs GD, Beatty PG, Klebanoff SJ, Harlan JM: Monoclonal antibody-defined functional epitopes on the adhesion-promoting glycoprotein complex (CDw18) of human neutrophils. *Blood* 67:1007, 1986.

18. Altieri DC, Edgington TS: The Mac-1 molecule on myeloid cells has a receptor recognition specificity for fibrinogen. *FASEB J* 2:A1461, 1988.

19. Altieri DC, Edgington TS: The saturable high affinity association of factor X to ADP-stimulated monocytes defines a novel function of the Mac-1 receptor. *J Biol Chem* 263:7007, 1988.

20. Schwartz BR, Ochs HD, Beatty PG, Harlan JM: A monoclonal antibody-defined membrane antigen complex is required for neutrophil-neutrophil aggregation. *Blood* 65:1533, 1985.

21. Dana N, Styrt B, Griffin JD, Todd III RF, Klempner MS, Arnaout MA: Two functional domains in the phagocyte membrane glycoprotein Mo1 identified with monoclonal antibodies. *J Immunol* 137:3259, 1986.

22. Wallis, WJ, Beatty PG, Ochs HD, Harlan JM: Human monocyte adherence to cultured vascular endothelium: monoclonal antibody-defined mechanisms. *J Immunol* 135:2323, 1985.
23. Harlan JM, Killen PD, Senecal FM, Schwartz BS, Yee EK, Taylor RF, Beatty PG, Price TH, Ochs HD: The role of neutrophil membrane glycoprotein GP-150 in neutrophil adherence to endothelium in vitro. *Blood* 66:167, 1985.
24. Dienner AM, Beatty PG, Ochs HD, Harlan JM: The role of neutrophil membrane glycoprotein 150 (GP-150) in neutrophil-mediated endothelial cell injury in vitro. *J Immunol* 135:537, 1985.
25. Simon RH, Dehart PD, Todd III RF: Neutrophil-induced injury of rat pulmonary alveolar epithelial cells. *J Clin Invest* 78:1375, 1986.
26. Ismail G, Morganroth ML, Todd III RF, Boxer LA: Prevention of pulmonary injury in isolated perfused rat lungs by activated human neutrophils preincubated with anti-Mo1 monoclonal antibody. *Blood* 69:1167, 1987.
27. Wright SD, Reddy PA, Jong MTC, Erickson BW: C3bi receptor (complement receptor type 3) recognizes a region of complement protein C3 containing the sequence Arg-Gly-Asp. *Proc Natl Acad Sci USA* 84:1965, 1987.
28. Wright SD, Jong MTC: Adhesion-promoting receptors on human macrophages recognize *Eschericia coli* by binding to lipopolysaccharide. *J Exp Med* 164:1876, 1986.
29. Lo SK, Wright SD: CF3 mediates binding of PMN to endothelial cells (EC) via its RGD binding, not the LPS binding site. *FASEB J* 2:A1236, 1988.
30. Todd III RF, Freyer DR: The CD11/CD18 leukocyte glycoprotein deficiency. *Hematol/Oncol Clin N Am* 2:13, 1988.
31. Malech HL, Gallin JI: Neutrophils in human diseases. *N Eng J Med* 317:687, 1987.
32. Braunwald E, Kloner RA: Myocardial reperfusion: a double-edged sword? *J Clin Invest* 76:1713, 1985.
33. Nayler WG, Elz JS: Reperfusion injury: Laboratory artifact or clinical dilemma? *Circulation* 74:1713, 1985.
34. Engler R: Granulocytes and oxidative injury in myocardial ischemia and reperfusion. *Fed Proc* 46:2395, 1987.
35. Romson JL, Hook BG, Kunkel SL, Abrams GD, Schork MA, Lucchesi BR: Reduction of the extent of ischemic myocardial injury by neutrophil depletion in the dog. *Circulation* 67:1016, 1983.
36. Mullane KM, Read N, Salmon JA, Moncada S: Role of leukocytes in acute myocardial infarction in anesthetized dogs: Relationship to myocardial salvage by anti-inflammatory drugs. *J Pharmacol Exp Ther* 228:510, 1984.
37. Romson JL, Hook BG, Rigot VH, Schork MA, Swanson DP, Lucchesi BR: The effect of ibuprofen on accumulation of 111-indium labelled platelets and leukocytes in experimental myocardial infarction. *Circulation* 66:1002, 1982.
38. Bednar M, Smith B, Pinto A, Mullane KM: Nafazatrom-induced salavage of ischemic myocardium in anesthetized dogs is mediated through inhibition of neutrophil function. *Circ Res* 57:131, 1985.
39. Simpson PJ, Mitsos SE, Ventura A, Gallagher KP, Fantone JC, Abrams GC, Schork MA, Lucchesi BR: 1987. Prostacyclin protects ischemic reperfused myocardium in the dog by inhibition of neutrophil activation. *Am Heart J* 113:129, 1987.

40. Simpson PJ, Mickelson JK, Fantone JC, Gallagher KP, Lucchesi BR: Iloprost inhibits neutrophil function in vitro and in vivo and limits experimental infarct size in the canine heart. *Circ Res* 60:666, 1987.
41. Lucchesi BR, Burmeister W, Lomas T, Abrams G: Ischemic changes in the canine heart as affected by the dimethylquaternery analog of propranolol UM 272 (SC27761). *J Pharmacol Exp Ther* 199:310, 1976.
42. Letvin NL, Todd III RF, Palley LS, Schlossman SF, Griffin JD: Conservation of myeloid surface antigens on primate granulocytes. *Blood* 61:408, 1983.
43. Giger U, Boxer LA, Simpson PJ, Lucchesi BR, Todd III RF: Deficiency of leukocyte surface glycoproteins Mo1, LFA-1, and Leu M5 in a dog with recurrent bacterial infections. An animal model. *Blood* 69:1622, 1987.
44. Wright SD, Jong MTC, Levin SM: CR3 expresses two binding sites, one for RGD-peptide, and one for bacterial LPS. *FASEB J* 2:A1236, 1988.
45. Simpson PJ, Todd III RF, Fantone JC, Mickelson JK, Griffin JD, and Lucchesi BR: Reduction of experimental canine myocardial reperfusion injury by a monoclonal antibody (anti-Mo1, anti-CD11b) that inhibits leukocyte adhesion. *J Clin Invest* 81:624, 1988.
46. Simpson PJ, Todd III RF, Mickelson JK, Fantone JC, Gallagher KP, Tamura Y, Lee KA, Kitzen JM, Lucchesi BR: Sustained limitation of myocardial reperfusion injury by a monoclonal antibody that inhibits leukocyte adhesion. *FASEB J* 2:A1237, 1988.
47. Simpson PJ, Fantone JC, Mickelson JK, Gallagher KP, Lucchesi BR: Identification of a time window for therapy to reduce experimental canine myocardial injury: suppression of neutrophil activity during 72 hours of reperfusion. *Circ Res* 63:1070, 1988.
48. Reimer KA, Lowe JE, Rasmussen MM, Jennings RB: The wavefront of ischemic cell death. 1. Myocardial infarct size vs. duration of coronary occlusion. *Circulation* 56:786, 1977.
49. Arfors K-E, Lundberg C, Lindbom L, Lundberg K, Beatty PG, Harlan JM: A monoclonal antibody to the membrane glycoprotein complex CD18 inhibits polymorphonuclear leukocyte accumulation and plasma leakage in vivo. *Blood* 69:338, 1986.
50. Price TH, Beatty PG, Corpuz SR: In vivo inhibition of neutrophil function in the rabbit using monoclonal antibody to CD18. *J Immunol* 139:4174, 1987.
51. Rosen H, Gordon S. Monoclonal antibody to the murine type 3 complement receptor inhibits adhesion of myelomonocytic cells in vitro and inflammatory cell recruitment in vivo. *J Exp Med* 166:1685, 1987.
52. Hernandez, LA, Grisham MB, Twohig B, Arfors K-E, Harlan JM, Granger DN: Role of neutrophils in ischemia-reperfusion-induced microvascular injury. *Am J Physiol* 253:H699, 1987.
53. Vedder NB, Winn RK, Rice CL, Chi EY, Arfors K-E, Harlan JM: A monoclonal antibody to the adherence-promoting leukocyte glycoprotein, CD18, reduces organ injury and improves survival from hemorrhagic shock and resuscitation in rabbits. *J Clin Invest* 81:939, 1988.

10

CDw18 Dependent Adhesion of Leukocytes to Endothelium and Its Relevance for Cardiac Reperfusion

Elke Seewaldt-Becker, Robert Rothlein, and Jürgen W. Dämmgen

Introduction

Mortality following myocardial infarction is critically related to the amount of myocardium that becomes infarcted and thus becomes non-functional (1). Myocardial damage is determined by the duration of is-chemia, the myocardial oxygen demand, and the amount of residual coronary blood flow.

Salvage of myocardial tissue by early reperfusion is the primary goal of various therapeutic regimens in acute myocardial infarction. In the clinical setting, coronary flow is restored by enzymatic thrombolysis, by mechanical removal of the clot (ptca), or by surgically implanted bypass grafts. Moreover, coronary blood flow is frequently reestablished by spon-taneous clot lysis or by relief of coronary spasm.

There is, however, increasing evidence that reperfusion, though ab-solutely necessary for myocardial salvage, by itself may induce cardiac damage (2). Animal experiments, in which depletion of leukocytes re-sulted in a dramatic reduction of infarct size (3), strongly suggested that leukocytes may be responsible for this additional cell damage called *re-perfusion injury*. The chemical signal directing granulocytes into the is-chemic myocardium and the pathway of signal transduction in the gran-ulocyte activation process are far from clear. More information is available about the different functional consequences resulting from the activation of granulocytes such as: increased adhesion of granulocytes to the endothelium, production of highly reactive oxygen radicals leading to oxidative damage, activation of the lytic system, and production of chemotactic factors.

Several different compounds interfering with granulocyte functions have been shown to reduce infarct size in experimental models of acute myocardial infarction. In vitro some of them, such as BW 755, nafaza-trom, and AA 861 inhibit the production of chemotactic factors; others, such as iloprost, inhibit the production of free-oxygen radicals in granulocytes (4).

The fact that these compounds reduce infarct size as well as the accumulation of granulocytes in myocardial tissue can be taken as indirect evidence for the important role of granulocytes in mediating damage during ischemia and reperfusion. Another key event that has been described to take place during ischemia and reperfusion is an impaired reperfusion of the previously ischemic region ("no reflow" phenomenon) which seems to be caused by entrapped granulocytes (5). This obstruction of capillaries by granulocytes may attenuate the restoration of blood flow in the ischemic region and may, thereby, augment and extend the area of ischemia. The entrapment of granulocytes in the capillaries is suggested to be the result of increased adhesion during ischemia and no washout during reperfusion (5).

In summary, inhibiting the adhesion of granulocytes to endothelium seems to be a possible therapeutic target for limiting granulocyte-dependent injury during myocardial infarction. If the adhesion of granulocytes to the endothelium is indeed an active process mediated by activated granulocytes or endothelial cells and if adhesion is a critical early step in the sequence of events leading to granulocyte-dependent cellular injury, agents blocking the adhesion process may be expected to reduce infarct size.

Recently, cell surface glycoproteins on granulocytes such as the CDw18 complex (Mac-1,LFA-1pl50,95 family) have been characterized. This family has been implicated in a variety of leukocyte adhesion-related functions. Additionaly, some adhesion-promoting glycoproteins expressed on the surface of endothelial cells such as the intercellular adhesion molecule (ICAM-1) or endothelial-leukocyte adhesion molecule (ELAM-1) have been described (6–8).

The objective of this study was to examine whether monoclonal antibodies binding to different glycoproteins of the CDw18 complex (MAb R3.3 and MAb R3.1) and a monoclonal antibody directed against ICAM-1 (MAb R6.5) can reduce infarct size in an ischemia-reperfusion model in the rabbit.

Results

In Vitro Effect of the Monoclonal Antibodies R3.3 and R3.1 on the Adherence of Granulocytes to Endothelium

We evaluated the abilities of two murine MAbs to inhibit the adherence of granulocytes to endothelial cells (EC) in vitro, in response to phorbol myristate acetate (PMA) as a stimulator that directly enhances the adhesion of granulocytes.

Monoclonal antibody R3.3 is an IgG3 directed against an epitope on the common β chain of the CDw18 cell-surface glycoprotein complex and MAb R3.1 is a IgG1 that binds to an epitope on the LFA-1 α-chain of the CDw18 complex.

As illustrated in Figure 10.1, MAb R3.3 (anti-common β chain) significantly ($p < 0.05$) reduced the adherence of human granulocytes to human umbilical vein endothelial cells (HUVEC). No statistically significant reduction of adherence could be seen using MAb R3.1 (anti-LFA-1 α chain).

A pronounced inhibition with R3.3 and no effect with R3.1 have also been found measuring the adherence of granulocytes to gelatin as a cell-free surface (Table 10.1).

Comparing the adherence of granulocytes from different species, we found that MAb R3.3 inhibited the adherence of rabbit granulocytes as much as the adherence of human granulocytes. No crossreactivity of MAb R3.3 was found with rat and mouse granulocytes in this in vitro assay (results not shown).

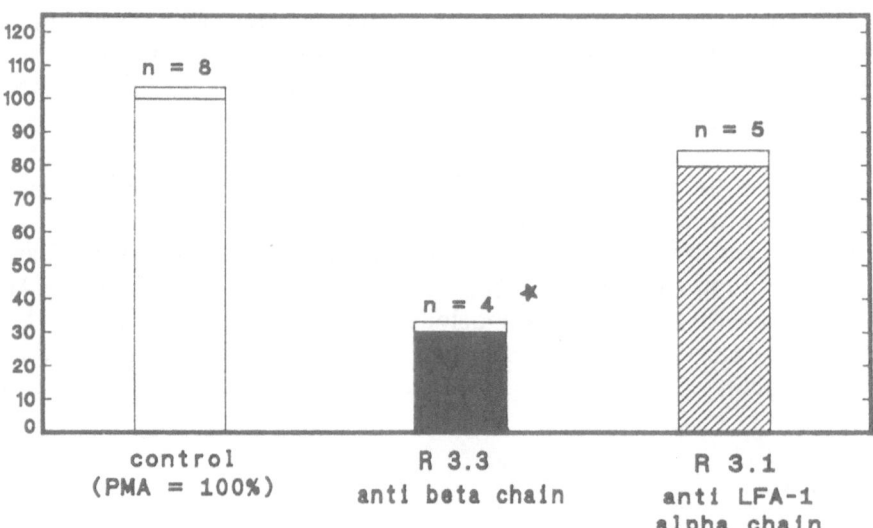

FIGURE 10.1. In vitro adherence of human granulocytes to human umbilical vein endothelial cells (HUVEC) with phorbol myristate acetate (PMA) as stimulator. The mean ± SE percent adherence (n = number of different experiments with four to eight replicates) is shown as percent adherence of the control (PMA = 100% adherence).

TABLE 10.1. Inhibition of Granulocyte Adherence by Monoclonal Antibodies

	% Adherence	
In Vitro Assay	MAb R 3.3	MAb R 3.1
1. Adherence of human granulocytes to HUVEC	30.3 ± 2.7 (4)	77.9 ± 3.4 (5)
2. Adherence of human granulocytes to cell-free surfaces	21.2 ± 2.6 (5)	102.2 ± 5.5 (5)
3. Adherence of rabbit granulocytes to cell-free surfaces	27.0 ± 2.4 (3)	83.5 ± 3.1 (3)

Human umbilical vein endothelial cells were plated in medium 199 with 10% fetal calf serum (FCS) into 16 mm microtiter wells and grown to confluence (1. passage). Granulocytes were preincubated with MAbs for 2 hours at 4 °C and added to washed EC monolayers or to cell-free surfaces (coated with gelatin).
After adding phorbol myristate acetate (PMA) (10 ng/ml) as a stimulator, the cells were incubated in the presence or absence of saturating concentrations of monoclonal antibody for 15 minutes at 37 °C. Nonadherent granulocytes were removed by gently washing each well (three times). Adherent granulocytes were stained with crystal violet (Merck) and quantitated photometrically.
The results represent the mean ± SE percent adherence of the control (PMA = 100%) resulting from the treatment with MAB with the number of experiments in parentheses. In each experiment (with four to eight replicates) different donors for granulocytes and endothelial cells were used.

Effect of Monoclonal Antibodies on Myocardial Infarct Size in Rabbits

Two MAbs binding to distinct molecules on the CDw18 glycoprotein complex (R3.3 and R3.1) and one MAb (R6.5) recognizing the intercellular adhesion molecule (ICAM-1) have been used to determine their effect on the resulting myocardial infarct size in an ischemia-reperfusion model in the rabbit. An unspecific murine MAb was prepared in a similar way and used as a control MAb.

Regional myocardial ischemia was produced by occluding the left anterior descending coronary artery (LAD) for 60 minutes. The period of occlusion was followed by 5 hours of reperfusion. In the first set of experiments either MAb or saline were administered by infusing them slowly (over 2–3 minutes) 15 minutes before occluding the LAD.

Treatment groups:

Group 1, control: Received the vehicle for the
     (n = 15)   antibody (0.9% saline), i.v.
Group 2, MAb R3.1: Received 0.5 mg/kg MAb R3.1
     (n = 10)   anti-LFA-1 α-chain (IgG1), i.v.
Group 3, MAb R3.3: Received 0.5 mg/kg MAb R3.3
     (n = 7)   anti-common beta chain (IgG3), i.v.

Group 4, MAb R6.5: Received 1 mg/kg MAb R6.5
   (n = 9)   anti-ICAM-1 (IgG 2a), i.v.
Group 5, MAb GZ 4: Received 1 mg/kg of a control
   (n = 10)   murine MAb (IgG1), i.v.

The area at risk was similar in all five treatment groups. As illustrated in Figure 10.2, the infarct sizes of rabbits subjected to 1-hour occlusion and 5 hours of reperfusion treated with MAb R3.3, R3.1, and R6.5 were significantly reduced compared to the saline control group as well as compared to the MAb control group. Resulting infarct sizes expressed as a percentage of risk area were: control (saline): 64.37 ± 3.06; MAb R3.1: 32.54 ± 7.40; MAb R3.3: 20.43 ± 4.10; MAb R6.5: 31.21 ± 6.3, and control MAb: 51.36 ± 6.85.

R3.1 reduced the infarct size by 50%, R3.3 by 68%, and R6.5 by 51% compared to the control (saline). The infarct size of the saline control group was not significantly different compared to the MAb control group.

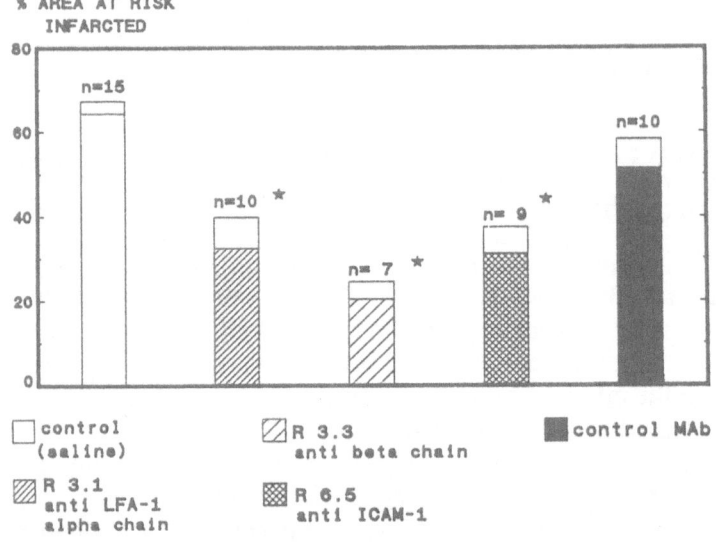

FIGURE 10.2. Rabbits of either sex with a body weight of ± 2 kg were anaesthetized, the chest and pericardium were opened and the animals were allowed to equilibrate for 30 to 60 minutes. Rabbits were then subjected to 60 minutes of regional myocardial ischemia (occluding the LAD) and 5 hours of reperfusion.
The area at risk, the nonischemic area and the infarcted area were demarcated by a dual staining method (9) and the quantification of these areas was made planimetrically.
Myocardial infarct size is expressed as percent of area at risk ± SE (n = number of animals in the treatment group).
Statistical comparison between groups were made by Student's t-test. Differences were considered significant when $p < 0.05$.

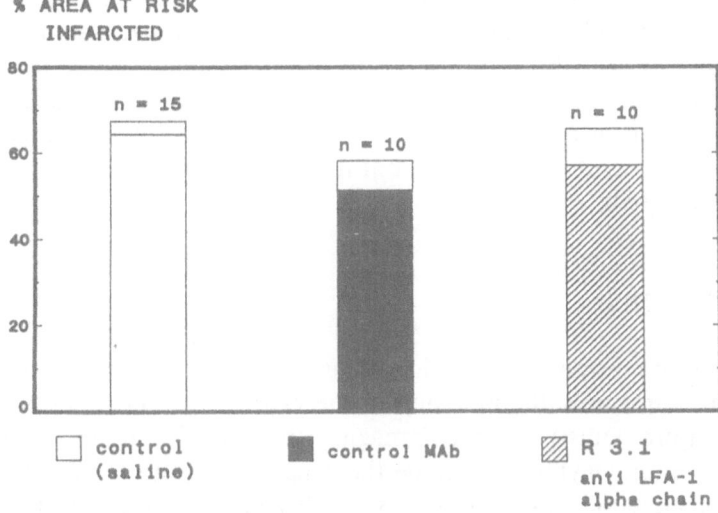

FIGURE 10.3. Myocardial infarct size is expressed as a percentage of area at risk. Regional myocardial ischemia was produced by occluding the LAD for 60 minutes. MAb R3.1 (anti-LFA-1 α chain) was administered at 0.5 mg/kg, 3 minutes before beginning the reperfusion period of 5 hours.

These results suggest that granulocytes play a central role in the development of myocardial ischemia-reperfusion injury.

To get more information about the critical time period in which granulocytes exert their damaging effect, we altered the experimental protocol giving MAb R3.1 not before occlusion, but at the end of the occlusion period (3 minutes prior to reperfusion).

In contrast to the pronounced reduction of infarct size when administered before occlusion MAb R3.1 does not reduce infarct size significantly compared to either saline control or compared to the MAb control group when given only during reperfusion (Figure 10.3).

Hemodynamic Parameters

Mean aortic pressure, heart rate, LVP system, and dp/dt_{max} were monitored at different times throughout the experiment. No significant differences in hemodynamic parameters were observed between the treatment groups and control. Therefore, the observed beneficial effects of MAbs concerning the infarct size cannot be explained on the basis of hemodynamic differences.

Circulating Leukocyte Counts

To see if the beneficial effects of MAb R3.3, MAb R3.1, and MAb R6.5 result from a depletion of circulating granulocytes, we measured the circulating leukocyte counts during the period of ischemia and reperfusion.

As illustrated in Figure 10.4, treatment with control MAb and R3.3 did not alter significantly circulating leukocytes at any measurement point compared to the saline control group.

Leukocyte counts in rabbits treated with R3.1 were significantly less in only one measurement point. In the R6.5 treatment group, leukocyte counts were significantly reduced at five measurement points compared to control MAb.

Conclusions

Taking into account the fact that adhesion of granulocytes to the endothelium is not only the first step in the process of infiltration, but may also be a prerequisite for the damaging potency of granulocytes and the fact that granulocytes seem to be the cause of the "no reflow" phenomenon, blocking the adhesion process has been suggested to be a promising target for therapeutical intervention in myocardial infarction.

Monoclonal antibodies recognizing adhesion-promoting molecules on granulocytes and on endothelial cells, which are recently available, represent important tools and enable us to get more information about the relative importance of the adhesion process itself and cellular participants playing a role in myocardial ischemia-reperfusion injury.

Our study shows that ischemia-reperfusion damage can be reduced by either blocking adhesion-promoting molecules on granulocytes or blocking the ligand molecule located on endothelial cells.

R3.3 as a MAb binding to the common β chain and R3.1 as a MAb recognizing to the LFA-1 α chain of the CDw18 complex reduce infarct size in the rabbit when administered before occlusion. In vivo experiments concerning reperfusion injury using antibodies recognizing molecules of the CDw18 complex have already been published.

Studies in which MAb 60.3 was used as a highly selective monoclonal antibody binding to a function-related idiotype of the CDw18 complex demonstrated a marked reduction of multiple organ injury and an improvement in survival rates in a model of hemorrhagic shock in the rabbit (10).

Also in accordance to our study are results found in an ischemia-reperfusion model using MAb 904, which binds to Mol, a glycoprotein of the Cdw18 complex (11) and reduces infarct size in the dog.

Our study extends these observations to indicate that, in the rabbit, reduction of infarct size can also be achieved by interfering with ICAM-1, the ligand for LFA-1, which is located on the surface of endothelial cells.

Eliminating leukocytes from the circulation is not likely because there is no correlation between reduction of infarct size and reduction of circulating leukocytes.

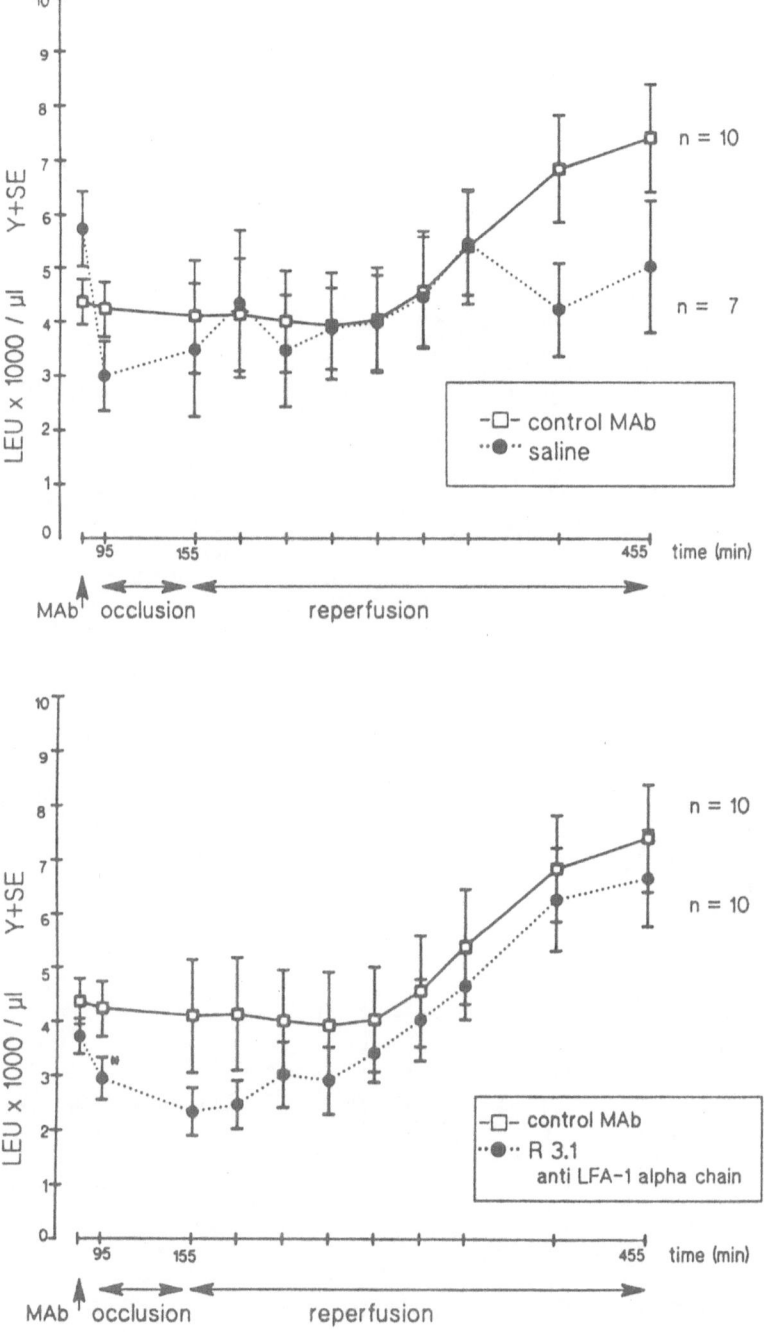

FIGURE 10.4. Time course of circulating leukocyte counts during ischemia and reperfusion. Antibodies or saline were administered 15 minutes prior to occlusion. Leukocyte counts are expressed as mean ± SE.

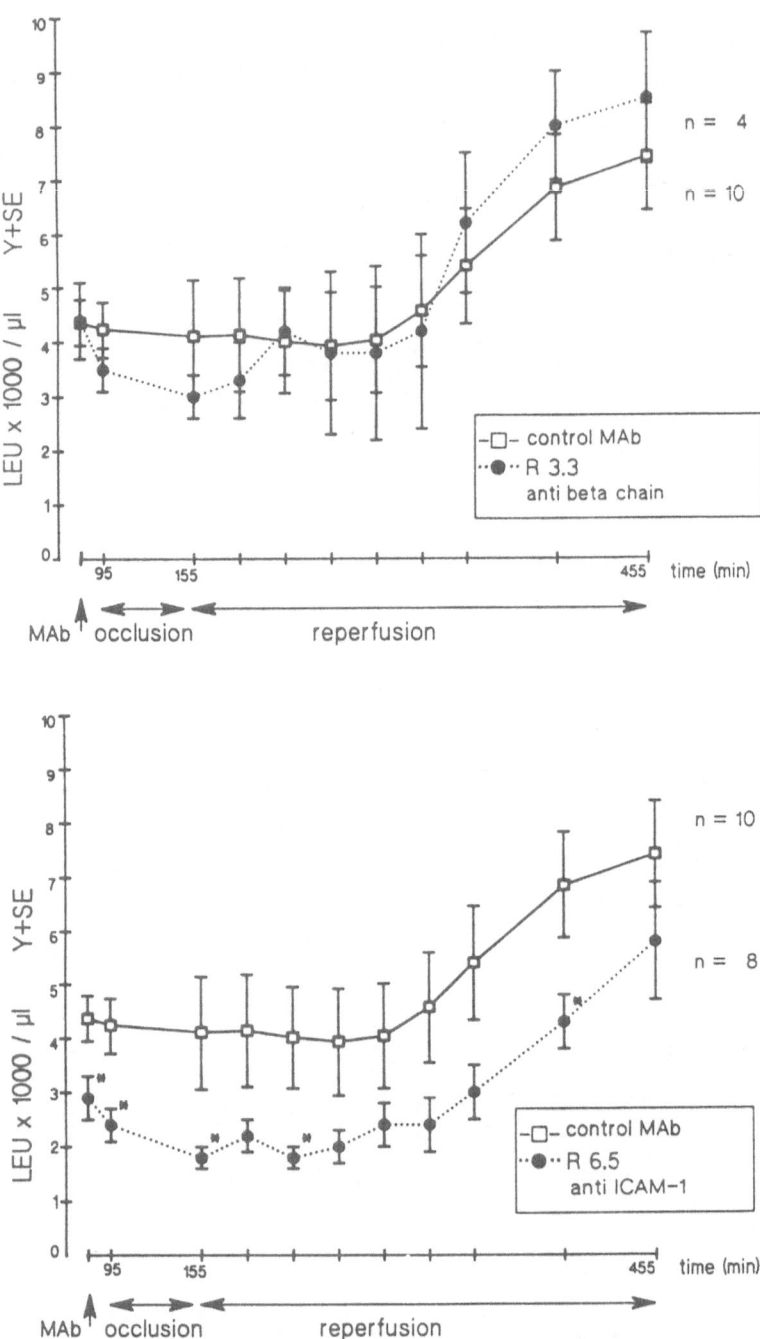

FIGURE 10.4. Continued

R3.3, showing the most prominent reduction of infarct size, does not affect circulating leukocytes at any measurement point during ischemia and reperfusion.

R3.1 and R6.5 which both reduce infarct size have lower endotoxin levels (R3.1:4.8 EU/ml; R6.5:1.2 EU/ml) compared to the control MAb (control MAb:19 EU/ml) which did not reduce infarct size significantly.

Therefore the possibility that reduction of circulating leukocytes, which could theoretically be responsible for the beneficial effect of the antibodies, is the result of a higher endotoxin level of that antibody preparation can also be ruled out.

Our results suggest that adhesion, either mediated by granulocytes or endothelial cells, plays an important role in the pathogenesis of ischemia-reperfusion in the rabbit. Our experiments show that R3.1, though effective if given before occlusion, does not reduce infarct size when administered prior to reperfusion. This suggests that at least a part of the granulocyte-mediated injury occurs during the ischemic phase. If this observation applies also to other adhesion-blocking antibodies or even to agents inhibiting other granulocyte functions, the resultant necessity of administering such agents to an early time point would have important clinical implications.

References

1. Henderson AH: The problem and the question: A clinical viewpoint, in Hearse DJ, Yellon DM (ed.): *Therapeutic Approaches to Myocardial Infarct Size Limitation*. New York, Raven Press, 1984.
2. Jolly SR, Kane WJ, Bailie MB, Abrams GD, Lucchesi BR: Canine myocardial reperfusion injury: Its reduction by the combined administration of superoxide dismutase and catalase. *Circ Res* 54:277–285, 1984.
3. Romson JL, Hook BG, Kunkel SL, Abrams GD, Schork MA, Lucchesi BR: Reduction of the extent of ischemic myocardial injury by neutrophil depletion in the dog. *Circulation* 67:1016–1023, 1983.
4. Simpson PJ, Lucchesi BR: Free radicals and myocardial ischemia and reperfusion injury. *J Lab Clin Med* 110(1):13–30, 1987.
5. Schmid-Schönbein GW, Engler RL: Granulocytes as active participants in acute myocardial ischemia and infarction. *Am J Cardiovasc Pathol* 1:15–30, 1986.
6. Anderson DC, Springer TA: Leukocyte adhesion deficiency: An inherited defect in the Mac-1,LFA-1 and pl50,95 glycoproteins. *Ann Rev Med* 38:175–194, 1987.
7. Rothlein R, Dustin ML, Marlin SD, Springer TA: An intercellular adhesion molecule (ICAM-1). *J Immunol* 137:1–5, 1986.
8. Bevilacqua MP, Pober JS, Mendrick DL, Cotran RS, Gimbrone, Jr. MA: Identification of an inducible endothelial-leukocyte adhesion molecule (E-LAM 1) using monoclonal antibodies. *Fed Proc* 46:405a, 1987.
9. Warltier DC, Zyvoloski MG, Gross GJ, Hardman HL, Brooks HI: Determination of experimental myocardial infarct size. *J Pharmacol Methods* 6:199–210, 1981.

10. Vedder NB, Winn RK, Rice CL, Chi EY, Arfors KE, Harlan JM: A monoclonal antibody to the adherence promoting leukocyte glycoprotein CD18, reduces organ injury and improves survival from hemorrhagic shock and resuscitation in rabbits. *J Clin Invest* 81:939–944, 1988.
11. Simpson PJ, Todd III RF, Fontane JC, Mickelson JK, Griffin JD, Lucchesi BR: Reduction of experimental canine myocardial reperfusion injury by a monoclonal antibody that inhibits leukocyte adhesion. *J Clin Invest* 81:624–629, 1988.

11

Role of Anti-Adhesion Monoclonal Antibodies in Rabbit Lung Inflammation

RANDALL W. BARTON, ROBERT ROTHLEIN, JOHN KSIAZEK, AND CHARLES KENNEDY

Introduction

The molecules on the leukocyte cell surface that are partially responsible for the adherence to cellular substrates have been identified as members of the LFA-1/MAC-1 family of adhesion molecules or the CD18 complex (reviewed in 1–3). A group of patients has been identified (leukocyte adherence deficiency, LAD) whose leukocytes do not express the CD18 complex on their surfaces and are unable to perform normally in in vitro, adhesion-dependent, leukocyte functional assays (4–9). The role of the CD18 complex in mediating leukocyte adhesion is confirmed when one observes that the same in vitro adhesion-related defects found in neutrophils from LAD patients are seen in leukocytes from normal individuals when they are assayed in the presence of anti-CD18 complex antibodies (3,10–15).

An inducible ligand for at least one member of the CD18 complex has been identified. This molecule, called *intercellular adhesion molecule-1* (ICAM-1) (16,17), is inducible on many cell types including vascular endothelium and keratinocytes by inflammatory mediators such as IL-1, TNF, IFNγ (18,19). Recently, it was reported that ICAM-1 mediated the CD18-dependent adhesion of neutrophils to endothelium in vitro (20). Here we show that antibodies to either CD18 or a ligand for the CD18 complex, ICAM-1, are able to inhibit granulocyte infiltration into lungs during a phorbol-ester-induced inflammatory lung model in rabbits.

Results

In Vitro Neutrophil Migration

The ability of the anti-adhesion MAbs to inhibit rabbit neutrophil migration in vitro was examined. The MAbs designated R3.1, R3.3, and R6.5 are directed against the human LFA-1α chain, the LFA-1β chain, and the ICAM-1 molecule, respectively. An additional MAb, M1/70, that binds to the MAC-1α chain (21) was also studied. As shown in Figure 11.1, both random migration and chemokinesis induced by zymosan-activated rabbit serum were inhibited >50% by anti-β (R3.3) and anti-MAC-1α (M1/70). Both anti-LFA-1α (R3.1) and anti-ICAM-1 (R6.5) were comparatively ineffective.

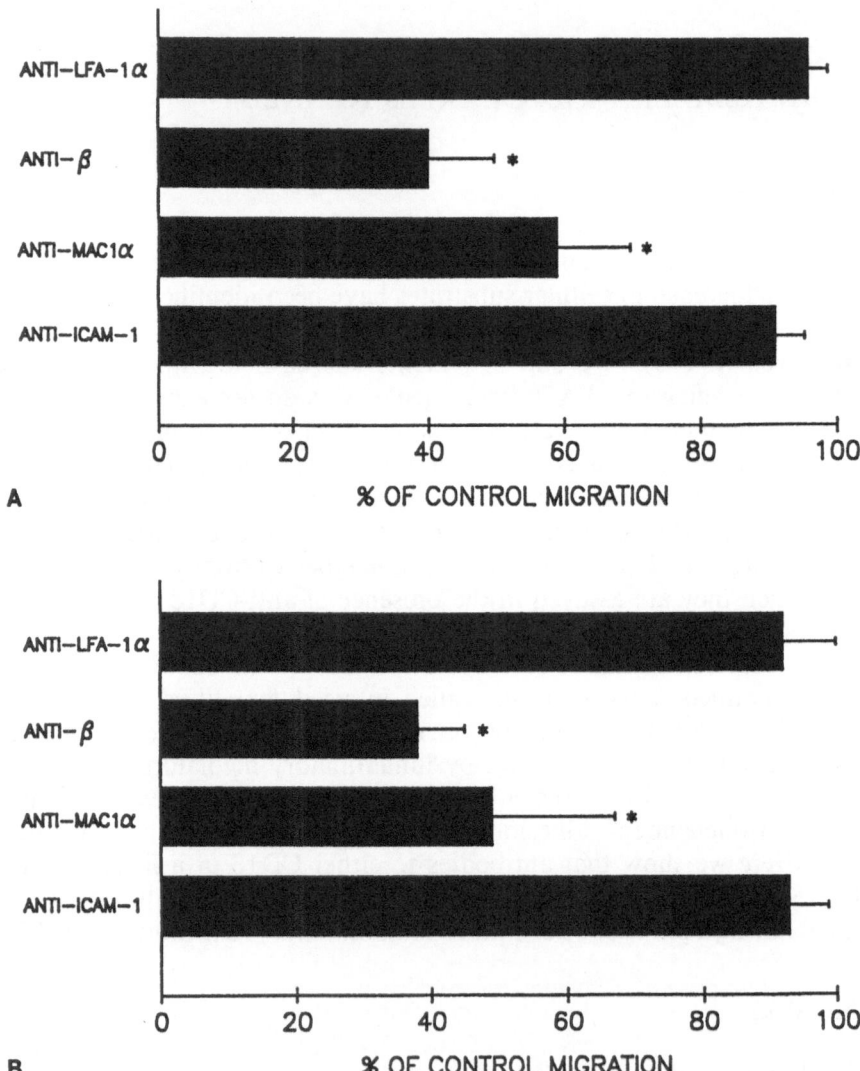

FIGURE 11.1. Effect of anti-human adhesion MAbs on in vitro neutrophil migration. (A) Random migration; (B) Chemokinesis. * $p < 0.05$

In Vivo Neutrophil Migration

In rabbits, a single intravenous injection of PMA, 40 µg/kg, induces an acute pulmonary inflammatory reaction that is characterized by neutrophil influx peaking at 20 hours after PMA injection (22,23). The efficacy of the R3.1, R3.3, and R6.5 MAbs to inhibit this neutrophil influx was compared (Fig. 11.2). One hour pretreatment of the rabbits with anti-β (R3.3), 1.25 mg/kg, and anti-ICAM-1 (R6.5), 1.5 mg/kg, resulted in significant inhibition (74 and 64%, respectively) of neutrophil influx at 20 hours post-PMA. Pretreatment with anti-LFA-1α (R3.1), 1.5 mg/kg, did not produce a statistically significant reduction. Peripheral blood cell differential analysis showed no differences between any of the antibody-treated groups and the PMA only groups.

Dose response experiments were conducted to compare the efficacy of the three anti-adhesion MAbs (Fig. 11.3 a-c). As seen in the previous experiments, anti-LFA-1α pretreatment did not produce a significant reduction of neutrophils at any dose tested. Anti-ICAM-1 pretreatment produced a significant reduction (67%) of neutrophil influx only at the highest dose whereas anti-β pretreatment produced a statistically significant reduction at both 1.25 and 0.125 mg/kg (75 and 69% reduction, respectively).

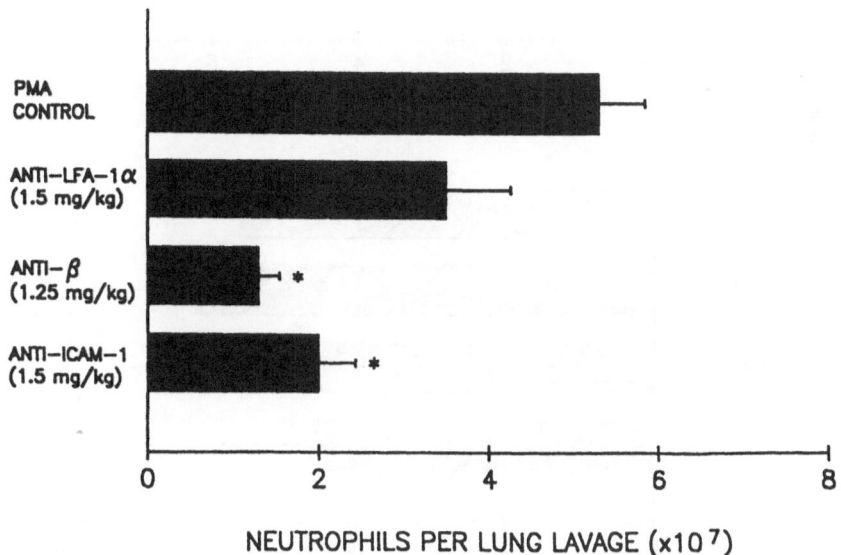

NEUTROPHILS PER LUNG LAVAGE (x10⁷)

FIGURE 11.2. Effect of anti-human adhesion MAbs on pulmonary neutrophil content 20 hours after a single intravenous injection of PMA. * $p < 0.05$.

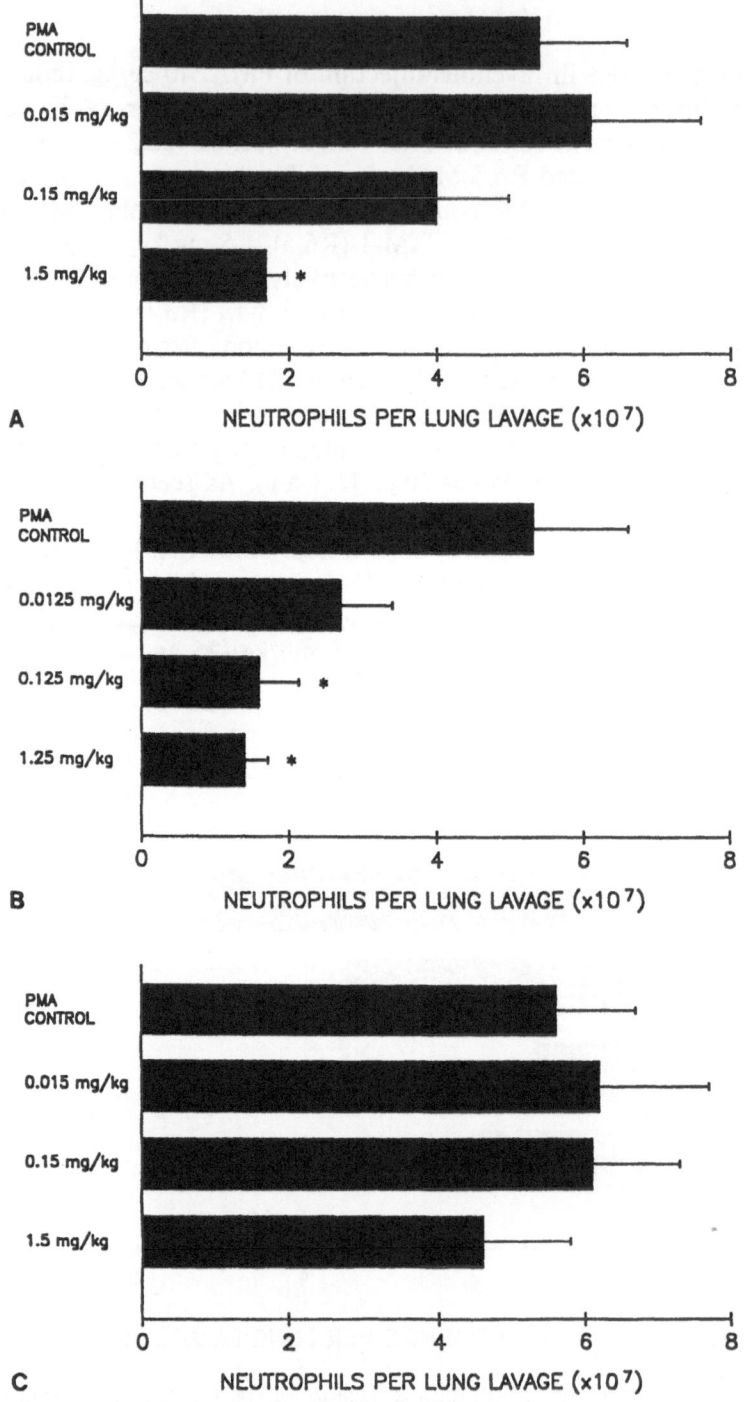

FIGURE 11.3. Dose response of anti-human adhesion MAbs on pulmonary neutrophil content 20 hours after a single intravenous injection of PMA. (A) anti-LFA-1α; (B) anti-β; (C) anti-ICAM-1.

Discussion

In the present study, we have shown that both anti-CD18 and anti-CD11b, but not anti-CD11a MAbs, inhibit rabbit neutrophil migration in vitro, an adhesion-dependent function. This effect is similar to previous findings with human neutrophils; anti-CD11b and anti-CD18, but not anti-CD11a, inhibited human neutrophil migration in vitro (1). Another anti-CD18 has been previously shown to inhibit rabbit neutrophil migration in vitro (24). We have also identified other anti-human CD18 MAbs that inhibit rabbit neutrophil migration. Thus, rabbit neutrophils express a cell surface molecule that is both antigenically and functionally similar to the human CD11b/CD18 molecule.

The ability of anti-CD18 MAb to inhibit in vitro neutrophil migration was reflected in vivo. Pretreatment with anti-CD18 resulted in a significant inhibition of lung neutrophil content 20 hours after PMA injection. In dose response experiments, this effect was even more striking; approximately similar inhibition was seen at 0.125 mg/kg as at 1.25 mg/kg. The in vivo efficacy of anti-CD18 MAb has also been reported (24–26) in inflammation models in rabbits and in a myocardial ischemia-reperfusion model in dogs (27).

The effect of anti-ICAM-1 MAb on pulmonary neutrophil influx after PMA injection was equally striking and has important implications. Specifically, that ICAM-1 mediates, at least in part, the migration of neutrophils to inflammatory sites.

ICAM-1 was originally identified by an MAb prepared against CD18-deficient human lymphocytes (16). The use of this MAb revealed that ICAM-1 mediates adherence interactions of lymphocytes with various cell types (17). Expression of ICAM-1 is enhanced in vivo on vascular endothelium and keratinocytes at sites of inflammation (17) and in vitro on fibroblasts, tumor cell lines and endothelial cells by inflammatory mediators (17,18).

Recently, ICAM-1 has been shown to be important in the CD18-dependent adherence of human neutrophils to human endothelial cells in vitro (20). The adherence of human neutrophils to human umbilical vein endothelial cells was enhanced by pretreatment of the neutrophils with f-Met-Leu-Phe (fMLP) or by pretreatment of the endothelial cells with LPS or IL-1. Both control adherence and adherence enhanced by fMLP, LPS, or IL-1 was equally blocked by the R6.5 anti-ICAM-1 MAb pretreatment of endothelial cells or anti-CD18 MAb pretreatment of granulocytes.

In summary, our results indicate that ICAM-1 mediates neutrophil migration into an inflammatory site. Moreover, since ICAM-1 expression is enhanced by inflammatory mediators in vitro and at sites of inflammation in vivo, then agents that antagonize ICAM-1 should act selectively at inflammatory sites to modulate neutrophil influx.

References

1. Martz E: Mechanism of specific tumor cell lysis by alloimmune T-lympho-cytes: Resolution and characterization of discrete steps in the cellular inter-actions. *Human Immunol* 18:3, 1977.
2. Springer TA, Dustin ML, Kishimoto TK, Marlin SD: The lymphocyte func-tion-associated LFA-1, CD2, and LFA-3 molecules: Cell adhesion receptors of the immune system. *Ann Rev Immunol* 5:223, 1987.
3. Anderson DC, Miller LJ, Schmalstieg FC, Rothlein R, Springer TA: Contri-butions of the Mac-1 glycoprotein family to adherence-dependent granulocyte functions: Structure-function assessment employing subunit-specific mono-clonal antibodies. *J Immunol* 137:15, 1986.
4. Anderson DC, Springer TA: Leukocyte adhesion deficiency: An inherited de-fect in the Mac-1, LFA-1, and p150,95 glycoproteins. *Ann Rev Med* 38:175, 1987.
5. Crowley CA, Curnutte JT, Rosin RE, Andre-Schwartz J, Gallin JI, Klempner M, Synderman R, Southwick FS, Stossel TP, Babior BM: An inherited ab-normality of neutrophil adhesion: Its genetic transmission and its association with a missing protein. *New Eng J Med* 302:1163, 1980.
6. Anderson DC, Schmalstieg FC, Arnaout MA, Kohl S, Tosi MF, Dana N, Buffone GJ, Hughes BJ, Brinkley BR, Dickery WD, Abrahamson JS, Springer T, Boxer LA, Hollers JM, Smith CW: Abnormalities of polymorphonuclear leukocyte function associated with a heritable deficiency of high molecular weight surface glycoproteins (GP138): Common relationship to diminished cell adherence. *J Clin Invest* 74:536, 1984.
7. Dana N, Todd III RF, Pitt J, Springer TA, Arnaout MA: Deficiency of a surface membrane glycoprotein (Mo1) in man. *J Clin Invest* 73:153, 1984.
8. Beatty PG, Ochs HD, Harlan JM, Price TD, Rosen H, Taylor RF, Hansen JA, Klebanoff SJ: Absence of monoclonal-antibody-defined protein complex in a boy with abnormal leukocyte function. *Lancet* 1:535, 1984.
9. Fischer A, Seger R, Durandy A, Grospierre B, Virelizier JL, LeDeist F, Gris-celli C, Fischer E, Kazatchkine M, Bohler MC, Descamps-Latscha B, Trung PH, Springer TA, Olive R, Mawas C: Deficiency of the adhesive protein complex lymphocyte function antigen 1, complement receptor 3, glycoprotein p150,95 in a girl with recurrent bacterial infection. *J Clin Invest* 76:2385, 1985.
10. Davignon D, Martz E, Reynolds T, Kurzinger K, Springer TA: Lymphocyte function-associated antigen 1 (LFA-1): Surface antigen distinct from Lyt-2,3 that participates in T lymphocyte-mediated killing. *Proc Natl Acad Sci USA* 78:4535, 1981.
11. Kohl S, Springer TA, Schmalstieg FC, Loo LS, Anderson DC: Defective nat-ural killer cytotoxicity and polymorphonuclear leukocyte antibody-dependent cellular cytotoxicity in patients with LFA-1/OKM-1 deficiency. *J Immunol* 133:2972, 1984.
12. Buescher ES, Gaither T, Nath J, Gallin JI: Abnormal adherence-related func-tions of neutrophils, monocytes, and Epstein-Barr virus-transformed B cells in a patient with C3bi receptor deficiency. *Blood* 65:1382, 1985.
13. Rothlein R, Springer TA: The requirement for lymphocyte function-associated antigen 1 in homotypic leukocyte adhesion stimulated by phorbol ester. *J Exp Med* 163:1132, 1986.

14. Tonnesen, MG, Anderson DC, Springer TA, Knedler A, Avdi N, Henson PM: Mac-1 glycoprotein family mediates adherence of neutrophils to endothelial cells stimulated by chemotactic peptides. *Clin Res* 34:419a, 1986.
15. Harlan JM, Killen PD, Senecal FM, Schwartz BR, Yee EK, Taylor RF, Beatty PG, Price TH, Ochs HD: The role of neutrophil membrane glycoprotein GP-150 in neutrophil adherence to endothelium in vitro. *Blood* 6:167, 1985.
16. Rothlein R, Dustin ML, Marlin SD, Springer TA: A human intercellular adhesion molecule (ICAM-1) distinct from LFA-1. *J Immunol* 137:1270, 1986.
17. Dustin ML, Rothlein R, Bahn AK, Dinarello CA, Springer TA: Induction by IL-1 and Interferon-gamma: Tissue distribution, biochemistry, and function of a natural adherence molecule (ICAM-1). *J Immunol* 137:245, 1986.
18. Pober JS, Gimbrone MA, Lapierre LA, Mendrick DL, Fiere W, Rothlein R, Sprinter TA: Overlapping patterns of activation of human endothelial cells by interleukin-1, tumor necrosis factor and immune interferon. *J Immunol* 137:1893, 1986.
19. Lange Vejlsgaard G, Ralfkiaer E, Avnstorp C, Czajkowski M, Marlin SD, Rothlein R: Kinetics and characterization of intercellular adhesion molecule-1 (ICAM-1) expression on keratinocytes in various inflammatory skin lesions and malignant cutaneous lymphomas. *J Am Acad Dermatol* 20:782, 1989.
20. Smith CW, Rothlein R, Hughes BJ, Mariscalco MM, Schmalsteig FC, Anderson DC: Recognition of an endothelial determinant for CD18-dependent human neutrophil adherence and transendothelial migration. *J Clin Invest* 82:1746, 1988.
21. Galfre AG, Secher DS, Milstein C: Mac-1: A macrophage differentiation antigen identified by monoclonal antibody. *Eur J Immunol* 9:301, 1979.
22. McCall CE, Taylor RG, Cousart SL, Woodruff RD, Lewis JC, O'Flaherty JT: Pulmonary injury induced by phorbol myristate acetate following intravenous administration in rabbits: Acute respiratory distress followed by pulmonary interstitial pneumonitis and pulmonary fibrosis. *Am J Pathol* 111:258, 1983.
23. Taylor RG, McCall CE, Thrall RS, Woodruff RD, O'Flaherty JT: Histopathologic features of phorbol myristate acetate-induced lung injury. *Lab Invest* 52:61, 1985.
24. Price TH, Beatty PG, Corpuz SR: In vivo inhibition of neutrophil function in the rabbit using monoclonal antibody to CD18. *J Immunol* 139:4174, 1987.
25. Arfors K-E, Lundberg C, Lindbom L, Lundberg K, Beatty PG, and Harlan JM: A monoclonal antibody to the membrane glycoprotein complex CD18 inhibits polymorphonuclear leukocyte accumulation and plasma leakage in vivo. *Blood* 69:338, 1987.
26. Vedder NB, Winn RK, Rice CL, Chi EY, Arfors K-E, Harlan JM: A monoclonal antibody to the adherence-promoting leukocyte glycoprotein CD18, reduces organ injury and improves survival from hemorrhagic shock and resuscitation in rabbits. *J Clin Invest* 81:939, 1988.
27. Simpson PJ, Todd III RF, Fantone JC, Mickelson JK, Griffen JD, Lucchesi BR: Reduction of experimental canine myocardial reperfusion injury by a monoclonal antibody (anti-Mo1, anti-CD11b) that inhibits leukocyte adhesion. *J Clin Invest* 81:624, 1988.

Part 3
Myeloid Cell Adhesion

12

The Role of LFA-1 and Related Antigens in Adhesion-Mediated Functions of Human Monocytes

CARL G. FIGDOR, ANJE TE VELDE, AND
JAN E. DE VRIES

Introduction

Both monocytes and granulocytes constituting the myeloid cell lineage play a pivotal role in the defense against foreign pathogens. Bacterial products released upon infection contain chemotactic compounds (1) that can cause activation of granulocytes, monocytes, and macrophages. This activation results in a rapid migration to the site of inflammation, thereby initiating an inflammatory response (2,3). Despite their common ancestor, major differences can be observed between monocytes and granulocytes. Granulocytes have a short half-life compared to monocytes which are destined to become long-lived tissue macrophages. In addition, they differ significantly in their kinetics. For example, monocytes accumulate in skin windows much slower than granulocytes, but are the predominant cell type after 12 to 24 hours (4). Furthermore, monocytes/macrophages fulfill an important function in the regulation of the immune response, as becomes apparent from their capacity to process and present antigen and to release immunomodulatory products such as interleukin-1 (IL-1) (3,5).

Most functions exerted by myeloid cells are dependent on cell adhesion. Adhesion to vascular endothelium prior to emigration to inflammatory sites (6–8), adhesion to microorganisms (9,10), tumor cells (11), or to other cells of the immune system (12, 13) is an essential first step in triggering the various functions exerted by these cells. Adhesion is mediated by a number of glycoprotein molecules embedded in the cell membrane. Recent studies demonstrated that the LFA-1 family of cell surface molecules, consisting of three structurally highly related glycoproteins (LFA-1, CR3, and p150,95) are of particular interest since they support numerous adhesion-mediated functions of leukocytes (14,15). The discovery that certain patients suffering from severe chronic infections lack these molecules on the cell surface of their leukocytes has significantly contributed to our understanding of their role in the inflammatory/immune response.

In the present study, we attempted to delineate the specific function of each of the molecules comprising the LFA-1 family on human monocytes, during different stages of activation/differentiation.

Results

Expression of LFA-1 Antigens on Human Monocytes and U937 Cells

Human monocytes become rapidly activated by manipulation. Adherence to plastic surfaces, which is often used to isolate these cells, causes activation as can readily be observed by an increase in the expression of HLA class II antigens (16). Therefore, we employed a different cell separation technique to purify these cells from peripheral blood. This method has been shown not to affect the functional properties of monocytes. Mononuclear cells were obtained by density centrifugation in a blood component separator, in which exposure of cells to body foreign substances was omitted (17). Subsequently, monocytes were separated from lymphocytes by means of centrifugal elutriation, which yielded >95% pure monocytes as judged by non-specific esterase staining. Table 12.1 shows the expression of LFA-1 family antigens on freshly isolated monocytes compared to monocytes cultured in the presence or absence of the phorbol ester PMA. Furthermore, we studied the effect of PMA, which can induce monocyte like differentiation (18), on the expression of these antigens on cells of the monocytic cell line U937.

TABLE 12.1. Expression of LFA-1 Family Antigens by Human Monocytes and U937 Cells

| | Fluorescence Intensity | | | |
	LFA-1α	CR3α	p150,95α	Common β
Monocytes				
Freshly isolated	15	22	11	24
Cultured (72 h) −PMA	15	23	26	26
+PMA	28	57	38	58
U937 cells				
Cultured (96 h) −PMA	5	3	3	9
+PMA	23	16	15	30

Data are expressed as fluorescence intensity analyzed with a facscan.
Monoclonal antibodies used: Anti-LFA-1: SPV-L7, anti-CR3: bear-1, anti-p150,95: S HCl-3 and anti-common β: CLB-LFA-1/1 (11,18).

The ratio in expression of LFA-1, CR3, and p150,95 on freshly isolated monocytes varies considerably among different donors. However, the expression of CR3 was always high compared to that of LFA-1, whereas the expression of p150,95 was low (18). Upon culture a rapid increase in the expression of p150,95 could be observed, which is in accordance with findings of Hogg et al. (19) that the expression of p150,95 on monocytes and granulocytes is low compared to that of tissue macrophages. A most prominent increase in the expression of all three LFA-1 family antigens can be observed on U937 cells, after exposure to PMA, which was also accompanied with an increase of the adherent properties of these cells. Similarly, treatment of monocytes with PMA generally caused a twofold increase in the expression of LFA-1 family antigens in comparison to monocytes cultured in the absence of PMA (Table 12.1). In addition, PMA induced homotypic cell aggregation (see below).

Biological Function of LFA-1 Family Molecules on Monocytes

Many of the physiological functions of monocytes depend on cell adhesion. MAbs against the common β chain of the LFA-1 family antigens inhibit adhesion, migration/chemotaxis, and phagocytosis of human monocytes (6,18,20). A more detailed analysis with a panel of anti-α chain MAbs revealed that LFA-1 α, CR3 α, and p150,95 α were also associated with adhesion-related processes, but that their individual contribution varied, and was also dependent on the function tested. The results of these studies are summarized in Table 12.2. The differences in the capacity of anti-α chain MAbs to block adhesion and adhesion related processes are not related to the level of expression of LFA-1, CR3, or p150,95 since our results demonstrate (Table 12.2; 18) that in spite of its relative low expression, p150,95 is involved in most monocyte functions tested. Adhesion to artificial substrates or to monolayers of cells, migratory or chemotactic responses to FMLP, as well as phagocytosis, are all inhibited by anti-p150,95 α antibodies. The role of CR3 seems to be limited, since anti-CR3 antibodies have only a moderate inhibitory effect on the chemokinetic and chemotactic response of monocytes to FMLP. This is in contrast to the observations with granulocytes, where CR3 seems to play a primary role in adhesion to endothelium, chemotaxis, and phagocytosis (21–24). Similarly, antibodies directed against the α chain of LFA-1 were less efficient in their capacity to block monocyte adhesion-mediated functions, although the same antibodies effectively blocked the cytolytic activity of cloned cytotoxic T lymphocytes (25,26). We consistently observed a reduced random migration of monocytes in the presence of anti-LFA-1 α antibodies. However, we were unable to correlate this observation with an impaired chemokinetic or chemotactic response as obtained with anti-p150,95 α antibodies. Antigen presenta-

TABLE 12.2. Role of LFA-1 Family Antigens in Adhesion Mediated Monocyte (U937 cell) Function

Biological Activity Tested	% Inhibition of Monocyte (U937)[a] Function with MAbs Against:			
	LFA-1α	CR3α	p150,95α	Common β
Adhesion				
To plastic surface	<5 (<5)	<5 (<5)	55 (70)[b]	65 (70)[b]
To endothelial cell monolayer	<5	<5	35[b]	35[b]
Migration				
Random	50 (50)	<5 (<5)	60 (40)[b]	70 (90)[b]
Chemokinesis	<5	10	15	35[b]
Chemotaxis	<5	20[b]	25[b]	55[b]
Phagocytosis				
Of latex beads	<5	<5	60[b]	80[b]
Antigen presentation				
HPH-induced lymphocyte proliferation	90[b]	<5	nt	nt
PWM-drived nonspecific antibody production	95[b]	10	nt	nt
TT-induced lymphocyte proliferation	50[b]	<5	<5	80[b]

See legend to Table 12.1 for MAbs used.
[a]Values between parenthesis are obtained with PMA stimulated U937 cells (see also ref. 18).
[b]p values from 0.01 to 0.005.
nt, not tested.

tion seems to be primarily mediated by LFA-1 since antibodies directed against this molecule block both antigen specific (*Helix pomatia* hemocyanin and tetanus toxoid; 12, unpublished) lymphocyte proliferation as well as polyclonal (PWM) antibody production.

Although LFA-1 family molecules play an important role in adhesion and adhesion-mediated processes, blocking of these functions by antibodies to the individual α chains or the common β chain is never complete. Also, combinations of antibodies against the different LFA-1 family molecules or of antibodies directed against distinct epitopes on one of the glycoproteins (18,21,23, unpublished) do not lead to a complete inhibition, suggesting that also other molecules are involved. This notion is compatible with observations indicating that also other members of the integrin supergene family, homing receptors, and Fc receptors may contribute to the adhesive properties of human monocytes (27–29).

Homotypic and Heterotypic Adhesion of Human Monocytes

EBV transformed B cell lines tend to aggregate spontaneously in culture in an LFA-1 dependent manner (30). The phorbol ester PMA has been shown to stimulate homotypic aggregation of these B cell line cells and induce also aggregation of CTL clones, which do not aggregate spontaneously (14,15,31). We studied the LFA-1 dependence of homotypic adhesion of human monocytes cultured in the presence of PMA or γ-IFN. Both γ-IFN and the PMA induced aggregation of human monocytes, which became maximal after 24 hours of incubation. This aggregation could be inhibited by anti-common β and anti-LFA-1 α antibodies but not significantly by anti-CR3 α or p150,95 α antibodies (Table 12.3). These results indicate that in spite of the fact that monocytes express all three members of the LFA-1 family, this type of adhesion seems to depend solely on LFA-1, which is in line with observations of others (32). In contrast, FMLP induced homotypic aggregation of granulocytes is mainly mediated by CR3 (33).

Furthermore, we investigated heterotypic cell adhesion. Two different methods were employed: (i) adhesion to monolayers of melanoma or endothelial cells; (ii) adhesion by means of single cell conjugate formation (11). The results which are summarized in Table 12.3 show a clear discrepancy between the two different assays. Adhesion to monolayers of cells, or to artificial substrates (Table 12.2), is only mediated by p150,95 since only anti-p150,95 α and anti-common β antibodies inhibit binding.

TABLE 12.3. Role of LFA-1 Family Antigens in Homotypic and Heterotypic Adhesion of Monocytes

Cell Type and Treatment	% Inhibition of Cell Adhesion with MAbs Against:			
	LFA-1α	CR3α	p150,95α	Common β
Homotypic adhesion				
γ-IFN aggregation	>80	10	10	>90
PMA-induced aggregations	>80	10	10	>90
Heterotypic adhesion:				
To cell monolayer of				
Endothelial cells	<5	<5	35	35
Melanoma cells	<5	<5	30	30
To single cell suspension of				
Endothelial cells	70	50	60	85
Melanoma cells	65	50	55	90

See legend to Table 12.1 for MAb used. The heterotypic and homotypic adhesion assays were carried out as described previously (11, 31).

On the other hand, all three members of the LFA-1 family are involved in single cell conjugate formation of human monocytes with endothelial cells or with melanoma tumor cells, irrespective of the fact that almost twice as many melanoma cells adhered to the monocytes in comparison with the endothelial cell/monocyte conjugates (11). All anti-α and anti-common β antibodies were effective in inhibiting conjugate formation. Collectively, these data indicate that different interaction mechanisms are involved in the various assays. The observation that binding to monolayers is only mediated by p150,95 suggests that the putative counter-structures to which LFA-1 or CR3 bind are absent or less available on the cell surface. This notion is supported by observations which show that cells growing in monolayers display a reorganization of cellular structures during adhesion, whereby molecules required for adhesion move to the interface substrate/cell surface (34).

The single cell experiments demonstrate that all LFA-1 family molecules can participate in monocyte mediated cell adhesion, provided that the ligands on the target cell to which they bind are available. In addition, the results show that the antibodies used in this study all recognize functionally active domains on the different members of the LFA-1 family. Homotypic adhesion of monocytes is mainly mediated by LFA-1 despite the presence of all three LFA-1 family molecules on the cell surface. This cannot be explained by a specific upregulation of LFA-1 upon stimulation with PMA, since the expression of all three LFA-1 antigens is increased (Table 12.1). It should be noted that homotypic adhesion is bi-directional, since both cells express LFA-1. However, this does not explain why p150,95 and CR3, which are also expressed on monocytes, do not participate in homotypic adhesion. On the other hand, there is no direct correlation between expression of LFA-1 family molecules and homotypic aggregations (31,33). This is illustrated by the observation that PMA rapidly induces (within 30 minutes) homotypic aggregation of CTL clones without a concurrent increase in the expression of LFA-1 (31). These results support the hypothesis that LFA-1 can be expressed on the cell membrane in different forms, and that PMA or other signalling molecules (31) can shift LFA-1 from an inactive to an active state, thereby inducing cell aggregation.

Modulation of LFA-1 Family Molecules by Interleukin-4

Recently, we observed that human recombinant IL-4 not only acts on lymphocytes, but also affects the function and morphology of human peripheral blood monocytes (35). Monocytes, which are cultured for 24 hours in vitro secreted chemotactic factors, but culturing in the presence of 50 U rIL-4 inhibited the release of these chemotactic substances. Furthermore, we found that IL-4 also inhibited the release of cytostatic factors which impaired the growth of A375 melanoma cell line cells (35). We now have preliminary evidence that the reduced chemotactic and cytostatic activities of supernatants of monocytes cultured in the presence of IL-4 may at least, in part, be attributed to a reduced secretion of IL-1. Interleukin-1 has been shown to act as a chemotactic factor for monocytes (36) and has cytostatic/cytotoxic effects on certain human tumor cell lines (37). Inhibition of secretion of IL-1 by IL-4 was demonstrated by means of the classical thymocyte proliferation assay and at the IL-1 mRNA level (35, unpublished).

Furthermore, we observed major changes in the morphology of monocytes during culture in the presence of IL-4. After 3 to 5 days, the cells acquired a macrophage/dendritic like appearance with long processes. This was accompanied by a major change in the phenotype of the monocytes as shown in Table 12.4. Monocytes cultured in the presence of IL-4 express a two- to threefold enhanced expression of CR3 and p150,95, whereas the number of LFA-1 molecules expressed is not altered (Table 12.4). The upregulation of CR3 and p150,95 became apparent within 20 hours after incubation and remained elevated throughout the whole culture period (7 days). γ-IFN has been shown also to increase the expression of LFA-1 family antigens. Interestingly, γ-IFN and IL-4 seem to regulate the expression in a different manner. Interleukin-4 induces primarily the expression of CR3 and p150,95, whereas incubation in the presence of γ-IFN results in an increased LFA-1 expression and the levels of CR3 and p150,95 remained unchanged (32). These data indicate that the three members of the LFA-1 family are independently regulated. This notion was supported by the finding that IL-4 is able to induce the expression of LFA-1 (Table 12.4) and LFA-3 (not shown) on Burkitt lymphoma cell line cells and a UD61 EBV-transformed B-cell line which express only

TABLE 12.4. Modulation of the Expression of LFA-1 Family Antigens on Monocytes and B Cell Line Cells by IL-4

Cells		LFA-1α	CR3α	p150,95α	Common β	CD23	HLA DR
Monocytes							
Cultured	−IL-4	99 (15)	99 (23)	99 (26)	100 (23)	0 (0)	99 (26)
(72 h)	+IL-4	99 (16)	100 (78)	99 (40)	100 (60)	45 (5)	99 (91)
B cell line cells							
BL2 cultured	−IL-4	1	0	0	0	0	nt
(48 h)	+IL-4	31	3	0	30	47	nt
UD61 cultured	−IL-4	17	0	0	14	25	nt
(48 h)	+IL-4	50	1	0	51	63	nt

See legend to Table 12.1 for anti-LFA-1 family MAb used. CD23 was detected by MAb 25 and HLA-DR by Q5/13 (see ref. 35). Data are expressed as percentage positive cells and fluorescence intensity (in parentheses) analyzed with a facscan.

low levels of LFA-1. Interleukin-4 induced LFA-1 and LFA-3 in a dose dependent fashion. Induction of LFA-1 and LFA-3 could be blocked by anti-IL-4 antiserum (38). Although recent studies indicate that certain B cells can express p150,95 on the cell membrane (39) we found no induction of CR3 or p150,95 after incubation of these B cell line cells in the presence of IL-4, indicating that IL-4 acts differently on B cells in comparison with monocytes.

Whether the upregulation of p150,95 and CR3 by IL-4 on monocytes has functional consequences is currently under investigation. The increased expression of these two antigens was also accompanied by an induction of CD23 (FcεR) and of HLA class II antigens (Table 12.4). It cannot be excluded that the macrophage/dendritic morphology and the increased HLA class II expression may affect the antigen-presenting capacity of human monocytes. More detailed functional studies are required to dissect whether IL-4 accelerates normal macrophage maturation or that it leads to the generation of cells with more specialized functions, like dendritic cells.

Conclusion

Monocytes express all three members of the LFA-1 family on their cell surface. CR3 is the most abundant antigen followed by LFA-1 and p150,95. Despite the relative low expression of p150,95 it plays a major role in the various monocyte functions. Exposure of monocytes or U937 cells to PMA results in a rapid upregulation of p150,95 and also of CR3 and LFA-1. The function of CR3 is limited and LFA-1 seems primarily of interest between interactions of cells which all express LFA-1 on their cell surface; antigen presentation and homotypic cell aggregation. The relative contribution of the three members constituting the LFA-1 family in heterotypic cell adhesion depends on the assay system applied. The results support the notion that p150,95 fulfills a pivotal role in monocyte adhesion dependent functions. Experiments with IL-4 and γ-IFN demonstrate that (i) IL-4 and γ-IFN modulate the expression of the LFA-1 family molecules, (ii) LFA-1, CR3, and p150,95 can be independently regulated (iii) modulation of the expression of the LFA-1 family antigens is different on monocytes in comparison with B cell line cells. Interleukin-4 and γ-IFN seem to act as antagonists on human monocytes, not only with respect to the monocyte phenotype, but also secretion of monokines are differently regulated (unpublished results). It is tempting to speculate that IL-4 induces macrophage/dendritic cell differentiation. However, it remains to be determined to what extent the morphological and phenotypical changes induced by IL-4 lead to altered functions such as antigen processing and presentation.

Acknowledgments. We thank Dr. G.D. Keizer for his contribution in this study and Marie Anne van Halem for secretarial help. This study was partly supported by the Koningin Wilhelmina Fonds (Netherlands Cancer Foundation), grant no. NKI 86-1.

References

1. Snyderman R, Pike MC: Chemoattractant receptors on phagocytic cells. *Ann Rev Immunol* 2:257, 1984.
2. Harlan JM: Leukocyte-endothelial interactions. Blood 65:513, 1985.
3. Nathan CF: Secretory productions of macrophages *J Clin Invest* 79:319, 1987.
4. Van Furth R, Raeburn JA, van Zwet TL: 1979. Characteristics of human mononuclear phagocytes. *Blood* 54:485, 1979.
5. Unanue ER, Rosenthal AS, (Eds): *Macrophage Regulation of Immunity.* New York, Academic Press, 1980.
6. Wallis WJ, Beatty PG, Ochs HD, Harlan JM: Human monocyte adherence to cultured vascular endothelium: Monoclonal antibody-defined mechanisms. *J Immunol* 135:2323, 1985.
7. Pohlman TH, Stanness KA, Beatty PC, Ochs HD, Harlan JM: An endothelial cell surface factor(s) induced in vitro by lipopolysaccharide, interleukin 1, and tumor necrosis factor-α increases neutrophil adherence by a CDw18-dependent mechanism. *J Immunol* 136:4548, 1986.
8. Bevilacqua MP, Pober JS, Wheeler ME, Contra RS, Gimbrone Jr. MA: Interleukin 1 acts on cultured human vascular endothelium to increase the adhesion of polymorphonuclear leukocytes, monocytes, and related leukocyte cell lines. *J Clin Invest* 76:2003, 1985.
9. Wright SD, Jong MT: Adhesion-promoting receptors on human macrophages recognize *Escherichia coli* by binding to lipopolysaccharide. *J Exp Med* 164:1876, 1986.
10. Bullock WE, Wright SD: Role of the adherence-promoting receptors, CR3, LFA-1, and p150,95 in binding of histoplasma capsulatum by human macrophages. *J Exp Med* 165:195, 1987.
11. Te Velde AA, Keizer GD, Figdor CG: Differential function of LFA-1 family molecules (CD11 and CD18) in adhesion of human monocytes to melanoma and endothelial cells. *Immunology* 61:261, 1987.
12. Keizer GD, Borst J, Figdor CG, Spits H, Miedema F, Terhorst C, de Vries JE: Biochemical and functional characteristics of the human leukocyte membrane antigen family LFA-1, Mo-1 and p150,95. *Eur J Immunol* 15:1142, 1985.
13. Dougherty GJ, Hogg N: The role of monocyte lymphocyte function-associated antigen 1 (LFA-1) in accessory cell function. *Eur J Immunol* 17:943, 1987.
14. Martz E: LFA-1 and other accessory molecules functioning in adhesions of T and B lymphocytes. *Human Immunol* 18:3, 1987.
15. Springer TA, Dustin ML, Kishimoto TK, Marlin SD: The lymphocyte function-associated LFA-1, CD2 and LFA-3 molecules: cell adhesion receptors of the immune system. *Ann Rev Immunol* 5:223, 1987.
16. Figdor, CG, te Velde AA, Leemans J, Bont WS: Differences in functional, phenotypical and physical properties of human peripheral blood monocytes

(Mo) reflect their various maturation stages, in *Leukocytes and Host Defense.* New York, Alan R. Liss, section V, p. 283–289, 1986.

17. Figdor, CG, Bont WS, Touw I, de Roos J, Roosnek EE, de Vries JE. Isolation of functionally different human monocytes by counterflow centrifugation elutriation. *Blood* 60:46, 1982.

18. Keizer, GD, te Velde AA, Schwarting R, Figdor CG, de Vries JE: Role of p150,95 in adhesion, migration, chemotaxis and phagocytosis of human monocytes. *Eur J Immunol* 17:1317, 1987.

19. Hogg N, Takacs L, Palmer DG, Selvendran Y, Allen C: The p150,95 molecule is a marker of human mononuclear phagocytes: comparison with expression of class II molecules. *Eur J Immunol* 16:240, 1986.

20. Hildreth JEK, August JT: The human lymphocyte function-associated (HLFA) antigen and a related macrophage differentiation antigen (HMac-1): Functional effects of subunit-specific monoclonal antibodies. *J Immunol* 134:3272, 1985.

21. Ross GD, Thompson RA, Walport MJ, Springer TA, Watson JV, Ward RHR, Lida J, Newman SL, Harrison RA, Lachman PJ: Characterization of patients with an increased susceptibility to bacterial infections and genetic deficiency of leukocyte membrane complement receptor type 3 and the related membrane antigen LFA-1. *Blood* 66:882, 1985.

22. Huu, TP, Chollet-Martin S, Perianin A, Marquetty C, Sourbier P, Babin-Chevaye C, Olive D, Gougerot-Pocidalo M-A, Debre P, Hakim J: Comparison of blocking effects of monoclonal antibodies anti-MO1-α and anti-LFA1-α on human neutrophil functions. *Immunology* 62:61, 1987.

23. Dana N, Styrt B, Griffin JD, Todd III RF, Klempner MS, Arnaout MA: Two functional domains in the phagocyte membrane glycoprotein Mo1 identified with monoclonal antibodies. *J Immunol* 137:3259, 1986.

24. Hickstein, DD, Smith A, Fisher W, Beatty PG, Schwartz BR, Harlan JM, Root RK, Locksley RM: Expression of leukocyte adherence-related glycoproteins during differentiation of HL-60 promyelocytic leukemia cells. *J Immunol* 138:513, 1987.

25. Spits, H, Keizer G, Borst J, Terhorst C, Hekman A, de Vries JE: Characterization of monoclonal antibodies against cell surface molecules associated with cytotoxic activity of natural and activated killer cells and cloned CTL lines. *Hybridoma* 2:423, 1983.

26. Keizer GD, Borst J, Visser W, Schwarting R, de Vries JE, Figdor CG: Membrane glycoprotein p150,95 of human cytotoxic T-cell clones is involved in conjugate formation with target cells. *J Immunol* 15:3130, 1987.

27. Gallatin M, St. John TP, Siegelman M, Reichert R, Butcher EC, Weissman IL: Lymphocyte homing receptors. *Cell* 44:673, 1986.

28. Hynes RO: Integrins: A family of cell surface receptors. *Cell* 48:549, 1987.

29. Pals ST, Kraal G, Horst E, de Groot A, Scheper RJ, Meijer CJLM: Human lymphocyte-high endothelial venule interaction: Organ-selective binding of T and B lymphocyte populations to high endothelium. *J Immunol* 137:760, 1986.

30. Mentzer SJ, Gromkowski SH, Krensky AM, Burakoff SJ, Martz E: LFA-1 membrane molecule in the recognition of homotypic adhesions of human B lymphocytes. *J Immunol* 135:9, 1985.

31. Keizer GD, Visser W, Vliem M, Figdor CG: A monoclonal antibody (NKI-L16) directed against a unique epitope on the α-chain of human leukocyte

functional-associated antigen 1 induces homotypic cell-cell interactions. *J Immunol* 140:1393, 1988.

32. Mentzer SJ, Faller DV, Burakoff SJ: Interferon-γ induction of LFA-1 mediated homotypic adhesion of human monocytes. *J Immunol* 137:108, 1986.

33. Buyon JP, Abramson SB, Philips MR, Slade SG, Ross GD, Weissmann G, Winchester RG: Dissociation between increased surface expression of Gp165/95 and homotypic neutrophil aggregation. *J Immunol* 140:3156, 1988.

34. Chen W-T, Singer SJ: Immunoelectron microscopic studies of the sites of cell-substratum and cell-cell contacts in cultured fibroblasts. *J Cell Biol* 95:205, 1982.

35. Te Velde AA, Klomp JPG, Yard BA, de Vries JE, Figdor CG: Modulation of phenotypic and functional properties of human peripheral blood monocytes by IL-4. *J Immunol* 140:1548, 1988.

36. Luger TA, Charon JA, Colot M, Micksche M, Oppenheim JJ: Chemotactic properties of partially purified human epidermal cell-derived thymocyte-activating factor (ETAF) for polymorphonuclear and mononuclear cells. *J Immunol* 131:816, 1983.

37. Onozaki K, Matsushima K, Aggarwal BB, Oppenheim JJ: Human interleukin 1 is a cytocidal factor for several tumor cell lines. *J Immunol* 135:3962, 1985.

38. Rousset F, Billaud M, Figdor CG, Blanchard D, Lenoir GM, Spits H, de Vries JE: IL-4 induces LFA-1 and LFA-3 expression on Burkitt's lymphoma cell lines: requirement of LFA-1 activation for induction of homotypic cell adhesions. *J Immunol* in press.

39. De la Heral A, Alvaren-Mon M, Sanchez-Madrid F, Martinez C, Durantez A: Co-expression of Mac-1 and p150,95 on CD5⁺ B cells. Structural and functional characterization in a human chronic lymphocytic leukemia. *Eur J Immunol* 18:1131, 1988.

13

Role of ICAM-1 in the Adherence of Human Neutrophils to Human Endothelial Cells In Vitro

C. Wayne Smith, Steven D. Marlin,
Robert Rothlein, Michael B. Lawrence,
Larry V. McIntire, and Donald C. Anderson

Introduction

The adherence of human neutrophils to endothelial cells and protein-coated foreign surfaces in vitro is significantly increased by chemotactic stimulation or exposure of the cells to secretagogues (1–7). The CD11b/CD18 (Mac-1) heterodimer on the neutrophil's surface appears to play a role as shown by the inhibitory effect of some monoclonal antibodies reactive with either CD11b or CD18 (4,5,8,9). Stimulation of endothelial cells with bacterial endotoxin (LPS) (4,10,11), interleukin-1 (IL-1) (4,12,13), tumor necrosis factor-α (TNF-α) (4,5,12–14), or lymphotoxin (LT) (13) increases the adherence of unstimulated neutrophils. In contrast to the rapid response following chemotactic stimulation of neutrophils, this increase is not evident until >1 hour after stimulation, and protein synthesis is required. The specific CD11/CD18 heterodimers important to this cytokine-induced adherence have not been defined. We have recently obtained evidence that the CD18-dependent adherence of human neutrophils to cytokine-stimulated endothelial cells involves intercellular adherence molecule-1 (ICAM-1) expressed on the endothelial surface (7). In light of recent evidence that ICAM-1 (15,16) is a ligand for the CD11a/CD18 (LFA-1) heterodimer (17,18), consideration was given to the possibility that LFA-1 contributes to the adherence of neutrophils to cytokine stimulated endothelial cells. Though LFA-1 is apparently not involved in homotypic aggregation of neutrophils or their adherence to foreign surfaces (8), its role in the adherence of neutrophils to endothelial cells has not been determined.

In the present study, we evaluate the roles of the CD11/CD18 family of glycoproteins and ICAM-1 in the adherence of human neutrophils to human umbilical vain endothelial cells (HUVEC), and we define the conditions of wall shear stress under which these determinants are operative. We provide evidence that LFA-1 contributes to the adherence of unstimulated neutrophils to HUVEC, that LFA-1 is most likely interacting with ICAM-1 on the endothelial cell surface, and that LFA-1 is involved in the transendothelial migration induced by activation of HUVEC with cytokines. We also provide evidence that Mac-1 is important in the adhesion of chemotactically stimulated neutrophils to HUVEC, that Mac-1 on chemotactically stimulated cells may interact with ICAM-1, and that Mac-1 is involved in transendothelial migration.

Results

An in vitro model of leukocyte-endothelial cell adhesion was used in the current studies. Isolated human neutrophils were exposed to monolayers of HUVEC in two distinct settings. The first was a closed chamber in which neutrophils were allowed to settle onto the endothelial monolayer in the absence of shear stress using techniques previously described (6,7). The second was a parallel plate flow chamber into which the neutrophil suspension was infused over a 15 minute observation period, thus requiring the leukocytes to adhere in the presence of a wall shear stress determined by the flow of the cell suspension. The specific aspects of this technique have been published (19,20). Figure 13.1 is a schematic presentation of observations made in these two experimental settings. Neutrophils in suspension may adhere to the apical surface of the endothelial monolayer. After attachment, a portion of the cells become activated as evidenced by membrane ruffling and shape change, and a portion will migrate to a position beneath the monolayer (2,7,21–23). These cells become markedly flattened as they migrate in a space between the HUVEC monolayer and the substratum (Fig. 13.2). The percentage of neutrophils that attach, change shape, or exhibit transendothelial migration is greatly influenced by inflammatory mediators such as chemotactic factors or cytokines (e.g., interleukin-1) (7,21). In the absence of shear stress,

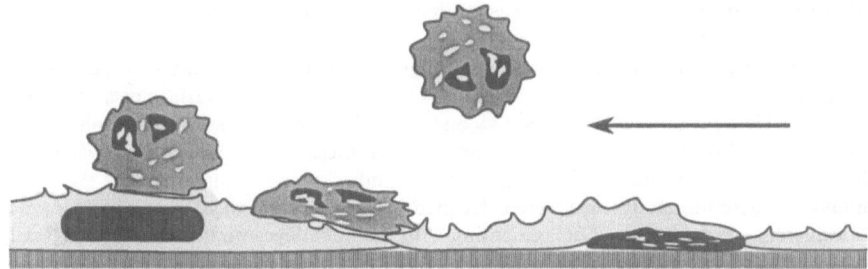

FIGURE 13.1. Schematic representation of neutrophil-endothelial cell interactions at defined wall shear stresses.

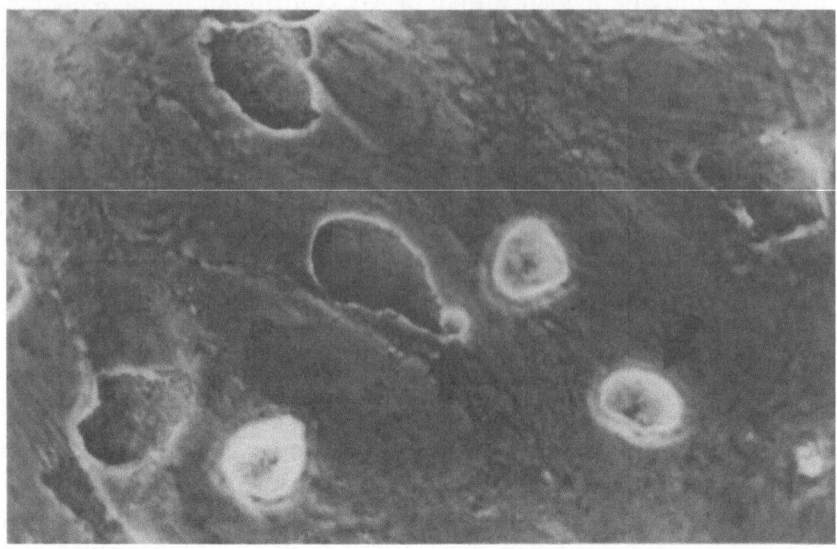

FIGURE 13.2. Phase contrast photomicrograph of neutrophils adherent to a confluent monolayer of human umbilical vein endothelial cells in vitro. Neutrophils on the apical surface of the monolayer are rounded with a surrounding halo (*arrow*). A neutrophil migrating through the monolayer is shown with the tip of its uropod remaining above the monolayer (*arrow*).

TABLE 13.1. Effects of IL-1 and fMLP on Adherence and Transendothelial Migration of Neutrophils In Vitro. Observations Made in the Absence of Shear Stress

Experimental Condition[a]	n[d]	Adherence[e]	Migration[e]
No stimulus	23	16.9 ± 6.4	None
IL-1 (0.3 U/ml, 4 hr.)[b]	25	89.6 ± 8.4**	64.9 ± 10.8
fMLP (10 nM, 5 min.)[c]	20	46.2 ± 6.9**	<2

[a]Treatment of cells prior to determining adherence and migration.
[b]Pretreatment of HUVEC. Monolayers were washed by dipping in two changes of PBS before being placed in the adherence chamber.
[c]Pretreament of neutrophils. fMLP was retained with the cells during adhesion assay.
[d]Number of separate experiments
[e]Expressed as percent adherence or percent migration ± 1 s.d. Neutrophil suspensions were injected into a closed chamber and allowed to settle onto the surface of a visually confluent HUVEC monolayer for 500 seconds at 37 °C. The number of neutrophils in 10 high power fields was determined using an inverted microscope. The chamber was then inverted for an additional 500 seconds and the number of cells remaining with the monolayer determined. Results were calculated in the following ways:
% Adherence = number of cells remaining with the monolayer/number of cells originally in contact with the monolayer × 100.
% Migration = number of cells beneath the monolayer/number of cells originally in contact with the monolayer × 100.
**$p < 0.01$ compared with values in the absence of stimulus.

IL-1 stimulation of the HUVEC markedly enhanced adherence and transendothelial migration of previously unstimulated neutrophils, and chemotactic stimulation significantly increased adherence of neutrophils to unstimulated HUVEC (Table 13.1). Interleukin-1 stimulation of HU-

TABLE 13.2. Adherence and Transendothelial Migration of Neutrophils In Vitro at Defined Wall Shear Stresses

Experimental Conditions[a]	n[b]	Adherence (PMNL/mm^2)[e]	Migration (PMNL/mm^2)[f]
No stimulus			
3.0	5	7 ± 4	<1
2.0	4	16 ± 9	<1
1.0	4	13 ± 7	<1
0.5	6	48 ± 14	<1
0.25	5	128 ± 23	<5
IL-1 (U/ml, 4 h)[c]			
3.0	5	10 ± 4	<1
2.0	5	371 ± 26	59 ± 22
1.0	—	—	—
0.5	4	637 ± 26	158 ± 40
0.25	—	—	—
fMLP (10 nM, 5 min.)[d]			
2.0	3	17 ± 15	<1
1.0	3	25 ± 21	<1
0.5	5	332 ± 71	<5
0.25	5	856 ± 39	<5

[a]Experiments were performed in a parallel plate flow chamber as previously described (19) at the wall shear stresses given, expressed in dynes/cm^2. When assembled, the two flat surfaces of the flow chamber were held by a silastic gasket approximately 250 microns apart, and a confluent HUVEC monolayer was on the lower glass plate. Neutrophil interactions with the monolayer were directly visualized using phase-contrast optics.
[b]Number of separate experiments.
[c]Pretreatment of HUVEC. Monolayers were washed for 5 minutes at 2.0 dynes/cm^2 with PBS after the flow chamber was assembled.
[d]Pretreatment of neutrophils. Neutrophils were exposed to fMLP immediately before the infusion into the flow chamber, and fMLP was retained with the cells during the observation period.
[e]Adherence occurred during flow. The number of cells associated with the monolayer over a 10 minute observation period was determined by evaluation videotape recording of each experiment. Cells were considered adherent if they were stationary on the monolayer surface, rolling at a velocity of <50 μm/sec., or engaged in transendothelial migration. Values given are numbers of neutrophils/mm^2 monolayer surface area.
[f]Transendothelial migration was determined by phase contrast microscopy. Neutrophils on the apical surface of the monolayer were phase bright with a surrounding halo. As they migrated through the monolayer they became phase dark. This conversion from phase bright to phase dark was observed on videotape recording of each experiment. Values given are number of migrated cells/mm^2 monolayer surface area.

VEC also significantly increased adherence and transendothelial migration of unstimulated neutrophils at wall shear stresses of 2.0 dynes/cm² or less (Table 13.2). However, chemotactic stimulation did not significantly increase adherence until the wall shear stress was reduced to 0.5 dynes/cm² or less (Table 13.2).

Role of CD18 in Adherence and Transendothelial Migration

In an effort to define the molecular determinants of neutrophil adherence to endothelium, blocking studies using monoclonal antibodies (MAbs) were performed. Confirmation that adherence of unstimulated neutrophils to IL-1 and LPS stimulated endothelium is partially dependent on CD18 is shown in Table 13.3. The anti-CD18 MAb, TS1/18, reduced adherence of normal neutrophils to the level exhibited by CD18-deficient cells. In addition, the transendothelial migration induced by LPS or IL-1 stimulation of the HUVEC was almost completely inhibited by the anti-CD18 MAb (Table 13.3). In contrast to these results, experiments performed with a wall shear stress of 2.0 dynes/cm² failed to show an

TABLE 13.3. Role of CD18 in Adherence and Transendothelial Migration of Human Neutrophils In Vitro. Observations in the Absence of Shear Stress

Pretreatment				
Neutrophils[a]	HUVEC[b]	n[d]	Adherence[e]	Migration[e]
PBS	Control	8	16.9 ± 5.4	<1
TS1/18	Control	4	3.5 ± 3.0	<1
PBS	LPS	8	72.4 ± 4.4	48.3 ± 8.9
TS1/18	LPS	8	39.8 ± 5.6*	<2*
PBS	IL-1	6	92.3 ± 7.5	69.4 ± 10.9
TS1/18	IL-1	6	42.3 ± 6.3**	4.7 ± 1.5**
CD18-Def[c]	Control	4	4.4 ± 2.0	<1
CD18-Def[c]	IL-1	4	44.9 ± 5.2**	<1**

[a]Neutrophils were suspended in PBS. Monoclonal antibody TS1/18 (anti-CD18) was added to a final concentration of 5 μg/ml, a level shown by flow cytometry to saturate binding sites on chemotactically stimulated neutrophils for 5 minutes prior to injecting the cell suspension with antibody into the adherence chamber.
[b]HUVEC monolayers were incubated with a complete medium alone (Control) or with LPS (10 ng/ml) or IL-1 (0.5 U/ml) for 3 to 4 hours, washed in PBS and placed into the adherence chamber.
[c]Neutrophils from two patients with CD18-deficiency (28).
[d]Number of separate experiments.
[e]See Table 13.1.
*$p < 0.01$ compared with LPS-treated monolayers without monoclonal antibody added.
**$p < 0.01$ compared with IL-1-treated monolayers without monoclonal antibody added.

inhibitory effect of the anti-CD18 MAb on the attachment of unstimulated neutrophils to IL-1 stimulated HUVEC (Table 13.4). In addition, CD18-deficient leukocytes attached as well as normal cells (Table 13.4). The only effect of the anti-CD18 MAb was on transendothelial migration, where the inhibition was >80%. Supporting a role for the CD18 family in this phenomenon were the observations that CD18-deficient neutrophils exhibited markedly low transendothelial migration (Table 13.4).

While attachment to cytokine-stimulated endothelium under static conditions was partially dependent on CD18, the elevated level of attachment produced by chemotactic stimulation of the neutrophil was inhibited by >90% by the anti-CD18 MAb (Table 13.5). Additional confirmation was obtained by attempts to enhance adherence of CD18-deficient neutrophils. Adherence was not increased (Table 13.5). CD18-dependent attachment appears to be more susceptible to shear stress than that induced by IL-1 stimulation of HUVEC. Figure 13.3 illustrates that CD18 is necessary for attachment at a wall shear stress of 0.25 dynes/cm², since the presence of the anti-CD18 MAb, adherence was as low as

TABLE 13.4. Role of CD18 in Adherence and Transendothlial Migration of Neutrophils In Vitro at Defined Wall Shear Stresses

Pretreatment				
A. 2.0 dynes/cm² Neutrophils[a]	HUVEC[b]	n[d]	Adherence (PMNL/mm²)[e]	Migration (PMNL/mm²)[e]
PBS	Control	8	3 ± 2	<5
PBS	IL-1	7	334 ± 63	59 ± 22
TS1/18	IL-1	7	305 ± 58	11 ± 5**
CD18-Def[c]	IL-1	1	323	2
B. 05. dynes/cm² Neutrophils[a]	HUVEC[b]			
PBS	Control	6	48 ± 34	6 ± 4
PBS	IL-1	4	637 ± 26	158 ± 40
TS1/18	IL-1	4	505 ± 22*	26 ± 8**
CD18-Def[c]	IL-1	1	368	0

[a]Neutrophils were suspended in PBS. Monoclonal antibody TS1/18 (anti-CD18) was added at 40 μg/ml, a level 8 times that shown by flow cytometry to sauturated binding sites on neutrophils, was added 15 minutes prior to beginning infusion of the cell suspension containing the antibody into the flow chamber.
[b]HUVEC monolayers were incubated in complete medium alone or with IL-1 (2 U/ml) for 4 hours prior to placement in the flow chamber. Monolayers were washed with PBS for 5 minutes at a shear stress of 2.0 dynes/cm² prior to infusion of the neutrophils.
[c]Neutrophils from a patient with CD18-deficiency (28).
[d]number of separate experiments.
[e]See Table 13.2.
*$p < 0.5$.
**$p < 0.01$.

TABLE 13.5. Role of CD18 in Chemotactic Enhancement of Neutrophil Adherence to Unstimulated Endothelial Monolayers. Observations Made in the Absence of Shear Stress

Pretreatment of Neutrophils[a]	n[c]	Adherence[d]*
PBS	8	16.9 ± 5.4
TS1/18	4	3.5 ± 3.0
fMLP	7	52.6 ± 6.3
fMLP, TS1/18	8	4.4 ± 3.9*
CD18-Def[b]	4	4.4 ± 2.0
CD18-Def[b], fMLP	4	3.4 ± 1.9

[a]Neutrophils were suspended in PBS. Monoclonal antibody TS1/18 (anti-CD18) was added to a final concentration ofd 5 μg/ml, a level shown by flow cytometry to saturate binding sites on chemotactically stimulated neutrophils, for 5 minutes prior to injecting the cell suspension with antibody into the adherence chamber.
[b]Neutrophils from two patients with CD18-deficiency (28).
[c]Number of separate experiments.
[d]See Table 1.
*$p < 0.01$ compared with fMLP stimulated adherence without monoclonal antibody added.

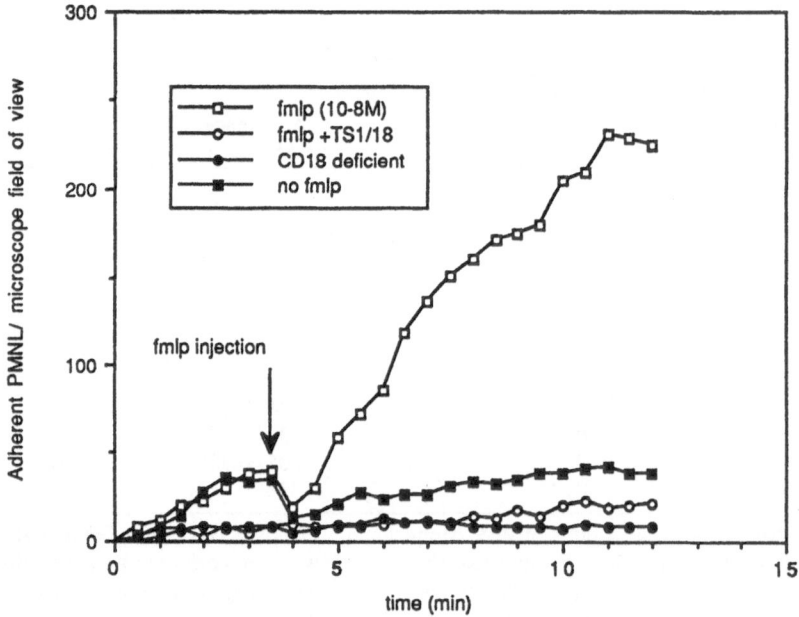

FIGURE 13.3. Kinetics of fMLP-stimulated increase in PMNL adherence to HU-VEC monolayers in vitro. These plots show addition of fMLP at a final concentration of 10 nM to the neutrophil suspension after a baseline level of unstimulated adherence had been established at 0.25 dynes/cm² for 4 minutes. In addition to fMLP stimulated normal neutrophils are experiments with neutrophils in the presence of 20 μg/ml TS1/18, neutrophils from a patient with CD18 deficiency, and normal neutrophils without fMLP stimulation.

that of CD18-deficient cells. As shown in Table 13.4, a CD18-dependent component to the adherence of neutrophils to IL-1 stimulated HUVEC is evident at a wall shear stress of 0.5 dynes/cm².

Role of ICAM-1 in Adherence and Transendothelial Migration

To evaluate the potential role of ICAM-1 in the adherence of neutrophils to endothelial cells, blocking experiments using a newly developed MAb (R6.5.D6, IgG2a) reactive with ICAM-1 were performed. The typical distribution of bound R6.5.D6 on the apical surface of endothelial cells is shown in Figure 13.4. Binding of R6.5.D6 to monolayers of endothelium is increased by stimulation of the HUVEC with IL-1 as shown in Figure 13.5.

Also shown in this figure is the coincidental increase in adherence of unstimulated neutrophils that occurs under static conditions and the partial inhibition resulting from preincubation of the HUVEC with R6.5.D6. The portion of stimulated adherence not inhibitable by R6.5.D6 peaks at 4 hours after stimulation and returns to baseline within 8 hours. The portion inhibited by R6.5.D6 remains relatively constant over the observation period after peaking at 4 hours.

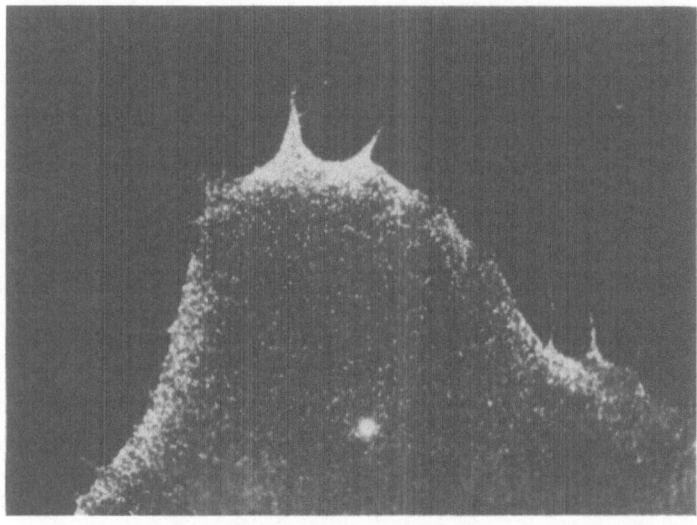

FIGURE 13.4. Immunofluorescent photomicrograph of a human umbilical vein endothelial cell exposed to monoclonal antibody R6.5.D6. The presence of bound R6.5.D6 was detected using FITC-labeled F(ab')₂ rabbit anti-mouse IgG antibodies.

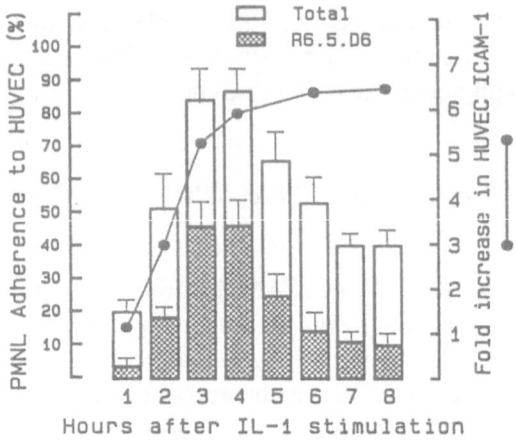

FIGURE 13.5. Kinetics of the effect of IL-1 stimulation of HUVEC monolayers on the adherence of neutrophils and the surface expression of ICAM-1. Monolayers were exposed to 0.3 U/ml IL-1 at 37 °C for the times indicated. Surface expression of ICAM-1 was determined using an enzyme immunoassay employing R6.5.D6 and alkaline phosphatase-labeled goat anti-mouse IgG antiserum. Adherence to stimulated monolayers or to stimulated monolayers preincubated with 10 μg/ml R6.5.D6 was determined in the absence of shear stress (expressed as mean percent adherence ±SD).

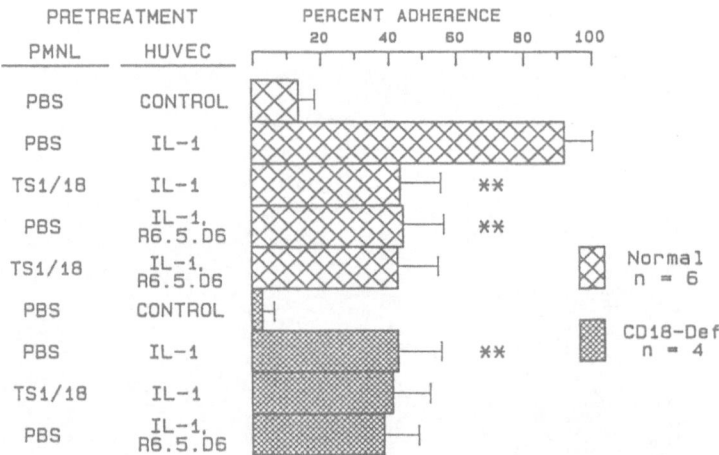

FIGURE 13.6. CD18 and ICAM-1 in the adherence of human neutrophils (PMNL) to HUVEC monolayers in vitro. Normal neutrophils or neutrophils from patients with CD18 deficiency were suspended in PBS alone or in PBS containing 5 μg/ ml TS1/18. HUVEC monolayers were unstimulated or stimulated for 4 hours with 0.5 U/ml IL-1, rinsed and placed in adherence chambers. The PMNL suspensions were injected and adherence was determined at room temperature without shear stress (7). **$p < 0.01$ compared to the adherence of untreated normal neutrophils to IL-1 stimulated HUVEC.

The relationship of ICAM-1 to CD18-dependent adherence is indicated by the results shown in Figure 13.6. Under static conditions, R6.5.D6 reduced adherence to the same degree as the anti-CD18 MAb, and when used together in the same experiment, they produced no additive or synergistic inhibition. Neither R6.5.D6 nor the anti-CD18 MAb reduced the attachment of CD18-deficient neutrophils to IL-1 stimulated HUVEC (Fig. 13.6). An isotype-matched control antibody, W6/32 (IgG2a), was without effect on neutrophil adherence (7).

Data presented above indicate that transendothelial migration is dependent on CD18. Data presented in Figure 13.7 indicate that ICAM-1 is also necessary for this migration. The reduction in adherence and migration caused by R6.5.D6 was comparable to that caused by anti-CD18

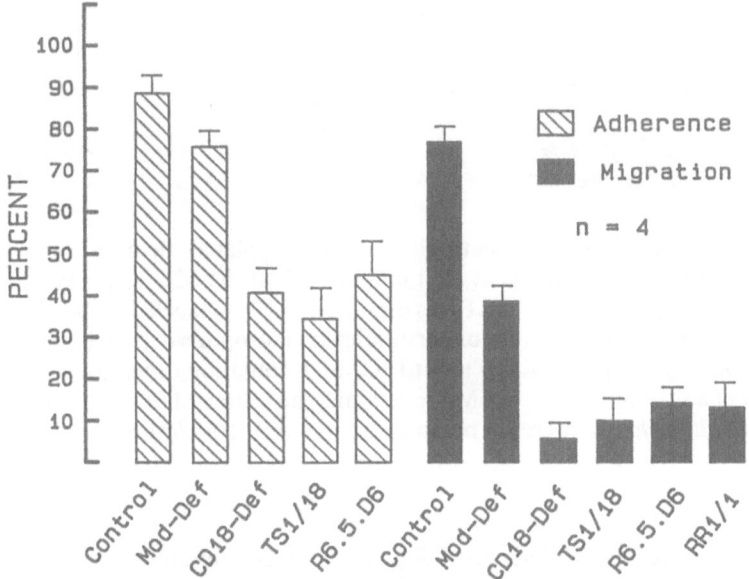

FIGURE 13.7. CD18 and ICAM-1 in the adherence and transendothelial migration of neutrophils. HUVEC monolayers were stimulated for 4 hours with 0.3 U/ml IL-1, rinsed and placed in adherence chambers. Neutrophils suspended in PBS were injected and adherence and transendothelial migration determined at 37 °C without shear stress. Mod-Def indicates the adherence of neutrophils from a patient with moderate deficiency of CD18. CD18-Def indicates the adherence of neutrophils from patients with severe CD18 deficiency. TS1/18 (5 μg/ml) was used to pretreat neutrophils before injection into the adherence chamber, and R6.5.D6 and RR1/1 (10μg/ml) were used to pretreat neutrophils before injection into the adherence chamber, and R6.5.D6 and RR1/1 (10 μg/ml) were used to pretreat the HUVEC monolayer prior to placing it into the adherence chamber. Results are expressed as the percent cells adhering or migrating ± 1 SD.

MAb, and that of CD18-deficient neutrophils. R6.5.D6 was equally effective in the form of F(ab')$_2$ fragments (7).

More direct evidence that neutrophils adhere to ICAM-1 was provided by studies with purified ICAM-1 in artificial planar membranes (17). As shown in Figure 13.8, unstimulated neutrophils adhered to ICAM-1 in-

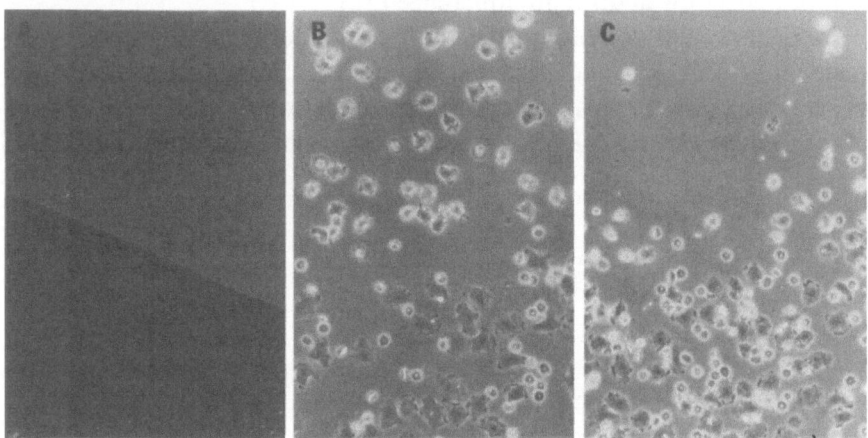

FIGURE 13.8. Adherence of neutrophils to artificial planar membranes containing ICAM-1. (*A*) Imunofluorescent photomicrograph of the edge of the planar membrane showing binding of R6.5.D6 as indicated by FITC-labeled rabbit anti-mouse IgG antiserum. (*B*) Adherence of unstimulated neutrophils of both untreated glass (*lower half of the photograph*) and to ICAM-1 containing membrane. (*C*) Inhibition of adherence to the ICAM-1-containing membrane but not untreated glass (*lower half of the photograph*) by pretreatment with 10µg/ml R6.5.D6.

TABLE 13.6. Role of ICAM-1 in Adherence and Transendothelial Migration of Neutrophils In Vitro. Observations Made at a Defined Wall Shear Stress of 2.0 dynes/cm^2

Pretreatment of HUVEC[a]	n[b]	Adherence (PMNL/mm^2)[c]	Migration (PMNL/mm^2)[c]
IL-1	6	287 ± 53	68 ± 40
IL-1, R6.5	6	246 ± 52	8 ± 4**

[a]HUVEC monolayers were exposed to IL-1 (2 U/ml) for 4 hours, placed in flow chambers and washed with PBS for 5 minutes at a shear stress of 2.0 dynes/cm^2. Unstimulated neutrophils suspended in PBS were infused throughout the observation period.
[b]Number of separate experiments.
[c]See Table 2.
**$p < 0.01$.

serted in planar membranes. This adherence was inhibited by pretreatment of the membrane with R6.5.D6.

The previous studies with R6.5.D6 were all performed in the absence of shear stress. Since CD18-dependent adherence of neutrophils to HUVEC was shown to be a function of shear stress and ICAM-1 appears to be involved in CD18-dependent adherence (7), we investigated the ability of R6.5.D6 to inhibit adherence under flow conditions. As shown in Table 13.6, this MAb was ineffective in reducing the adherence of unstimulated neutrophils to IL-1 stimulated HUVEC that occurs at a wall shear stress of 2.0 dynes/cm². However, like the anti-CD18 MAb, R6.5.D6 significantly inhibited transendothelial migration.

Cooperation of LFA-1 and Mac-1 in ICAM-1-dependent Neutrophil Adherence

The attachment of unstimulated neutrophils to purified ICAM-1 in artificial membranes was markedly inhibited by an anti-CD11a MAb, R3.1. As shown in Figure 13.9, the extent of inhibition was equivalent to that

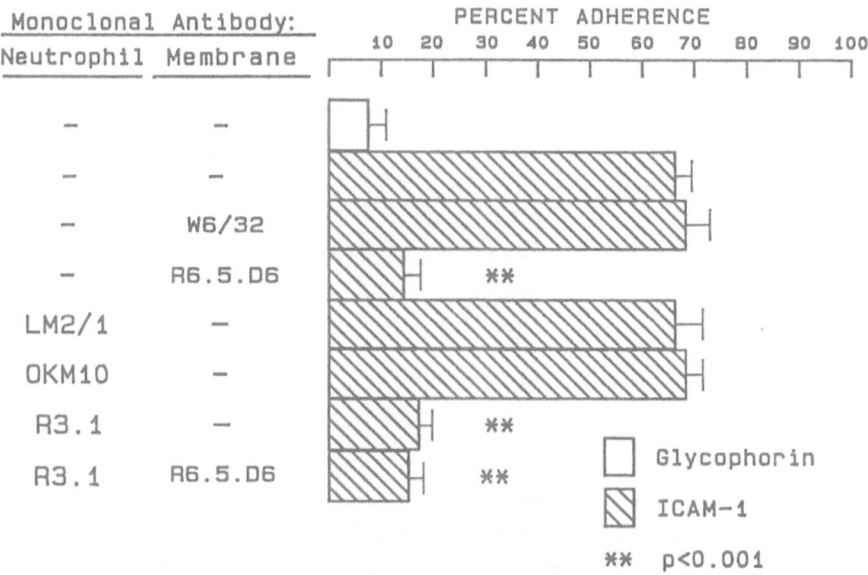

FIGURE 13.9. Adherence of neutrophils to artificial planar membranes containing either glycophorin or purified ICAM-1. Adherence was determined at room temperature in closed chambers in the absence of shear stress. Either the neutrophils or the membranes were pretreated with monoclonal antibodies as indicated. LM2/1 and OKM10 were used at 10μg/ml (over twice the saturating concentration as determined by flow cytometry) and R3.1 was used at 5μg/ml. W6/32 and R6.5.D6 were used at 10μg/ml. Preincubation with antibodies was 15 minutes at room temperature. Results are expressed as percentage adherent cells ± 1 SD.

produced by pretreatment of the artificial membrane with R6.5.D6, and the resulting level of adherence was only slightly greater than that of neutrophils to a control protein, glycophorin. The combination of R3.1 and R6.5.D6 was no more effective than either MAb alone. In clear contrast to the anti-CD11aMAb, two MAbs reactive with CD11b (LM2/1 and OKM10), were without inhibitory effect.

Since chemotactic stimulation increases the attachment of neutrophils to endothelial cells, we evaluated its effect on attachment of neutrophils to purified ICAM-1 in artificial membranes. While stimulation with fMLP did not increase adherence to the artificial membranes containing the control protein, glycophorin, it significantly increased adherence to membranes with ICAM-1 (Fig. 13.10). In contrast to the results with unstimulated neutrophils, MAbs reactive with both CD11a and CD11b significantly inhibited the adherence of stimulated neutrophils, and a

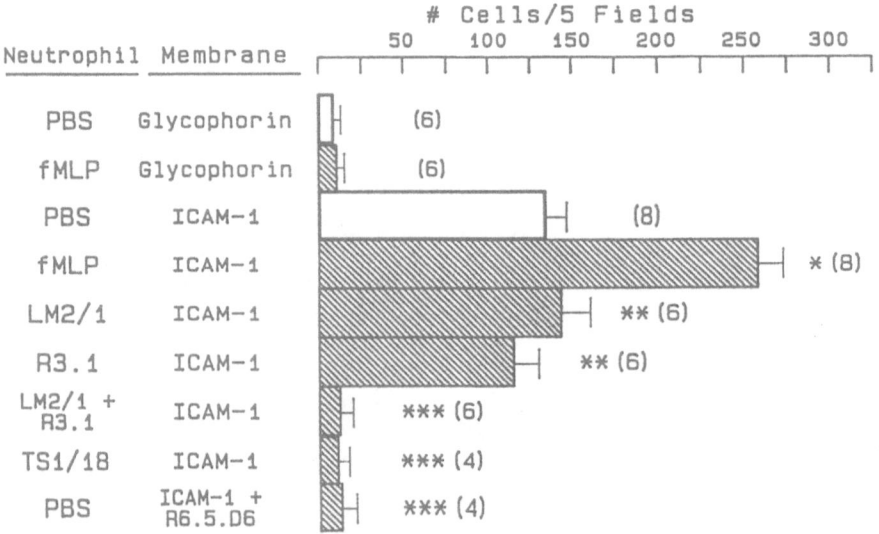

FIGURE 13.10. Adherence of neutrophils to artificial planar membranes containing either glycophorin or purified ICAM-1. Effects of chemotactic stimulation. Neutrophils were stimulated with fMLP (10 nM) for 2 minutes prior to being injected into the adherence chamber containing the membranes. Results with unstimulated cells are shown in open bars and results with stimulated cells (with and without monoclonal antibodies) are shown in fine hatched bars. Adherence was determined without shear stress. LM2/1, TS1/18, and R3.1 were used at 5μg/ml, and R6.5.D6 was used at 10μg/ml. Results are expressed as number of cells per 5 40X phase contrast fields ± 1 SD. *$p < 0.01$ compared to results of unstimulated cells on ICAM-1. **$p < 0.01$ compared to fMLP stimulated cells on ICAM-1 without antibody. ***$p < 0.01$ compared to fMLP stimulated cells on ICAM-1 with R3.1 alone.

combination of anti-CD11a and CD11b MAbs resulted in additive inhibition equivalent to that of the anti-CD18 MAb (Fig. 13.10).

Evidence of this cooperative effect of Mac-1 and LFA-1 was sought in the adherence of neutrophils to HUVEC monolayers. To reduce confounding effects of the ICAM-1-independent adhesion seen after 4 hours of IL-1 stimulation, we used HUVEC monolayers stimulated with IL-1 for 18 hours. At this time, ICAM-1-independent adhesion factors, e.g., ELAM-1 (12,24), are low, while ICAM-1 expression on the endothelial cell surface remains high (12). As shown in Figure 13.11, adhesion of unstimulated neutrophils to these monolayers was inhibited by the anti-CD11a MAb, R3.1, but not the anti-CD11b MAb, LM2/1. In contrast, the adherence of chemotactically stimulated neutrophils was inhibited by both anti-CD11a and anti-CD11b MAbs, and a combination of these MAbs was more inhibitory than either alone.

Cooperation of LFA-1 and Mac-1 appeared to occur in the phenomenon of transendothelial migration. R3.1 produced significant inhibition. The combination of R3.1 and LM2/1 was as effective as the anti-CD18 MAb in preventing migration (Fig. 13.12). Combination of R3.1 with

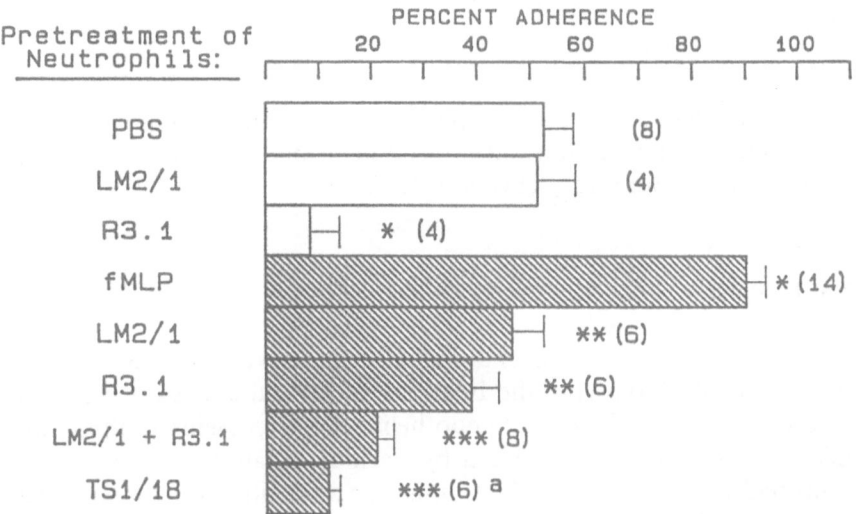

FIGURE 13.11. Adherence of neutrophils to IL-1 stimulated HUVEC monolayers 18 hours after the addition of IL-1. Unstimulated neutrophils (*open bars*) were suspended in PBS alone or in PBS with 5 μg/ml of the monoclonal antibodies indicated. Neutrophils were also suspended in PBS containing 10 nM fMLP (*fine hatched bars*) with or without the monoclonal antibodies indicated. Adherence was determined without shear stress. Results are expressed as percentage adherent cells $\pm$ 1 SD. *$p < 0.01$ compared to results of unstimulated cells without antibody. **$p < 0.01$ compared to results of fMLP stimulated cells without antibody. ***$p < 0.01$ compared to fMLP stimulated cells with R3.1 alone.

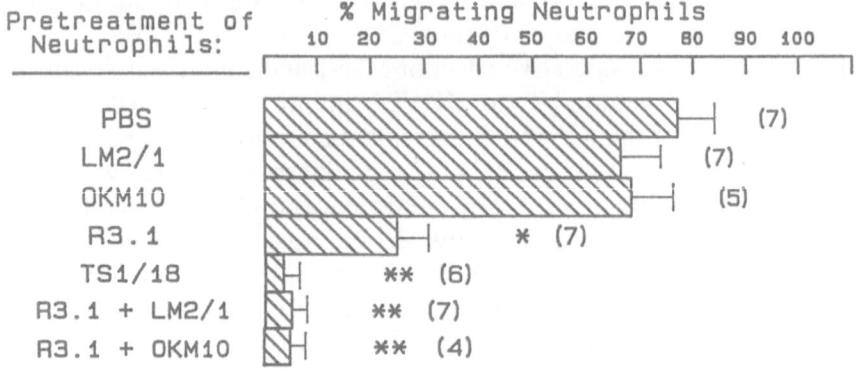

FIGURE 13.12. Effects of anti-CD11a and anti-CD11b monoclonal antibodies on transendothelial migration. HUVEC monolayers were stimulated with IL-1 (0.3 U/ml) for 4 hours, rinsed, and placed in adherence chambers. Unstimulated neutrophils were suspended in PBS with and without the monoclonal antibodies indicated (at 5 μg/ml), and injected into the chamber. Adherence was allowed without shear stress and migration was determined after a 1000-second incubation at 37 °C. Results are expressed as percentage cell migrating ± 1 SD. *$p < 0.01$ compared to results without antibody. **$p < 0.01$ compared to results with R3.1 alone.

4A5, a MAb that binds to neutrophils, is of the same isotype as LM2/1 (IgG1), but does not inhibit adherence under any conditions tested, did not enhance the inhibitory effects of R3.1.

Discussion

Adhesion of neutrophils to endothelial cells in vitro appears to be a complex process involving the interplay of several adhesion molecules on both the neutrophil and the endothelial cell. Expression of these molecules is differentially stimulated by various inflammatory mediators. Neutrophil-endothelial cell adhesion is also a sensitive function of shear stress. Under flow, the adhesive interaction can be separated into a CD18- and ICAM-1-dependent component which seems to be operative at wall shear stresses of 0.5 dynes/cm² and under, and CD18/ICAM-1-independent components expressed on the endothelial cell following cytokine stimulation which operate at wall shear stresses between 2.0 and 3.0 dynes/cm². These results suggest that circulating neutrophils may form attachments at significantly higher shear stresses with activated endothelium than would be possible if neutrophils alone were stimulated by a chemotactic factor diffusing from a focal tissue site. Localization of neutrophils at sites of inflammation may be primarily determined by endothelial cells and the local flow rates which determine the forces on the marginating neutrophils. When flow is within the physiologic range for venules, CD18/ICAM-1-independent factors may predominate, and when below the physiologic range, there may be participation of the CD11/CD18 family and ICAM-1.

Once attached, a portion of the neutrophils migrate through the endothelial monolayer (7,21,23,25). This phenomenon was found to be markedly increased by cytokine stimulation of endothelium, was seen after adhesion under static conditions and at shear stresses as high as 2.0 dynes/cm^2, and appeared in all cases to require the CD11/CD18 family and ICAM-1. Evidence supporting this requirement in vitro was provided by the lack of migration of CD18-deficient neutrophils and the pronounced inhibition of monoclonal antibodies specific for CD18 and ICAM-1. These results suggest that while margination may occur as a result of CD18/ICAM-1-independent factors, emigration requires the CD11/CD18 family and ICAM-1. Patients with CD18 deficiency appear to have a marginal pool of neutrophils, but neutrophils fail to emigrate into sites of inflammation (26). Rabbits treated with anti-CD18 monoclonal antibody exhibit marginal neutrophil pools and markedly reduced neutrophil emigration into inflammatory lesions (27).

The specific heterodimers of the CD11/CD18 family involved in the attachment of neutrophils to ICAM-1 has been unclear. Our results demonstrate that unstimulated neutrophils specifically adhere to purified ICAM-1 in an artificial membrane, and that this adherence requires LFA-1 on the surface of the neutrophil. The data supporting this conclusion come from experiments showing anti-CD11a monoclonal antibody completely inhibits adherence to ICAM-1 while anti-CD11b monoclonal antibodies fail to show significant inhibition, that CD18-deficient neutrophils do not adhere to purified ICAM-1, and an anti-ICAM-1 monoclonal antibody (R6.5.D6) completely inhibits adherence to purified ICAM-1. The results in this report further demonstrate that chemotactically stimulated neutrophils adhere to purified ICAM-1 in significantly greater numbers than unstimulated neutrophils. The mechanism of this increase is suggested by the results of experiments utilizing normal cells pretreated with MAbs, and results of experiments with CD18-deficient cells. Chemotactic stimulation may enable Mac-1 to interact with ICAM-1. The failure of the anti-LFA-1 MAb to completely inhibit the adhesion of fMLP stimulated neutrophils suggests that involvement of molecules other than LFA-1. Mac-1 is implicated by the demonstration that anti-CD11b MAbs reduced the level of adhesion of stimulated neutrophils to that of unstimulated neutrophils. The cooperative involvement of both heterodimers in neutrophil-HUVEC adhesion is shown by the additive inhibition of combinations of anti-CD11a and CD11b MAbs, essentially reducing adherence to the level induced by the anti-CD18 or anti-ICAM-1 MAbs (i.e., the level of adherence to the control protein, glycophorin). The involvement of ICAM-1 in this process is further supported by the finding that fMLP did not increase adherence of neutrophils to glycophorin containing planar membranes. This latter observation discounts the possibility that stimulated neutrophils release factors that enhance the adhesivity of the artificial membrane in some manner independent of ICAM-1. The recruitment of Mac-1 in adhesion of neutrophils to ICAM-1 is indicated by the failure of anti-CD11b to inhibit adhesion of unstimulated neutrophils. Whether this apparent recruitment is the result of newly upregulated molecules (8,28–30), or altered function or distribution of existing molecules (31–34) remains to be evaluated.

Chemotactic stimulation also leads to increased adhesion of neutrophils to HUVEC monolayers stimulated in vitro for 18 hours with IL-1, i.e., conditions that result in high expression of ICAM-1 and very low expression of ELAM-1 (12,24). As in the experiments with purified ICAM-1, Mac1 appears to play an important role. Anti-CD11b MAbs are inhibitory, and in combination with an anti-CD11a MAb, reduce adherence to the same low level induced by the anti-ICAM-1 MAb (R6.5.D6). Thus, the current evidence indicates that Mac-1 (CD11b/CD18) is recruited by chemotactic stimulation to interact with ICAM-1 on the surface of the endothelial cell. Recent investigations from several laboratories have provided evidence that Mac-1 may recognize ligands in addition to iC3b (8,34–38).

References

1. Gimbrone MA, Brock AF, Schafer AI: Leukotriene B_4 stimulates polymorphonuclear leukocyte adhesion to cultured vascular endothelial cells. *J Clin Invest* 74:1552–1555, 1984.
2. Tonnesen MG, Smedley LA, Henson PM: Neutrophil-endothelial cell interactions: Modulation of neutrophil adhesiveness induced by complement fragments C5a and $C5a_{desArg}$ and formyl-methionyl-leucyl-phenylalanine *in vitro*. *J Clin Invest* 74:1581–1592, 1984.
3. Charo IF, Yuen C, Perez HD, Goldstein IM: Chemotactic peptides modulate adherence of human polymorphonuclear leukocytes to monolayers of cultured endothelial cells. *J Immunol* 136:3412–3419, 1986.
4. Pohlman TH, Stanness KA, Beatty PG, Ochs HD, Harlan JM: An endothelial cell surface factors(s) induced *in vitro* by lipopolysaccharide, interleukin-1, and tumor necrosis factor increases neutrophil adherence by a CDw18 (LFA)-dependent mechanism. *J Immunol* 136:4548–4553, 1986.
5. Gamble JR, Harlan JM, Klebanoff SJ, Vadas MA: Stimulation of the adherence of neutrophils to umbilical vein endothelium by human recombinant tumor necrosis factor. *Proc Natl Acad Sci USA* 82:8667–8674, 1985.
6. Smith CW, Hollers JC, Patrick RA, Hassett C: Motility and adhesiveness in human neutrophils: Effects of chemotactic factors. *J Clin Invest* 63:221–229, 1979.
7. Smith CW, Rothlein R, Hughes BJ, Mariscalco MM, Schmalstieg FC, Anderson DC: Recognition of an endothelial determinant for CD18-dependent

human neutrophil adherence and transendothelial migration. *J Clin Invest* 82:1746–1756, 1988.

8. Anderson DC, Miller LJ, Schmalstieg FC, Rothlein R, Springer TA: Contributions of the Mac-1 glycoprotein family to adherence-dependent granulocyte functions: Structure-function assessments employing subunit-specific monoclonal antibodies. *J Immunol* 137:15–27, 1986.

9. Zimmerman GA, McIntyre TM: Neutrophil adherence to human endothelium in vitro occurs by CDw18 (Mo1, MAC-1/LFA-1/GP150,95) glycoprotein-dependent and independent mechanisms. *J Clin Invest* 81:531–537, 1988.

10. Pohlman TH, Munford RS, Harlan JM: Deacylated lipopolysaccharide inhibits neutrophil adherence to endothelium induced by lipopolysaccharide in vitro. *J Exp Med* 165:1393–1402, 1987.

11. Dunn CJ, Fleming WE: Increased adhesion of polymorphonuclear leukocytes to vascular endothelium by specific interaction of endogenous (interleukin-1) and exogenous (lipopolysaccharide) substances with endothelial cells "in vitro." *Eur J Rheum Inflam* 7:80–86, 1984.

12. Pober, JS, Gimbrone MA, Lapierre LA, Mendrick DL, Fiers W, Rothlein R, Springer TA: Activation of human endothelium by lymphokines: Overlapping patterns of antigenic modulation by interleukin 1, tumor necrosis factor and immune interferron. *J Immunol* 137:1893–1896, 1986.

13. Pober JS, Lapierre LA, Stolpen AH, Brock TA, Springer TA, Fiers W, Bevilacqua MP, Mendrick Jr. DL, Gimbrone MA: Activation of cultured human endothelial cells by recombinant lymphotoxin: comparison with tumor necrosis factor and interleukin 1 species. *J Immunol* 138:3319–3324, 1987.

14. Broudy VC, Harlan JM, Adamson JW: Disparate effects of tumor necrosis factor-a/cachectin and tumor necrosis factor-b/lymphotoxin on hematopoietic growth factor production and neutrophil adhesion molecule expression by cultured human endothelial cells. *J Immunol* 138:4298–4302, 1987.

15. Rothlein R, Dustin ML, Marlin SD, Springer TA: An intercellular adhesion molecule (ICAM-1) distinct from LFA-1. *J Immunol* 137:1270–1275, 1986.

16. Dustin ML, Rothlein R, Bhan AK, Dinarello CA, Springer TA: Induction by IL-1 and interferon-gamma: Tissue distribution, biochemistry, and function of a natural adherence molecule (ICAM-1). *J Immunol* 137:245–254, 1986.

17. Marlin SD, Springer TA: Purified intercellular adhesion molecule-1 (ICAM-1) is a ligand for lymphocyte function-associated antigen 1 (LFA-1). *Cell* 51:813–819, 1987.

18. Makgoba MW, Sanders ME, Ginther Luce GE, Dustin ML, Springer TA, Clark EA, Mannoni P, Shaw S: ICAM-1 a ligand for LFA-1-dependent adhesion of B, T and myeloid cells. *Nature* 331:86–88, 1988.

19. Lawrence MB, McIntire LV, Eskin SG: Effect of flow on polymorphonuclear leukocyte/endothelial cell adhesion. *Blood* 70:1284–1290, 1987.

20. Lawrence MB, McIntire LV, Eskin SG, Smith CW: Involvement of CD18 in human neutrophil adhesion to endothelium under flow conditions. *FASEB* 2(5):A1237–5469a, 1988.

21. Furie MB, McHugh DD: Stimulation of neutrophil transendothelial migration by interleukin-1. *J Cell Biol* 105:276a (Abstract), 1987.

22. Beesley JE, Pearson JD, Carleton JS, Hutchings SA, Gordon JL: Interactions of leukocytes with vascular cells in culture. *J Cell Sci* 33:85–101, 1978.

23. Beesley JE, Pearson JD, Hutchings A, Carleton JS, Gordon JL: Granulocyte migration through endothelium in culture. *J Cell Sci* 38:237–248, 1979.

24. Bevilacqua MP, Pober JS, Mendrick DL, Cotran RS, Gimbrone Jr. MA: Identification of an inducible endothelial-leukocyte adhesion molecule. *Proc Natl Acad Sci USA* 84:9238–9242, 1987.
25. Hoover RL, Robinson M, Karnovsky MJ: Adhesion of polymorphonuclear leukocytes to endothelium enhances the efficiency of detoxification of oxygen-free radicals. *Am J Pathol* 126:258–268, 1987.
26. Anderson DC, Schmalstieg FC, Finegold MJ, Hughes BJ, Rothlein R, Miller LJ, Kohl S, Tosi MF, Jacobs RL, Waldrop TC, Goldman AS, Shearer WT, Springer TA: The severe and moderate phenotypes of heritable Mac-1, LFA-1, p150,95 deficiency: Their quantitative definition and relation to leukocyte dysfunction and clinical features. *J Infec Dis* 152:668–689, 1985.
27. Arfors KE, Lundberg C, Lindbom L, Lundberg K, Beatty PG, Harlan JM: A monoclonal antibody to the membrane glycoprotein complex CD18 inhibits polymorphonuclear leukocyte accumulation and plasma leakage in vivo. *Blood* 69:338–340, 1987.
28. Anderson DC, Schmalstieg FC, Kohl S, Arnaout MA, Hughes BJ, Tosi MF, Buffone GJ, Brinkley BR, Dickey WD, Abramson JS, Springer TA, Boxer LA, Hollers JM, Smith CW: Abnormalities of polymorphonuclear leukocyte function associated with a heritable deficiency of high molecular weight surface glycoproteins (GP138): Common relationship to diminished cell adherence. *J Clin Invest* 74:536–551, 1984.
29. O'Shea JJ, Brown EJ, Seligmann BE, Metcalf JA, Frank MM, Gallin JI: Evidence for distinct intracellular pools or receptors for C3b and C3bi in human neutrophils. *J Immunol* 134:2580–2587, 1985.
30. Petty HR, Francis JW, Todd III RF, Petrequin PR, Boxer LA: Neutrophil C3bi receptors: Formation of membrane clusters during cell triggering requires intracellular granules. *J Cell Physiol* 133:235–242, 1987.
31. Vedder NB, Harlan JM: Increased surface expression of CD11b/CD18 (Mac-1) is not required for stimulated neutrophil adherence to cultured endothelium. *J Clin Invest* 81:676–682, 1988.
32. Philips M, Buyon J, Winchester R, Weissmann G, Abramson S: Upregulation of iC3b receptors (CR3) is neither necessary nor sufficient to promote neutrophil aggregation. *J Clin Invest* 82:495–501, 1988.
33. Detmers PA, Wright SD, Olsen E, Kimball B, Cohn ZA: Aggregation of complement receptors on human neutrophils in the absence of ligand. *J Cell Biol* 105:1137–1145, 1987.
34. Altieri DC, Edgington TS: The saturable high affinity association of Factor X to ADP-stimulated monocytes defines a novel function of the Mac-1 receptor. *J Biol Chem* 263:7007–7015, 1988.
35. Wright SD, Jong MTC: Adhesion-promoting receptors on human macrophages recognize *Escherichia coli* by binding to lipopolysaccharide. *J Exp Med* 164:1876–1888, 1986.
36. Ross, GD, Thompson RA, Walport MJ, Springer TA, Watson JV, Ward RHR, Lida J, Newman SL, Harrison RA, Lachmann PJ: Characterization of patients with an increased susceptibility to bacterial infections and a genetic deficiency of leukocyte membrane complement receptor type 3 and the related membrane antigen LFA-1. *Blood* 66:882–890, 1985.

37. Wright SD, Jong MTC, Levin SM. 1988. CR3 expresses two binding sites, one for RGD-peptide, and one for bacterial LPS. *FASEB* 2:a1237 (Abstract), 1988.
38. Lo SK, Wright SD: CR3 mediates binding or PMN to endothelial cells (EC) via its RGD binding, not the LPS binding site. *FASEB* 2:A1236 (Abstract), 1988.

14

Specificity and Regulation of CD18-Dependent Adhesions*

SAMUEL D. WRIGHT, SIU KONG LO, AND
PATRICIA A. DETMERS

Introduction

The ability of polymorphonuclear leukocytes (PMN) to adhere to other cells and substrates is under strict physiological regulation. This is best illustrated by considering PMN circulating within the vasculature. These cells show negligible affinity for the endothelial cells (EC) which surround them. When PMN encounter chemotactic factors this situation is rapidly altered, and the adhesion of PMN to EC is dramatically enhanced. In normal circumstances, stimulated PMN soon break their adhesion to EC and either return to the circulation or leave the vasculature by diapedesis. Thus, PMN possess the means of transiently making then breaking adhesion to EC and other substrates in response to specific stimuli.

Human phagocytes express a family of cell surface glycoproteins, CD11a/CD18 (LFA-1), CD11b/CD18 (CR3), and CD11c/CD18 (p150,95), collectively known as the *CD18 complex.* Each of these receptors is a dimer composed of a β chain (CD18) which is identical in all three members, and a noncovalently associated α chain that is unique to each molecule. Several lines of evidence indicate that the CD18 complex mediates adhesion of stimulated PMN to several substrates including EC. Monoclonal antibodies (MAbs) directed against CD18 block the adhesion of PMN to surfaces coated with C3bi (1), fibrinogen (2), serum proteins (3), as well as adhesion of PMN to EC both in vitro (4, 5) and in vivo 6–8. Further, PMN from patients deficient in CD18 fail to adhere to all of these substrates (9). A principal role for CD18 in the adhesion of PMN is also suggested by the observation that the capacity of CD18 to mediate adhesion is regulated: Its binding activity can be rapidly turned

*Supported by USPHS grants AI22003 and AI24775 and a Grant-in-Aid from the American Heart Association, New York City Affiliate. SDW is an Established Investigator of the American Heart Association, SKL is an Irvington Institute Fellow, and PAD is an Investigator of the American Heart Association, New York City affiliate.

on, then back off again to mediate transient adhesions (10). Here we discuss several adhesion events mediated by CD18 on phagocytes. First, we describe evidence that CD18 can recognize a variety of specific ligands and that these ligands fall into two structurally distinct categories. The two categories, peptide and carbohydrate, appear to be recognized by two distinct binding sites on CD18. Then we review the evidence that the adhesion mediated by the peptide binding site of CD18 can be rapidly turned on and off, and we discuss the mechanism of this form of regulation. We will focus primarily on CD11b/CD18, the CD18 member of greatest abundance on the PMN.

CD11b/CD18 Binds Several Proteins by Recognizing a Linear Sequence of Amino Acids

C3bi

CD11b/CD18 was first recognized as a receptor for the surface-bound complement protein C3bi (1). This was demonstrated by down-modulating CD11b/CD18 from the dorsal surface of human macrophages by allowing the cells to spread on tissue culture plastic derivatized with specific anti-receptor MAbs. The receptors diffuse in the plane of the membrane and are trapped by antibody at the basal surface of the macrophage, leaving the apical surface devoid of a particular receptor. Substrates coated with the MAbs OKM1, OKM9, or OKM10, all directed against CD11b/CD18, completely inhibited the binding of C3bi-coated erythrocytes (EC3bi) to adherent phagocytes. Direct evidence of the interaction of CD11b/CD18 with C3bi was provided by the observation that surfaces coated with purified CD11b/CD18 bound specifically to EC3bi (1). These observations established that CD11b/CD18 mediates the adhesion of C3bi-coated particles to phagocytes by physically binding to C3bi.

The portion of the C3bi molecule recognized by CD11b/CD18 was defined through the use of synthetic peptides modelled on the deduced amino acid sequence of human C3 (11). CD11b/CD18 bound specifically to erythrocytes coated with a synthetic peptide based on amino acids 1383 to 1403, a region near the carboxy terminus of the α chain of C3bi. This peptide contains the sequence Arg-Gly-Asp (RGD, see Fig. 14.1). Several other adhesion-promoting receptors, such as the fibronectin receptor and the vitronectin receptor, have been shown to recognize the RGD triplet in their target ligands (12). Thus our observations cemented the emerging notion that CD11b/CD18 is a member of a superfamily of adhesion-promoting receptors termed *integrins* (13). The subsequent cloning of the β (14, 15) and α (16, 17) chains of the CD18 complex has

Protein	Sequence

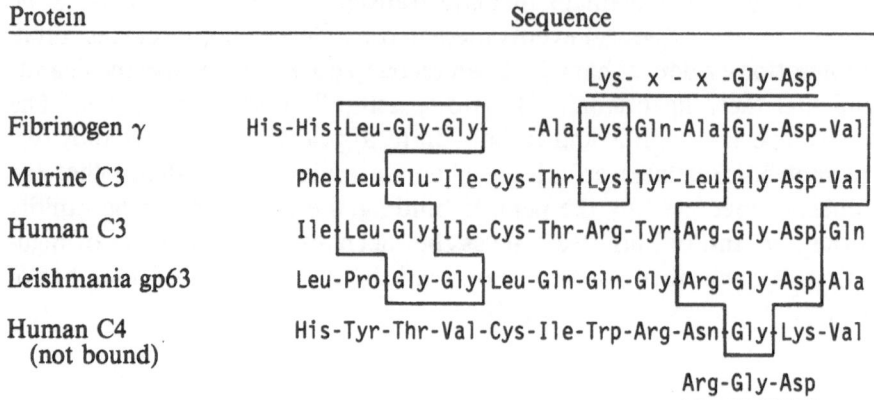

FIGURE 14.1. Sequences of proteins recognized by CD11b/CD18.

amply confirmed the relationship between CD11b/CD18 and the other integrins.

In its role as a receptor for the opsonin C3bi, CD11b/CD18 functions in the binding, phagocytosis, and killing of complement-coated microbes. Phagocytosis of particles opsonized with IgG is accompanied by the release of toxic products such as proteases and reactive oxygen intermediates (O_2^- and H_2O_2) and by the release of inflammatory lipids such as leukotrienes and prostaglandins. To determine whether these products are released during phagocytosis mediated by CD11b/CD18, particles were prepared bearing only IgG or only C3. Phagocytosis of IgG-coated particles induced brisk release of H_2O_2 (18) and arachidonic acid metabolites (19). In contrast, phagocytosis of particles coated with C3 caused no release of either mediator (18, 19). These studies emphasize the functional difference between Fc receptors, which mediate adhesion of host cells and microbes for lethal purposes, and CD11b/CD18, which also functions in the cooperative interaction between host cells. The capacity of CD11b/CD18 to mediate adhesion without triggering secretion of toxic compounds may reflect its evolutionary origins as a receptor that functioned in the movement and positioning of cells in embryogenesis. Its use in the immune system may have appeared much later in evolution.

Leishmania gp63

The ability of CD11b/CD18 to mediate phagocytic uptake of particles without initiating cytolytic activity makes it an advantageous portal for the entry of intracellular parasites. *Leishmania* promastigotes enter a human host during the bite of an infected sandfly. The parasite is quickly phagocytosed by a macrophage and replication occurs only within the phagosome. Recently we have shown that CD11b/CD18 recognizes a dominant component on the surface of *Leishmania,* a protein called *gp63* (20), and may therefore be involved in the tropism of the parasite for macrophages and for their subsequent phagocytosis. Beads coated with purified gp63 bind avidly to macrophages, and binding is completely inhibited by down-modulating CD11b/CD18 with surface bound MAbs.

gp63 contains an RGD triplet (20, and see Fig. 14.1), and this region constitutes the recognition sequence (21). Synthetic peptides based on this sequence competitively inhibit the binding of purified gp63, and competitively inhibit the binding of EC3bi to macrophages. Thus, *Leishmania* appears to have mimicked a short sequence in C3bi to gain entry into macrophages. These observations confirm the importance of the RGD triplet in the recognition specificity of CD11b/CD18.

Fibrinogen

During C3bi-mediated phagocytosis, macrophages appress the leading edge of their advancing pseudopod against the ligand-coated target. The contact between pseudopod and target is sufficiently tight to exclude soluble protein molecules in the medium from the zone of contact (22). In this way, a phagocyte may generate a sealed compartment between itself and an opsonized target that is protected from proteins in the medium. Recent evidence suggests that PMN also form such a compartment when they bind to surfaces coated with fibrinogen, and that the compartment serves to protect secreted elastase from protease inhibitors in plasma (23). We have shown that CD11b/CD18 is responsible for forming the protected compartment by binding to fibrinogen. Monoclonal antibodies directed against CD11b/CD18 specifically blocked the adhesion of PMN to fibrinogen-coated surfaces and thereby prevented formation of the protected compartment (2).

The region of fibrinogen recognized by CD11b/CD18 lies at the carboxy terminus of the γ chain (2). A synthetic decameric peptide (Fig. 14.1) based on this sequence effectively inhibited the binding of PMN to fibrinogen-coated surfaces and blocked the formation of a sealed compartment as assayed by protection of elastase activity from protease inhibitors in the medium. Interestingly, this region of fibrinogen does not contain an RGD triplet. Instead, the sequence KxxGD is observed. The positively charged lysine (K) group of KxxGD appears critical for recognition since peptides in which this group was modified by acetylation did not competitively inhibit binding of PMN to fibrinogen (2). Thus it appears that the positively charged lysine group (K) in this sequence substitutes for the positively arginine (R) group in RGD. The capacity of CD11b/CD18 to recognize both RGD and KxxGD is confirmed by the observation that murine C3, a good ligand for both murine and human CD11b/CD18, contains KxxGD at the region homologous to the RGD site in human C3 (Fig. 14.1).

The fibrinogen receptor of platelets, gpIIb/IIIa, shares extensive sequence homology with CD11b/CD18, and, not surprisingly, binds to the same region of the γ chain of fibrinogen (24). gpIIIb/IIIa binds RGD-containing sequences in fibronectin and in von Willibrand factor, and a KxxGD sequence in fibrinogen (24). Thus the binding specificity of gpIIb/IIIa appears similar to that of CD11b/CD18. An important difference between CD11b/CD18 and gpIIb/IIIa, however, is that CD11b/CD18 does not recognize fibronectin (25), and binding of C3bi (25, 2), gp63 (20), or fibrinogen (2) to CD11b/CD18 cannot be inhibited by peptides based on the sequence of fibronectin.

A Single Site on CD11b/CD18 Binds C3bi, gp63, and Fibrinogen

Monoclonal antibody OKM10, directed against an epitope on the α chain of CD11b/CD18, effectively blocks binding of C3bi (1), gp63 (unpublished), and fibrinogen (2) to PMN, while other anti-α chain MAbs against different epitopes (such as OKM1 and 904) are without effect on these interactions. Thus, a single epitope on CD11b/CD18 appears to be involved in the recognition of all three ligands. This point is confirmed by the observation that peptides from the γ chain of fibrinogen that block the binding of PMN to fibrinogen also block the binding of EC3bi (2) and gp63 (20) to CD11b/CD18. Thus, CD11b/CD18 recognizes three different ligands, C3bi, gp63, and fibrinogen, via the same binding site which we shall call the peptide binding site.

Sequences Recognized by the Peptide Binding Site of CD11b/CD18

Several sequences recognized by CD11b/CD18 are shown in Fig. 14.1, and structural features may be described as follows. The core contains the zwitterionic sequences RGD or KxxGD, and both charged groups are necessary for recognition. The negative charge of D (asp) is critical since substitution of a neutral residue abolishes recognition of synthetic peptides (unpublished results). Further, in the complement protein C4, which is highly homologous to C3, a positively charged group (K) appears in place of the negatively charged D at this position, and C4 is not recognized by CD11b/CD18. The positive charge of K or R is also critical to recognition since acetylation of K (2, 20) or substitution of arg with an uncharged residue (unpublished) abolishes the recognition of peptides.

While the RGD or KxxGD core is part of the recognition sequence, it is clear that the core in itself is not sufficient for recognition. Some proteins that have an RGD sequence, such as fibronectin or vitronectin, are not recognized by CD11b/CD18 (25). Short RGD-containing peptides

(GRGDSP [25] or YRGDQ [11]) do not inhibit binding of ligands to C3bi even when used at extremely high concentrations. Thus it appears likely that the identity of the flanking amino acids provides the necessary additional recognition specificity. The residues C terminal to RGD were first suspected to contribute to recognition because weak sequence homology is observed up to 30 amino acids to the carboxyl side of the RGD site in several ligands for integrins (fibronectin, vitronectin, and C3) (11). However, this portion of C3 is unlikely to be of prime importance in recognition by CD11b/CD18 since the sequence of the γ chain of fibrinogen, a good ligand for CD11b/CD18, terminates one residue after the KxxGD sequence. The region on the amino terminal side of RGD is, therefore, the most likely to be recognized by CD11b/CD18, but critical residues other than RGD have not yet been identified.

CD18 Molecules Also Recognize Ligands Not Composed of Protein

Several workers have noted that antibodies against CD11b/CD18 not only inhibited the binding of C3bi-coated particles, but also partially inhibited binding of particles such as zymosan (26, 27) and *Staphlyococcus* (28). Since these binding experiments were performed in the absence of a source of complement, it was proposed that CD11b/CD18 interacts not with C3bi, but the components of the particles themselves. Further support for this idea came from the observation that PMN from LAD patients failed to bind zymosan (28) and *Escherichia coli* (29), again in the absence of serum opsonins.

More detailed experiments have indicated that CD11a/CD18, CD11b/CD18, and CD11c/CD18 are each able to directly bind the microbes, *Histoplasma capsulatum* (30) and *E. coli* (31). When all three receptors were cleared from the apical surface of the macrophage membrane by allowing the cells to spread on surfaces derivatized with MAbs directed against CD18, binding of microbes was completely abolished. Parallel experiments were performed on surfaces coated with MAbs directed against individual CD11 chains or pairs of CD11 chains such that two of the three receptors were down-modulated and a single CD11/CD18 receptor remained on the apical surface. Under these conditions, binding of microbes was still observed. These results indicate that each of the members of the CD18 family is individually capable of binding to *H. capsulatum* and *E. coli*.

CD11a/CD18, CD11b/CD18, and CD11c/CD18 Recognize Bacterial Lipopolysaccharide

Recent work indicates that the chemical structure of the ligands on *E. coli* recognized by the CD18 complex is different from that of the proteins, C3bi, gp63, or fibrinogen. Macrophages bind *E. coli* by recognizing lipopolysaccharide (LPS), the most prevalent molecule on the surface of the bacterium (31). Observations supporting this conclusion are that down-modulation of "LPS-receptors" caused by spreading of cells on an LPS-coated surface inhibited binding of E. coli to macrophages (31), and erythrocytes coated with purified LPS were bound by phagocytes (31). The binding of LPS is clearly mediated by the CD18 complex because binding of E. coli and LPS-coated erythrocytes did not occur with PMN from CD18-deficient patients (29). Further, soluble MAb 904, directed against CD11b, blocked the binding of LPS-coated erythrocytes to PMN, a cell which bears relatively little LFA-1, or p150,95 (29). Finally, down-modulation of CD11a/CD18, CD11b/CD18, and CD11c/CD18 with surface-bound MAbs inhibited binding of LPS-coated erythrocytes with a pattern identical to that seen for E. coli, i.e., down-modulation of any pair of the trio did not block binding, but modulation of all three CD18 molecules completely blocked binding of LPS-coated erythrocytes (31). Thus, all members of the CD18 family can recognize LPS.

The portion of the LPS molecule recognized by the CD18 complex lies within the "Lipid A" region. The smallest moiety of LPS thus far shown to be bound is the biosynthetic precursor of LPS termed Lipid IVa (31), consisting of a fatty-acylated diglucosamine bisphosphate (Fig. 14.2). Since the fatty acids of LPS are buried in the outer membrane of the bacterium and the fatty acids of Lipid IVa are buried in the outer leaflet of a Lipid IVa-coated erythrocyte, it is likely that the exposed hydrophilic diglucosamine phosphate provides the recognition structure bound by the CD18 complex. It is also possible that sugars or sugar phosphates on the surface of H. capsulatum and zymosan may account for their recognition by CD18.

Bacterial LPS (endotoxin) causes profound physiological effects in humans and animals. These include fever, shock, and the induction of the acute phase response (32). The cell type primarily responsible for these effects is the macrophage, which synthesizes large amounts of interleukin-1 and cachectin (TNFα) in response to LPS (33, 34). The studies described above indicate that CD11a/CD18, CD11b/CD18, and CD11c/CD18 can bind LPS, but whether these receptors directly mediate the many biological effects of LPS is not yet certain.

Lipopolysaccharide and Peptides Are Recognized by Distinct Sites on CD11b/CD18

Since peptides containing RGD are structurally dissimilar from LPS (Fig. 14.2), we sought to determine if these two ligands interact with the same or different binding sites on CD11b/CD18. We found that MAb OKM10 against CD11b, which blocks the binding of C3bi (1), gp63 (unpublished), and fibrinogen (2) does not block the binding of LPS-coated erythrocytes to PMN (29). Another MAb, 904, directed against a different epitope on

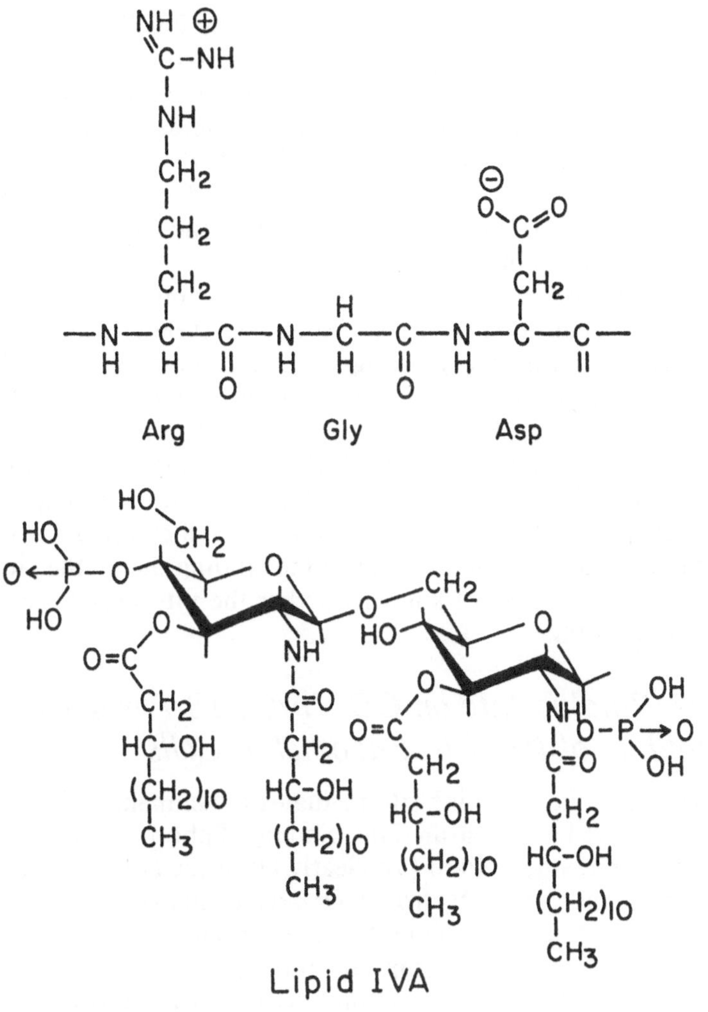

FIGURE 14.2. Chemical structure of Lipid IVa and RGD.

CD11b, blocks binding of LPS-coated erythrocytes to PMN (29), but does not block binding of C3bi (29) or fibrinogen (2). Thus it appears that OKM10 binds and blocks the binding site for peptides, and MAb 904 binds and blocks a separate binding site for LPS.

Further confirmation that the LPS binding site is separate from the peptide binding site derives from studies with low molecular weight synthetic peptides. Peptides based on the recognition sequences in C3bi or fibrinogen competitively inhibited the binding of EC3bi to PMN, but these competitors did not inhibit binding of LPS-coated erythrocytes (29). The binding site for LPS is thus physically distinct from that for peptide. We observed, however, that these peptides actually enhanced the binding of LPS-coated erythrocytes to PMN (29). Thus, although the binding sites for peptides and LPS are separate, they are functionally linked. The two binding sites on CD18 also differ in the contribution of the α chain to binding specificity. The specificity of the peptide binding site is strongly influenced by the α chain since the peptide ligands, C3bi and gp63, are not bound by CD11a/CD18 or CD11c/CD18. In contrast, the binding site for LPS on the CD18 family appears little affected by the identity of the α chain since CD11a/CD18, CD11b/CD18, and CD11c/CD18 exhibit equivalent ability to bind LPS (31).

The presence of two binding sites, one for peptide and a second for lipids, may be a common property of integrins. As described above, members of this superfamily recognize ligands that contain the sequences RGD or KxxGD. For example, the vitronectin receptor recognizes the triplet RGD in vitronectin (35). The vitronectin receptor is also capable of binding the ganglioside GD2 (36). Gangliosides are structurally similar to LPS in that both species are amphipathic with a strongly anionic hydrophilic group. The binding site on the CD18 complex for LPS may thus be the evolutionary homologue of the binding site on the vitronectin receptor for GD2. It is not known whether the vitronectin receptor can bind bacterial LPS.

Peptide Binding Site on CD11b/CD18 Participates in the Binding of PMN to Endothelial Cells

To determine which binding site mediates attachment of leukocytes to endothelium, we have measured the binding of phorbol-stimulated PMN to unstimulated endothelium (5). Nearly all of the adhesion observed is CD18-dependent since PMN from CD18-deficient patients exhibited negligible adherence under our conditions. Approximately 66% of the adhesion could be blocked by appropriate anti-CD11b/CD18 MAbs, 66% by anti-CD11a/CD18 MAbs, and a combination of these two types of antibodies blocked >90% of the adhesion. Thus, both CD11b/CD18 and CD11a/CD18 mediate adhesion to EC under these conditions. The anti-CD11b/CD18 reagent OKM10, which blocks the binding of the peptide ligands C3bi, gp63, and fibrinogen, successfully blocked adhesion to EC (66%). On the other hand, MAb 904, which blocks the binding of LPS, did not inhibit binding of PMN to EC at all. Therefore, the epitope on CD11b/CD18 involved in binding of peptide ligands is also involved in binding to EC.

Further evidence for a principal role of the peptide binding site comes from studies on the inhibition of adhesion of PMN to EC with synthetic peptides. An RGD-containing peptide based on the recognition sequence of gp63 competitively inhibited the binding of PMN to EC (37). This observation confirms the role of the peptide binding site in the recognition of EC by PMN, and suggests that an RGD or similar structure serves as a ligand on the endothelial surface.

Other Adhesive Interactions

The members of the CD18 complex mediate a variety of adhesive interactions that may be subserved by either the peptide or the LPS binding sites. For example, CD11a/CD18 mediates aggregation of B cells. This function is probably mediated by a peptide binding site since the ligand for LFA-1 in this system is the protein ICAM-1 (38). CD11b/CD18 mediates the binding of PMN to protein-coated surfaces, and this function, on the other hand, may be mediated by the LPS site. Lipopolysaccharide is a common contaminant of all laboratory solutions and binds avidly to tissue culture plastic (31). Thus, "protein-coated" plastic is likely to be LPS-coated as well, and the adhesion of PMN to these surfaces may be mediated by interaction of CD18 molecules with surface-bound LPS. This view is supported by the observation of Dana et al. (3) who showed that MAb904 strongly blocks binding of PMN to protein-coated plastic.

CD18 Molecules Mediate Transient Adhesion Events

CD18 and other integrins appear not to be involved in the strong, long-term adhesions found between cells in tissues. Thus, integrins appear unrelated to proteins of gap junctions, tight junctions, or desmosomes. Rather, integrins mediate relatively weak adhesions that may be readily broken.

The reversible nature of adhesions caused by integrins is illustrated by several findings. (i) Cells bearing fibronectin receptors can migrate rapidly across surfaces coated with fibronectin (39), indicating that sequential, transient attachments are made between fibronectin receptors and ligand. (ii) The CD11a/CD18-mediated binding of killer T cells to targets can be as short as 10 minutes, ending when the killer detaches and moves to another target (40). (iii) The CD18 complex mediates the binding of PMN to endothelial cells, which must be short lived to enable diapedesis to occur. (iv) Finally, high concentrations of synthetic peptides bearing the RGD sequence, which competitively block the action of several integrins (12), prevent normal migration of cells in avian, amphibian (41), and insect embryos (42), a process for which transient adhesions are obviously required.

One can imagine two models of receptor behavior that could be used to generate transient adhesions. In the first model, cells would deliver adhesion-promoting receptors to the cell surface only at the place and time required, and after they had functioned, either the receptor or the ligand would be proteolytically inactivated to break the adhesion. In a second model, adhesions are made or broken by allosteric alteration of the receptors that effectively raise or lower the binding affinity for ligand. An important prediction of this second model is that receptors may be left on the cell surface in an inactive state such that their activity can be altered when and where necessary. We present evidence below that both mechanisms operate together to control adhesivity of CD11/CD18 on PMN. In some instances, receptor expression is enhanced during upregulation of adhesiveness. However, enhanced adhesiveness may occur without changes in receptor number, and CD11b/CD18 may be expressed on the cell surface in a form that is unable to bind ligand.

Function of Adhesion-Promoting Receptors Is Regulated

Several observations suggest that integrins might exist on the cell surface in an inactive state. Transformed fibroblasts fail to bind fibronectin despite apparently normal levels of fibronectin receptors (43, 44). CD11a/CD18 may also exist in an inactive form, since the CD11a/CD18-dependent aggregation of B cells or lymphoblastoid cell lines is not observed unless the stimulant phorbol myristate acetate (PMA) is added (45). The most detailed studies of on/off behavior, however, are available for CD11b/CD18 on macrophages and PMN. We review here the evidence for regulation of binding by CD11b/CD18 in phagocytes and discuss the agents that mediate the regulation.

CD11b/CD18 on resting human monocytes and macrophages mediates avid binding of C3bi-coated erythrocytes with half-maximal binding at about 5,000 C3bi per erythrocyte (46). Culture of the phagocytes for 48 hours with IFNγ (but not IFNα) causes a striking decrease in the binding capacity of C3 receptors: half-maximal binding of erythrocytes is not obtained even with 120,000 C3bi per erythrocyte (47). This result contrasts with the behavior of Fc receptors in that IFNγ causes dramatically enhanced expression of Fc$_\gamma$R$_{p72}$ (48) and enhanced binding of IgG-coated erythrocytes (47).

The reduced binding activity of CD11b/CD18 in IFNγ-treated cells is not associated with changes in the number of cell surface receptor molecules, nor is it associated with proteolytic inactivation of the receptors (47). Rather, CD11b/CD18 appears to be reversibly disabled. This point is emphasized by the observation that the binding capacity of CD11b/CD18 can be fully restored in minutes by allowing the phagocytes to interact with surfaces coated with the extracellular matrix protein, fibronectin (see below).

The physiological significance of this "deactivation" of CD11b/CD18 by IFNγ is not currently clear, but IFNγ can be expected to diminish complement receptor activity on all macrophages except those in contact with the appropriate extracellular matrix components. Since IFNγ-treated macrophages possess extremely potent cytolytic activity, lowered capacity to adhere may control inappropriate cytolysis.

The binding activity of CD11b/CD18 on granulocytes is also regulated, but in a manner different from that seen in mononuclear cells (49). Resting PMN have *low* capacity to bind C3-coated particles. This binding activity is rapidly upregulated by several agents including PMA. Though PMA is not a physiologic stimulus, its activities are the best characterized and are described in detail below.

Stimulation of PMN with PMA produces a biphasic response (49). During the first 20 minutes, the capacity of CD11b/CD18 to bind and phagocytose C3bi-coated erythrocytes is dramatically enhanced. During the following 30 minutes, the capacity to bind and phagocytose is dramatically enhanced. During the following 30 minutes, the capacity to bind and phagocytose is depressed to levels below those shown by resting cells. Treatment of PMN with PMA causes a rapid rise in the expression of CD11b/CD18 on the cell surface (50, 51, 49). Specific granules (51, 52), granules identified by the presence of gelatinase (53), and granules identified by the presence of alkaline phosphatase (54) have all been implicated as a source of the newly expressed CD11b/CD18. Though increased expression of CD11b/CD18 is temporally associated with the enhanced capacity to bind and ingest C3-coated particles caused by PMA, an increased number of receptors is unlikely to be responsible for either the increased binding or phagocytosis. During activation of adherent PMN by PMA, the attachment of C3bi-coated erythrocytes increases 8- to 10-fold and phagocytosis increases 30- to 40-fold while the number of CD11b/CD18 per cell increases only 3-fold (49). More strikingly, the capacity of CD11b/CD18 to bind ligand and signal phagocytosis is eliminated during incubation with PMA from 20 to 65 minutes, but the expression of CD11b/CD18 does not change in this time. Increasing the expression of CD11b/CD18 on the cell surface is clearly not sufficient to induce ligand binding. Treating PMN with the chemotactic peptide n-formyl-methionyl-leucyl-phenylalanine (FMLP) causes a twofold increase in the amount of CD11b/CD18 on the cell surface (50), but does not affect the binding activity of this receptor (55). The capacity of CD11b/CD18 to bind C3bi must, therefore, be controlled in another way.

The capacity of CD11/CD18 to mediate adhesion to EC is also transiently enhanced by stimulation of PMN. Treatment of PMN with PMA causes first a dramatic increase, then a decrease in binding capacity with a time course identical with that shown for binding of EC3bi (5). The physiological stimulus, C5a, also causes a transient change in adhesion capacity but with a faster time course (5). As with binding of EC3bi, the changes in surface expression of CD11b/CD18 on PMN are not well correlated with changes in adhesive capacity toward EC suggesting an additional mechanism may be operative. This supposition is confirmed in experiments employing cytoplasts, fragments of PMN that are devoid of granules and cannot alter expression of surface receptors. Cytoplasts respond to PMA with a transient increase then a decrease in their binding to EC (5). These studies prove that changes in the nature, not the number, of CD11b/CD18 operate to control the CD18-dependent adhesiveness of PMN.

Several observations suggest that phosphorylation may be the biochemical event responsible for the transient changes in CD11b/CD18 activity. (i) The time course and reversibility of CD11b/CD18 activation are consistent with the hypothesis that a posttranslational modification like phosphorylation controls receptor activity. (ii) Phorbol myristate acetate, which enhances CD11b/CD18 activity, is a potent activator of a ubiquitous Ca^{2+}-activated, phospholipid-dependent protein kinase (56). (iii) Loading PMN with inorganic thiophosphate (thioP) allows irreversible activation of C3 receptors (49). ThioP resembles phosphate and is incorporated into nucleotides and phosphoproteins, but the resulting thiophosphoproteins are resistant to phosphatases. Thus one would expect that in a cell loaded with thioP, phosphorylation caused by stimulation of a kinase would result in a pool of thiophosphorylated proteins that are resistant to dephosphorylation. Therefore, it is likely that the irreversible activation of receptors observed in loaded cells is a consequence of irreversible thiophosphorylation. (iv) Finally, recent data show that CD11b/CD18 is a phosphoprotein, and its state of phosphorylation is regulated by PMA (C.B. Epstein and S.D.W., ms. in prep.).

Phosphorylation of CD11b/CD18 may alter adhesive capacity by affecting the distribution of the receptor in the plane of the membrane. Our recent work indicates that CD11b/CD18 aggregates in the plane of the membrane in response PMA, and such aggregation appears to be a prerequisite for ligand binding (55). When the surfaces of resting PMN are labelled with monoclonal anti-CD11b/CD18 (OKM1) and colloidal gold and viewed by transmission electron microscopy, the gold particles depicting CD11b/CD18 are present in a random distribution, with many as individuals. After PMN are treated with PMA for 25 minutes, which dramatically enhances binding activity, clusters of receptors are apparent. The time course of aggregation corresponds precisely with the time course of enhancement of binding: Aggregation is high at 25 minutes in PMA, and there is disaggregation by 50 minutes in PMA, by which time binding declines to levels comparable to those observed in resting cells. Other surface antigens on PMN (FcR, HLA) do not exhibit changes in their aggregation state in response to PMA, indicating that this is not a general property of membrane proteins. As mentioned above, treating PMN with the chemotactic peptide FMLP causes increased expression of CD11b/CD18 on the cell surface (50) but does not enhance the binding activity of the receptor (55). Also, FMLP does not cause aggregation of CD11b/CD18, again supporting the idea that clustering of CD11b/CD18 is required for binding ligand.

A possible mechanism by which clustering of receptors may endow them with enhanced binding activity is suggested by studies on the spatial distribution of the ligand, C3bi, on the surface of the ligand-coated particles (57). When C3 is deposited as random monomers, the resulting C3bi-coated erythrocytes are not bound by macrophages or PMN. However, if an equivalent number of C3bi molecules are deposited in clusters, binding of the EC3bi to phagocytes is avid. Even PMN stimulated with PMA, which express clustered CD11b/CD18 on the cell surface, are incapable of binding erythrocytes with randomly deposited C3bi, and avid binding is only observed when both the C3bi on the erythrocyte and the CD11b/CD18 on the PMN are clustered. Since the size of the clusters of C3bi (>5) is similar to the size of the clusters of CD11b/CD18 (6–10), it is likely that clusters of ligand interact with clusters of receptors to mediate effective binding between cells.

Why does clustering of receptors and ligand promote the interaction of cells? Though we cannot rule out the possibility that clustering causes conformational changes in the proteins, we prefer the hypothesis that the multivalent binding between clusters stabilizes cell-cell interactions. To disassociate a cluster of ligands from a cluster of receptors, each of the individual receptor-ligand interactions would need to be simultaneously broken. The unlikelihood of this event could prevent the detachment of a C3bi-coated cell from a phagocyte or the detachment of PMN from an endothelial cell.

The observation that clustering of CD11b/CD18 is required to mediate adhesion suggests a mechanism by which cells could "de-adhere" from a ligand-coated surface or cell. Receptors that are actively mediating adhesion are held in position by two types of bonds, those among components of a receptor cluster and those between receptor and the opposing ligand. To reverse adhesion, these two bonds may be broken sequentially. The bond tethering a receptor to other members of a cluster appears to involve phosphorylation of the receptor and may thus be broken by the action of a phosphatase. Since the binding affinity of the second bond (between an individual receptor and an individual ligand) is very low, the "off rate" will be high and receptors will release ligand often. Upon releasing ligand, untethered receptors would be free to diffuse in the plane of the membrane away from the cluster and away from ligand, thus reducing the chance of rebinding to ligand. Effective adhesion of a cluster of receptors with a cluster of ligands may thus be broken by sequentially breaking the individual receptor-ligand bonds and removing the receptors from the cluster.

Conclusion

The capacity of CD18 molecules to regulate adhesive interactions of cells makes them ideally suited for controlling movement of leukocytes across the endothelium. We believe that this type of regulated behavior will be a common feature of integrins that allows them to mediate cell movement and positioning. Thus studies on CD18 molecules may serve as a model for understanding a large class of adhesive interactions vital to multicellular organisms.

References

1. Wright SD, Rao PE, Van Voorhis WC, Craigmyle LS, Iida K, Talle MA, Westberg EF, Goldstein G, Silverstein SC: Identification of the C3bi receptor on human monocytes and macrophages by using monoclonal antibodies. *Proc Natl Acad Sci USA* 80:5699–5703, 1983.
2. Wright SD, Weitz JI, Huang AJ, Levin SM, Silverstein SC, Loike JD: Complement receptor type three (CR3, CD11b/CD18) of human polymorphonuclear leukocytes recognizes fibrinogen. *Proc Natl Acad Sci USA* 85:7734–7738, 1988.
3. Dana N, Styrt B, Griffin JD, Todd III RF, Klempner MS, Arnaout MA: Two functional domains in the phagocyte membrane glycoprotein Mol identified with monoclonal antibodies. *J Immunol* 137:3259–3263, 1986.
4. Harlan JM, Killen PD, Senecal FM, Schwartz BR, Yee EK, Taylor FR, Beatty PG, Price TH, Ochs HH: The role of neutrophil membrane protein GP-150 in neutrophil adhesion to endothelia in vitro. *Blood* 66:167–178, 1985.
5. Lo SK, Detmers PA, Levin SM, Wright SD: Transient adhesion of neutrophils to endothelium. *J Exp Med* 169:1779–1793, 1989.
6. Arfors K-E, Lunderg C, Lindblom L, Lundberg K, Beatty PG, Harlan JM: A monoclonal antibody to the membrane glycoprotein complex CD18 inhibits polymorphonuclear leukocyte accumulation and plasma leakage in vivo. *Blood* 69:338–340, 1987.
7. Rosen H, Gordon S: Monoclonal antibody to the murine type 3 complement receptor inhibits adhesion of myelomonocytic cells in vitro and inflammatory cell recruitment in vivo. *J Exp Med* 166:1685–1701, 1987.
8. Tuomanen EI, Saukkonen K, Sande S, Cioffe C, Wright SD: Reduction of inflammation, tissue damage, and mortality in bacterial meningitis in rabbits treated with monoclonal antibodies against adhesion-promoting receptors of leukocytes. *J Exp Med* (in press), 1989.
9. Todd III RF, Freyer DR: The CD11/CD18 leukocyte glycoprotein deficiency. *Hem /Onc Clin N Am* 2:13–31, 1988.
10. Wright SD, Detmers PA: Adhesion-promoting receptors on phagocytes. *J Cell Sci Suppl* 9:99–120, 1988.
11. Wright SD, Reddy PA, Jong MTC, Erickson BW: C3bi receptor (complement receptor type 3) recognizes a region of complement protein C3 containing the sequence Arg-Gly-Asp. *Proc Natl Acad Sci USA* 84:1965–1968, 1987.
12. Ruoslahti E, and Pierschbacher MD: Arg-Gly-Asp: A versatile cell recognition signal. *Cell* 44:517–518, 1986.
13. Hynes RO: Integrins: A family of cell surface receptors. *Cell* 48:549–554, 1987.
14. Law SKA, Gagnon J, Hildreth JE, Wells CE, Willis AC, Wong AJ: The primary structure of the beta-subunit of the cell surface adhesion glycoproteins LFA-1, CR3 and p150,95 and its relationship to the fibronectin receptor. *EMBO J* 6:915, 1987.
15. Kishimoto TK, O'Connor K, Lee A, Roberts TM, Springer TA: Cloning of the beta subunit of the leukocyte adhesion proteins: Homology to an extra-

cellular matrix receptor defines a novel supergene family. *Cell* 48:681–690, 1987.

16. Corbi AL, Miller LJ, O'Conner K, Larson RS, Springer TA: cDNA cloning and complete primary structure of the alpha subunit of a leukocyte adhesion glycoprotein, p150,95. *EMBO J* 6:4023–4028, 1987.

17. Arnaout MA, Gupta SK, Pierce MW, Tenen DG: Amino acid sequence of the alpha subunit of human leukocyte adhesion receptor Mo1 (Complement receptor type 3). *J Cell Biol* 106:2153–2158, 1988.

18. Wright SD, Silverstein SC: Receptors for C3b and C3bi promote phagocytosis but not the release of toxic oxygen from human phagocytes. *J Exp Med* 158:2016–2023, 1983.

19. Aderem AA, Wright SD, Silverstein SC, Cohn ZA: Ligated complement receptors do not activate the arachidonic acid cascade in resident peritoneal macrophages. *J Exp Med* 161:617–622, 1985.

20. Button LL, McMaster WR: Molecular cloning of the major surface antigen of Leishmania. *J Exp Med* 167:724, 1988.

21. Russell DG, Wright SD: Complement receptor type 3 (CR3) binds to an Arg-Gly-Asp-containing region of the major surface glycoprotein, gp63, of Leishmania promastigotes. *J Exp Med* 168: 279–292, 1988.

22. Wright SD, Silverstein SC: Phagocytosing macrophages exclude proteins from the zones of contact with opsonized targets. *Nature* 309:359–361, 1984.

23. Weitz JI, Huang AJ, Landman SL, Nicholson SC, Silverstein SC: Elastase-mediated fibrinogenolysis by chemoattractant-stimulated neutrophils occurs in the presence of physiological concentrations of antiproteinases. *J Exp Med* 166:1836–1850, 1987.

24. Lam SCT, Plow EF, Smith MA, Andrieux A, Ryckwaert JJ, Marguerie G, Ginsberg MH: Evidence that arginyl-glycyl-aspartate peptides and fibrinogen gamma chain peptides share a common binding site on platelets. *J Biol Chem* 262:947–950, 1987.

25. Wright SD, Meyer BC: The fibronectin receptor of human macrophages recognizes the amino acid sequence, Arg-Gly-Asp-Ser. *J Exp Med* 162:762–766, 1985.

26. Ezekowitz, RAB, Sim RB, Hill M, Gordon S: Local opsonization by secreted macrophage complement components. Role of receptors for complement in uptake of zymosan. *J Exp Med* 159:244–260, 1984.

27. Ross GD, Cain JA, Lachmann PJ: Membrane complement receptor type three (CR3) has lectin-like properties analogous to bovine conglutinin and functions as a receptor for zymosan and rabbit erythrocytes as well as a receptor for iC3b. *J Immunol* 134:3307–3315, 1985.

28. Ross GD, Thompson RA, Walport MJ, Springer TA, Watson JV, Ward RHR, Lida J, Newman SL, Harrison RA, Lachmann PJ: Characterization of patients with an increased susceptibility to bacterial infections and a genetic deficiency of leukocyte membrane complement receptor type 3 and the related membrane antigen LFA-1. *Blood* 66:882–890, 1985.

29. Wright SD, Levin SM, Jong MTC, Chad Z, Kabbash LG: CR3 (CD11b/CD18) expresses one binding site for Arg-Gly-Asp-containing peptides, and a second site for bacterial lipopolysaccharide. *J Exp Med* 169:175–183, 1989.

30. Bullock WE, Wright SD: The role of adherence-promoting receptors, CR3, LFA-1, and p150,95 in binding of *Histoplasma capsulatum* by human macrophages. *J Exp Med* 165:195–210, 1987.

31. Wright SD, Jong MTC: Adhesion-promoting receptors on human macrophages recognize *E. coli* by binding to lipopolysaccharide. *J Exp Med* 164:1867–1888, 1986.
32. Morrison DC, Ulevitch JR: The effects of bacterial endotoxins on host mediating systems. *Am J Pathol* 93:527–617, 1978.
33. Beutler B, Mahoney J, LeTrang N, Pekala P, Cerami A: Purification of cachectin, a lipoprotein lipase-supressing hormone secreted by endotoxin-treated RAW 264.7 cells. *J Exp Med* 161:984–995, 1985.
34. Durum SK, Schmidt JA, Oppenheim JJ: Interleukin-1: An immunological perspective. *Ann Rev Immunol* 3:263–287, 1985.
35. Suzuki S, Oldberg A, Hayman EG, Pierschbacher MD, Ruoslahti E: Complete amino acid sequence of human vitronectin deduced from cDNA. Similarity of cell attachment sites in vitronectin and fibronectin. *EMBO J* 4:2519–2524, 1985.
36. Cheresh, DA, Pytela R, Pierschbacher MD, Klier FG, Ruoslahti E, Reisfeld RA: An Arg-Gly-Asp-directed receptor on the surface of human melanoma cells exists in a divalent cation-dependent functional complex with disialoganglioside GD2. *J Cell Biol* 105:1163–1173, 1987.
37. Lo SK, Wright SD: CR3 mediates binding of PMN to endothelial cells (EC) via its RGD binding, not the LPS binding site. *FASEB J* 2:A991, 1988.
38. Marlin SD, Springer TA. 1987. Purified intercellular adhesion molecule-1 (ICAM-1) is a ligand for lymphocyte function-associated antigen (LFA-1). *Cell* 51:813–819, 1987.
39. Rovasio, RA, Delouvee A, Yamada KM, Timpl R, Thiery JP. Neural crest cell migration: Requirements for exogenous fibronectin and high cell density. *J Cell Biol* 96:462–473, 1983.
40. Sanderson CJ: 1976. The mechanism of T cell mediated cytotoxicity. II. Morphological studies of cell death by time-lapse microcinematography. *Proc R Soc Lond Ser B* 92:241–255, 1976.
41. Boucaut JC, Darribere T, Poole TJ, Aoyama H, Yamada KM, Thiery JP: Biologically active synthetic peptides as probes of embryonic development: A competitive peptide inhibitor of fibronectin function inhibits gastrulation in amphibian embryos and neural crest cell migration in avian embryos. *J Cell Biol* 99:1822–1830, 1984.
42. Naidet C, Semeriva M, Yamada KM, Thiery JP: Peptides containing the cell-attachment recognition signal Arg-Gly-Asp prevent gastrulation in Drosophila embryos. *Nature* 325:348–350, 1987.
43. Hirst R, Horwitz A, Buck C, Rohrschneider L: Phosphorylation of the fibronectin receptor complex in cells transformed by oncogenes that encode tyrosine kinases. *Proc Natl Acad Sci USA* 83:6470–6474, 1986.
44. Chen W-T, Wang J, Hasegawa T, Yamada SS, Yamada KM: Regulation of fibronectin receptor distribution by transformation, exogenous fibronectin, and synthetic peptides. *J Cell Biol* 103:1649–1661, 1986.
45. Rothlein R, Springer TA: The requirement for lymphocyte function-associated antigen 1 in homotypic leukocyte adhesion stimluated by phorbol esters. *J Exp Med* 163:1132–1149, 1986.
46. Wright SD, Silverstein SC: Tumor-producing phorbol esters stimulate C3b and C3b′ receptor-mediated phagocytosis in cultured human monocytes. *J Exp Med* 156:1149–1164, 1982.

47. Wright SD, Detmers PA, Jong MTC, Meyer BC: Interferon-gamma depresses binding of ligand by C3b and C3bi receptors on cultured human monocytes, an effect reversed by fibronectin. *J Exp Med* 163:1245–1259, 1986.
48. Guyre PM, Morganelli PM, Miller R: Recombinant immune interferon increases immunoglobulin G Fc receptors on cultured human mononuclear phagocytes. *J Clin Invest* 72:393–397, 1983.
49. Wright SD, Meyer BC: Phorbol esters cause sequential activation and deactivation of complement receptors on polymorphonuclear leukocytes. *J Immunol* 136:1759–1764, 1986.
50. Berger M, O'Shea J, Cross AS, Folks TM, Chused TM, Brown EJ, Frank MM. Human neutrophils increase expression of C3bi as well as C3b receptors upon activation. *J Clin Invest* 74:1566, 1984.
51. O'Shea J, Brown EJ, Seligmann BE, Metcalf JA, Frank MM, Gallin JI: Evidence for distinct intracellular pools of receptors for C3b and C3bi in human neutrophils. *J Immunol* 134:2580, 1985.
52. Todd III RF, Arnaout MA, Rosin RE, Crowley CA, Peters WA, Babior BM: Subcellular localization of the large subunit of Mo1 (Mo1 : formerly gp 110), a surface glycoprotein associated with neutrophil adhesion. *J Clin Invest* 74:1280, 1984.
53. Petrequin PR, Todd III RF, Devall LJ, Boxer LA, Curnutte III JT: Association between gelatinase release and increased plasma membrane expression of the Mo1 glycoprotein. *Blood* 69:605–610, 1987.
54. Borregaard N, Miller LJ, Springer TA: Chemoattractant-regulated mobilization of anovel intracellular compartment in human neutrophils. *Science* 237:1204–1206, 1987.
55. Detmers PA, Wright SD, Olsen E, Kimball B, Cohn ZA: Aggregation of complement receptors on human neutrophils in the absence of ligand. *J Cell Biol* 105:1137–1145, 1987.
56. Castagna M, Takai Y, Kaibuchi K, Sano K, Kikkawa U, Nishizuka Y: Direct activation of calcium-activated, phospholipid-dependent protein kinase by tumor-promoting phorbol esters. *J Biol Chem* 257:7847–7851, 1982.
57. Hermanowski-Vosatka A, Detmers PA, Goetze O, Silverstein SC, Wright SD: Clustering of binding sites on the plasma membrane enhances cell-cell adhesion. *J Biol Chem* 263:17,822–17,827, 1988.

15

CD18 Dependence of Primate Eosinophil Adherence In Vitro

CRAIG D. WEGNER, C. WAYNE SMITH, AND
ROBERT ROTHLEIN

Introduction

Evidence from several laboratories suggest that eosinophil migration into the lung interstitium and airway lumen plays a major role in the pathogenesis of airway hyperresponsiveness and asthma. In asthmatics, eosinophils in the airway lumen have been found to increase following allergen inhalation. This increase is correlated with the magnitude of the late phase response (1,2). Late phase responses have also been associated with increases in nonspecific airway responsiveness (3–5). We have reported that the presence of eosinophils in the airways of monkeys not only follows but appears to be predictive of an intense immediate response to inhaled antigen (6). Additionally, chronic airway eosinophilia in monkeys, either idiopathic (7) or induced by chronic antigen exposure (8), is associated with marked airway hyperresponsiveness to inhaled methacholine.

Once in the lung interstitium and airway lumen, eosinophils apparently generate mediators that have the unique ability to induce the pathologic characteristics of the airway inflammation found in asthmatics. The eosinophil granule major basic protein (MBP) causes epithelial exfoliation and impaired ciliary beating when added to guinea pig tracheal rings and human bronchial epithelium in vitro at concentrations similar to those found in sputum from asthmatics (9,10). Denudation of the airway epithelium has been found to increase the reactivity and maximal contraction of bronchial strips in vitro (11) and has been hypothesized to increase airway responsiveness by exposing sensory nerve endings in vivo (12,13). In studies of MBP immunofluorescence in human lung tissue obtained at autopsy, extracellular MBP was found associated with mucus plugs, damaged epithelium, eosinophil infiltration in the lamina propria, destruction of the basement membrane zone, and necrotic areas and amorphous debris below the basement membrane (14–16). Eosinophils also contain at least two other cytotoxic proteins, eosinophil cationic protein (17) and eosinophil-derived neurotoxin (18). Finally, eosinophils have a pronounced respiratory burst (18,19), release the potent broncho-constrictor and mucus producer, leukotriene C_4 (20), and generate the pro-inflammatory mediator, platelet activating factor (21).

The migratory and cytotoxic functions of eosinophils may require cell adhesion. To evaluate this possibility, we have initially investigated the properties of eosinophil adhesion in vitro. In particular, we have studied the role of the CD18 family of adhesion molecules (22–25) as well as one of its ligands, intercellular adhesion molecule-1 [ICAM-1 (26–29)], on primate eosinophil adhesion to protein-coated plastic and human endothelium.

Results

Eosinophils were obtained by bronchoalveolar lavage from adult male cynomolgus monkeys (*Macaca fascicularis*) with airway eosinophilia, purified (morphologically >93% pure) on a Percoll continuous density gradient, washed and added to 96 well flat bottom tissue culture plates (5×10^3 cells/well). After a 60-minute incubation at 37 °C, the nonadherent cells were removed by a plate washer. Adherent cells were quantitated visually (no aggregation or degranulation was observed) and by a colorimetric assay for eosinophil peroxidase, EPO (30).

The eosinophils were observed to spontaneously adhere and spread on the bottom of untreated wells or wells coated with immune complexes made of *Ascaris* extract and serum from an *Ascaris*-sensitive monkey. In contrast, eosinophils adhered poorly to wells coated with protein (*Ascaris* extract, bovine serum albumin or monkey serum). Of various soluble stimuli tested, platelet activating factor (PAF) was found to induce the most pronounced and consistent adherence of eosinophils to wells coated with protein (Fig. 15.1).

The role of the CD18 family of adhesion molecules as well as one of their ligands (intercellular adhesion molecule-1, ICAM-1) was investigated using monoclonal antibodies (MAbs) reactive with each of these membrane glycoproteins. The anti-CD18 MAbs were R3.3 and R15.7, the anti-CD11a MAb was R3.1, the anti-CD11b MAbs were M1/70 and LM2/1, and the anti-ICAM-1 MAbs were RR1/1 and R6.5.D6. The anti-HLA class I MAb W6/32 was also used.

The adhesion of primate lung eosinophils to wells coated with immune complexes appeared to be CD11b dependent, since adhesion was inhibited by MAbs against CD11b and CD18 but not CD11a, ICAM-1, or HLA (Fig. 15.2). Identical results were found when adhesion to wells coated with protein was induced by soluble stimuli including PAF.

On the other hand, PAF (10^{-7} M) induced adhesion of primate lung eosinophils to LPS (10 ng/ml) stimulated and glutaraldehyde-fixed human umbilical vein endothelial cells (HUVEC) (isolated as described in

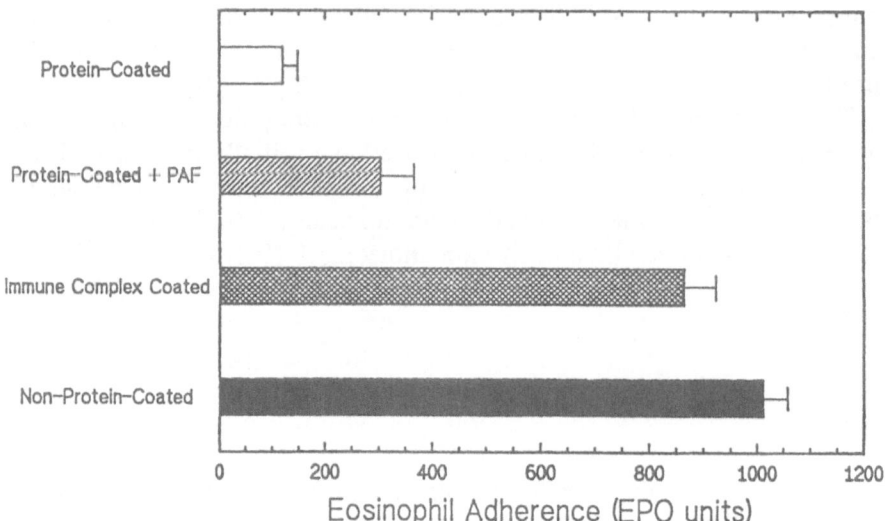

FIGURE 15.1. Eosinophil adherence to flat bottom tissue culture plate wells protein-coated with no stimulus, protein-coated with PAF (10^{-7} M) stimulus, immune complex (IC) coated and stimulus, and noncoated with no stimulus. Adhered cells were quantitated by a colormetric assay for eosinophil peroxidase (EPO units, mean ± SD).

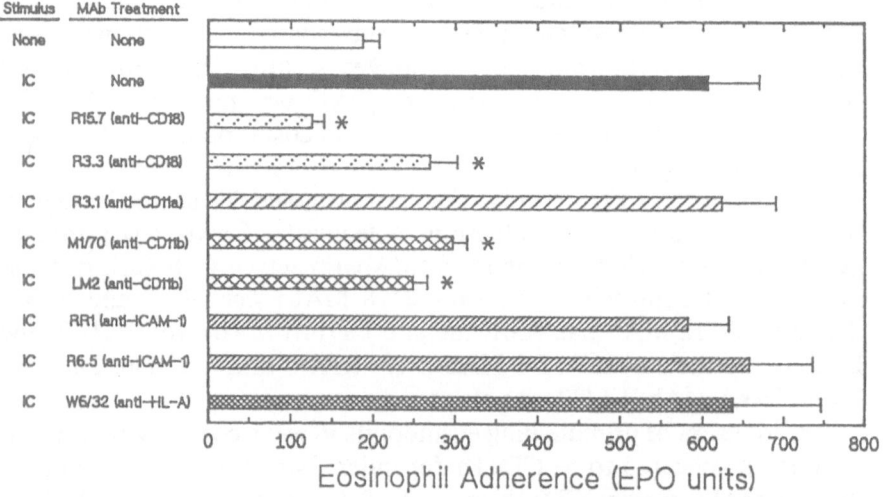

FIGURE 15.2. Effect of various monoclonal antibody (MAb) supernatants (1:4 dilution) on eosinophil adherence to flat bottom tissue culture plate wells coated with immune complexes (IC). Adhered cells were quantitated by a colormetric assay for eosinophil peroxidase (EPO units, mean ± SD). Statistically significant attenuation of adherence is signified by an asterisk (*).

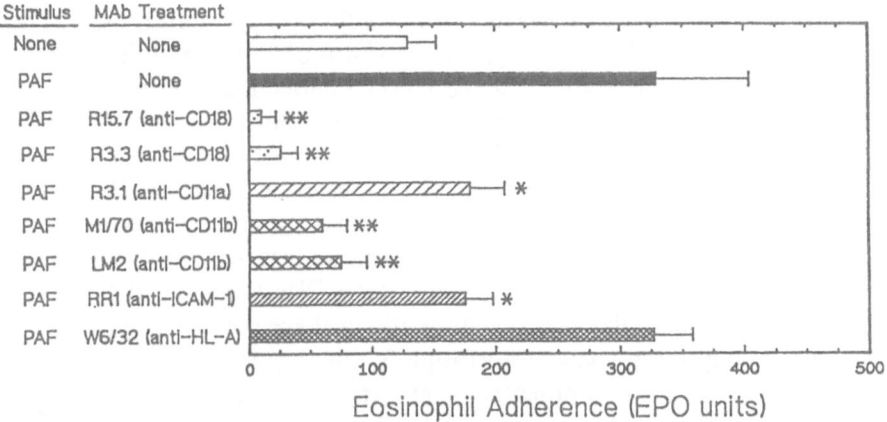

FIGURE 15.3. Effect of various monoclonal antibody (MAb) supernatants (1:4 dilution) on PAF ($10^{-7}M$) induced eosinophil adherence to LPS (10 ng/ml) stimulated and glutaraldehyde fixed endothelium. Adhered cells were quantitated by a colormetric assay for eosinophil peroxidase (EPO units, mean ± SD). Statistically significant attenuation of adherence is signified by an asterisk (*).

references 28 and 29) appeared to be partially dependent on CD11a, CD11b, and ICAM-1. That is, adhesion was completely inhibited by MAbs against CD18, partially inhibited by MAbs against CD11a, CD11b, and ICAM-1, and not inhibited by a MAb against HL-A (Fig. 15.3).

Discussion

The importance of leukocyte-cell and leukocyte-substrate adhesion in migration, inflammation, and immune defense is impressively demonstrated by a group of patients that are deficient in their expression of the CD18 family of adhesion molecules because of an inherited defect in the common β-chain (24,25). These *leukocyte adhesion deficiency* (LAD) patients present clinically with delayed umbilical cord separation, reoccurring and progressive soft tissue infections, and impaired pus formation, despite a striking blood leukocytosis. Histopathologic evaluations of infected tissues in LAD patients demonstrate a profound deficiency or total absence of neutrophilic granulocytes. However, relatively large numbers of eosinophils have been identified in these tissues (24). Thus, one might hypothesize that at least for migration, eosinophils differ from neutrophils in their dependence on the CD18 family of adhesion molecules.

In this communication, we have demonstrated that the adherence properties of primate lung eosinophils are, in the tests performed, nearly identical to those of previously reported for human neutrophils (28,29,31–

33) in their dependence on the membrane adhesion glycoproteins CD11a, CD11b, and ICAM-1. That is, adhesion of stimulated eosinophils to protein-coated plastic was found to be CD11b dependent and adhesion of stimulated eosinophils to stimulated endothelium was found to be CD11a, CD11b, and ICAM-1 dependent.

We conclude that the CD18 family of adhesion molecules play a major role in eosinophil adherence *in vitro* and that the selective presence of eosinophils in the infected tissues of LAD patients is not simply explained by a lack of dependence on these molecules. We speculate that the selective tissue accumulation of eosinophils in these patients is because of the much longer life span of eosinophils than neutrophils especially in the presence of lymphokines such as granulocyte-macrophage colony stimulating factor and interleukin-5 (34,35).

References

1. Durham SR, Kay AB: Eosinophils, bronchial hyperreactivity and late-phase asthmatic reactions. *Clin Allergy* 15:411, 1985.
2. De Monchy JGR, Kauffman HF, Venge P, Koeter GH, Jansen HM, Sluiter HJ, De Vries K: Bronchoalveolar eosinophilia during allergen-induced late asthmatic reactions. *Am Rev Respir Dis* 131:373, 1985.
3. Lam S, Tan F, Chan H, Chan-Yeung M: Relationship between types of asthmatic reaction, nonspecific bronchial reactivity, and specific IgE antibodies in patients with red cedar asthma. *Clin Allergy* 72:134, 1983.
4. Cockcroft DW, Ruffin RE, Dolovich J, Hargreave FE: Allergen-induced increase in non-allergic bronchial reactivity. *Clin Allergy* 7:503, 1977.
5. Cartier, A, Thomson NC, Frith PA, Roberts R, Tech M, Hargreave FE: Allergen-induced increase in bronchial responsiveness to histamine: relationship to the late asthmatic response and change in airway caliber. *J Allergy Clin Immunol* 70:170, 1982.
6. Wegner CD, Koker PJ, Sabo JP, Gundel RH: Interdependence of antigen-induced bronchoconstriction, airway inflammation and altered blood white cell composition in monkeys (abstract). *Am Rev Respir Dis* 135:A221, 1987.
7. Wegner CD, Koker PJ, Gundel RH: Relationship between eosinophils and airway reactivity in monkeys (abstract). *Am Rev Respir Dis* 135:A222, 1987.
8. Gundel RH, Gerritsen ME, Wegner CD: Increase in airway eosinophils is associated with increase in airway reactivity in monkeys (abstract). *Am Rev Respir Dis* 135:281, 1988.
9. Frigas E, Loegering DA, Gleich GJ: Cytotoxic effects of the guinea pig eosinophil major basic protein on tracheal epithelium. *Lab Invest* 42:35, 1980.
10. Frigas E, Loegering DA, Solley GO, Farrow GM, Gleich GJ: Elevated levels of the eosinophil granule major basic protein in the sputum of patients with bronchial asthma. *Mayo Clin Proc* 56:345, 1981.
11. Barnes PJ, Cuss FM, Palmer JB: The effect of airway epithelium on smooth muscle contractility in bovine trachea. *Br J Pharmac* 86:685, 1985.
12. Laitinen LA, Heino M, Laitinen A, Kava T, Haahtela T: Damage of the airway epithelium and bronchial reactivity in patients with asthma. *Am Rev Respir Dis* 131:599, 1985.

13. Barnes PJ: Asthma as an axon reflex. *Lancet* 1:242, 1986.
14. Filley WV, Holley KE, Kephart GM, GJ Gleich. Identification by immuno-fluorescence of eosinophil granule major basic protein in lung tissue of patients with bronchial asthma. *Lancet* 2:11, 1982.
15. Gleich GJ: The pathology of asthma: with emphasis on the role of the eosinophil. *NER Allergy Proc* 7:421, 1986.
16. Frigas E, Gleich GJ: The eosinophil and the pathophysiology of asthma. *J Allergy Clin Immunol* 77:527, 1986.
17. Ackerman SJ, Gleich GJ, Loegering DA, Richardson BA, Butterworth AE. Comparative toxicity of purified human eosinophil granule cationic proteins for schistosomula of *Schistosoma mansomi*. *Am J Trop Med Hyg* 34:735, 1985.
18. Gleich GJ, Adolphson CR: The eosinophilic leukocyte: Structure and function. *Adv Immunol* 39:177, 1986.
19. Learn DB, Brestel EP: A comparison of superoxide production by human eosinophils and neutrophils. *Agents Actions* 12:485, 1982.
20. Shaw RJ, Cromwell O, Kay AB: Preferential generation of leukotriene C_4 by human eosinophils. *Clin Exp. Immunol* 56:716, 1984.
21. Lee T, Lenihan DJ, Malone B, Roddy LL, Wasserman SI: Increased biosynthesis of platelet-activating factor in activated human eosinophils. *J Biol Chem* 259:5526, 1984.
22. Sanchez-Madrid F, Nagy J, Robbins E, Simon P, Springer TA: A human leukocyte differentiation antigen family with distinct alpha subunits and a common beta subunit: The lymphocyte function-associated antigen (LFA-1), the C3bi complement receptor (OKM1/Mac-1), and the p150,95 molecule. *J Exp Med* 158:1785, 1983.
23. Springer TA, Dustin ML, Kishimoto TK, Marlin SD: The lymphocyte function-associated LFA-1, CD2, and LFA-3 molecules: Adhesion receptors of the immune system. *Ann Rev Immunol* 5:223, 1987.
24. Anderson DC, Schmalsteig FC, Finegold MJ, Hughes BJ, Rothlein R, Miller LJ, Kohl S, Tosi MF, Jacobs RL, Waldrop TC, Goldman AS, Shearer WT, Springer TA: The severe and moderate phenotypes of heritable Mac-1, LFA-1 deficiency: Their quantitative definition and relation to leukocyte dysfunction and clinical features. *J Inf Dis* 152:668, 1985.
25. Anderson DC, Springer TA: Leukocyte adhesion deficiency: An inherited defect in the Mac-1, LFA-1, and p150,95 glycoproteins. *Ann Rev Med* 38:175, 1987.
26. Rothlein R, Dustin ML, Marlin SD, Springer TA: A human intercellular adhesion molecule (ICAM-1) distinct from LFA-1. *J Immunol* 137:1270, 1986.
27. Marlin SD, Springer TA: Purified intercellular adhesion molecule-1 (ICAM-1) is a ligand for lymphocyte function-associated antigen 1 (LFA-1). *Cell* 51:813, 1987.
28. Smith CW, Rothlein R, Hughes BJ, Mariscalco MM, Rudloff HE, Schmalsteig FC, Anderson DC: Recognition of an endothelial determinate for CD18-dependent human neutrophil adherence and transendothelial migration. *J Clin Invest* 82:1746, 1988.
29. Smith CW, Marlin SD, Rothlein R, Lawrence MB, McIntire LV, Anderson DC. 1989. The role of ICAM-1 in the adherence of human neutrophils to human endothelial cells in vitro. *This volume*, Chapter 13.

30. Strath M, Warren DJ, Sanderson CJ: Detection of eosinophils using an eosinophil peroxidase assay. Its use as an assay for eosinophil differentiation factors. *J Immunol Meth* 83:209, 1985.
31. Anderson DC, Miller LJ, Schmalsteig FC, Rothlein R, Springer TA: Contributions of the Mac-1 glycoprotein family to adherence-dependent granulocyte functions: structure-function assessment employing subunit-specific monoclonal antibodies. *J Immunol* 137:15, 1986.
32. Tonnesen MG, Anderson DC, Springer TA, Knedler A, Avdi N, Henson PM: Mac-1 glycoprotein family mediates adherence of neutrophils to endothelial cells stimulated by chemotactic peptides. *Clin Res* 34:419a, 1986.
33. Harlan JM, Killen PD, Senecal FM, Schwartz BR, Yee EK, Taylor RF, Beatty PG, Price TH, Ochs HD: The role of neutrophil membrane glycoprotein GP-150 in neutrophil adherence to endothelium in vitro. *Blood* 66:167, 1985.
34. Lopez AF, Williamson DJ, Gamble JR, Begley CG, Harlan JM, Klebanoff SJ, Waltersdorph A, Wong G, Clark SC, Vadas MA: Recombinant human granulocyte-macrophage colony-stimulating factor stimulates in vitro mature human neutrophil and eosinophil function, surface receptor expression, and survival. *J Clin Invest* 78:1220, 1986.
35. Begley CG, Lopez AF, Nicola NA, Warren DJ, Vadas MA, Sanderson CJ, Metcalf D: Purified colony-stimulating factors enhance the survival of human neutrophils and eosinophils in vitro: A rapid and sensitive microassay for colony-stimulating factors. *Blood* 68:162, 1986.

16

Identification and Characterization of Endothelial-Leukocyte Adhesion Molecule 1

M.P. BEVILACQUA AND M.A. GIMBRONE, JR.

Introduction

Increasing evidence suggests that the vascular endothelium actively contributes to a variety of proinflammatory processes, including leukocyte extravasation. In particular, recent in vitro studies have supported the hypothesis that inflammatory/immune mediators can act directly on vascular endothelial cells to increase the adhesion of blood leukocytes (1–14). In addition, in vivo studies have suggested that local activation of the vascular endothelium may contribute to accumulation of leukocytes at sites of inflammation in various pathophysiological contexts (15–18). Our laboratory has used standardized in vitro monolayer adhesion assays to study factors that alter endothelial-leukocyte adhesion and to investigate the mechanisms of these interactions. Initially, human monocyte derived IL-1 was found to act on cultured human endothelial cells (HEC) in a time- and protein-synthesis dependent fashion to increase the adhesion of isolated human blood polymorphonuclear leukocytes (PMN), monocytes and the related cell lines HL-60 and U937 (1). Subsequently, other cytokines including recombinant IL-1-α, IL-1-β, tumor necrosis factor (TNF) and lymphotoxin, as well as bacterial endotoxin (lipopolysaccharide, LPS) (2,3) were shown to induce endothelial-dependent mechanisms of leukocyte adhesion. Other investigators have made similar observations using various mediators and leukocyte cell types (4–14). Taken together, these studies suggested that activation of vascular endothelium increases the expression of cell surface adhesion molecules that can bind blood leukocytes. We have referred to these putative structures as *endothelial-leukocyte adhesion* molecules (ELAMs) (2,3). In this chapter, we review studies leading to the identification and characterization of an inducible endothelial cell surface adhesion molecule for PMN, designated ELAM-1 (19).

Results

Identification of ELAM-1

Our primary strategy for the identification of endothelial-leukocyte adhesion molecules has involved the development of monoclonal antibodies (MAb) against IL-1 and TNF-treated HEC. Initial efforts led to the generation of MAb H4/18, which recognized an endothelial cell surface protein expressed on cytokine- or endotoxin-stimulated HEC but not on unstimulated HEC in vitro (20). In addition, this MAb was found to be effective in identifying "activated" vascular endothelium in human tissues in situ (16). In functional assays, MAb H4/18 partially inhibited the adhesion of the promyelocytic cell line HL-60 to activated HEC monolayers (15–25%) (3), but failed to block the adhesion of isolated human

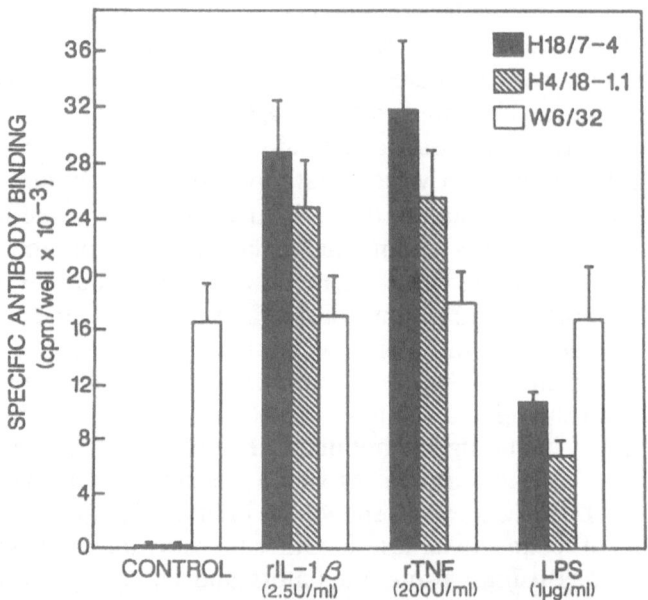

FIGURE 16.1. Endothelial cell surface radioimmunoassay. Serially passaged umbilical vein HECs were grown to confluence in 96-well tissue culture plates precoated with gelatin. Human recombinant IL-1 β, recombinant TNF, or LPS were added to the cultures in RPMI-1640 supplemented with 10% fetal calf serum. After incubation for 4 hours, the added mediators were removed, the HEC monolayers were washed three times, and then assessed for the binding of several monoclonal antibodies in a cell-surface RIA (19,20). "Specific antibody binding" was determined by subtracting nonspecific radioactivity that was observed in wells labeled with an isotype matched nonbinding MAb. Data are presented as a mean ± SD from quadruplicate wells.

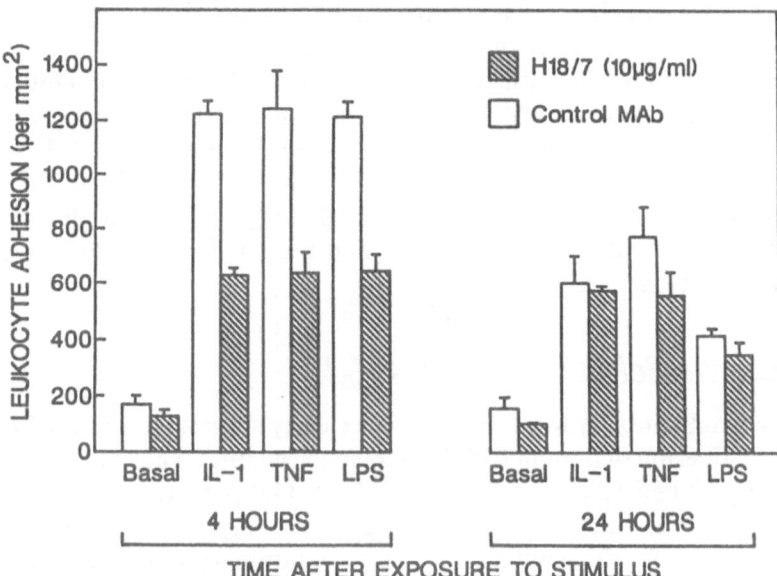

FIGURE 16.2. Inhibition of PMN adhesion to HEC monolayers by anti-ELAM-1 MAb H18/7. Human endothelium cells monolayers were activated for 4 hours (left panel) or for 24 hours (right panel) with IL-1, TNF, or LPS. Subsequently, the monolayers were washed and incubated for 45 minutes (room temperature) with a control MAb or with MAb H18/7. Coincidentally, radiolabelled PMN were incubated (45 minutes) with heat-aggregated rabbit IgG (20 mg/ml) to block their Fc receptors. The PMN preparation (100 μl per well; 1.0×10^6 leukocytes/ml, final concentration) was added to the HEC monolayers and an adhesion assay (10 minutes) was performed as described (19).

peripheral blood PMN. Subsequent efforts led to the generation of MAb H18/7 (19), which was found to recognize the same inducible endothelial cell surface molecule as H4/18. Monoclonal antibody H18/7 was effective in blocking both isolated human PMN (50%) and HL-60 cells (60%), and thus allowed the designation of ELAM-1 (19).

As depicted in Fig. 16.1, immunobinding studies with MAbs H4/18 and H18/7 have indicated that unstimulated vascular endothelial cells express little or no ELAM-1. However, exposure of HEC monolayers to IL-1, TNF, or endotoxin for four hours results in a dramatic increase in the specific binding of both anti-ELAM-1 MAbs. Continuous treatment of HEC monolayers with these mediators results in a time- and protein synthesis-dependent increase in the expression of ELAM-1, with onset at 30 to 60 minutes and peak activity at 4 to 6 hours, followed by a decline toward basal levels by 24 to 48 hours (19,20).

The role of ELAM-1 in endothelial-leukocyte adhesion is suggested by the effects of MAb H18/7 in monolayer adhesion assays. As shown in

Fig. 16.2, MAb H18/7 consistently blocks (by approximately 50%) the adhesion of isolated human PMN to HEC monolayers that have been activated for 4 hours by different cytokines or mediators (IL-1, TNF, and LPS). The adhesion of PMN to 24-hour activated HEC monolayers is also elevated, but this effect appears to primarily involve a non-ELAM-1 dependent mechanism. As described elsewhere in this book, the inducible cell adhesion molecule ICAM-1 (12–14) also contributes to endothelial-PMN adhesive interactions, and may, at least in part, explain this observation.

Leukocyte Selectivity of ELAM-1

Monoclonal antibody inhibition studies clearly suggest a role for ELAM-1 in the adhesion of human PMN and related cell lines (i.e., promyelocytic lines HL-60 and KG-1) to activated HEC. The potential contributions of ELAM-1 to the adhesion of other leukocyte cell types is currently under investigation. As depicted in Fig. 16.3, preliminary data indicate that MAb H18/7 fails to block the adhesion of several human T and B lymphoid cell lines. Similar results were obtained with isolated human mono-

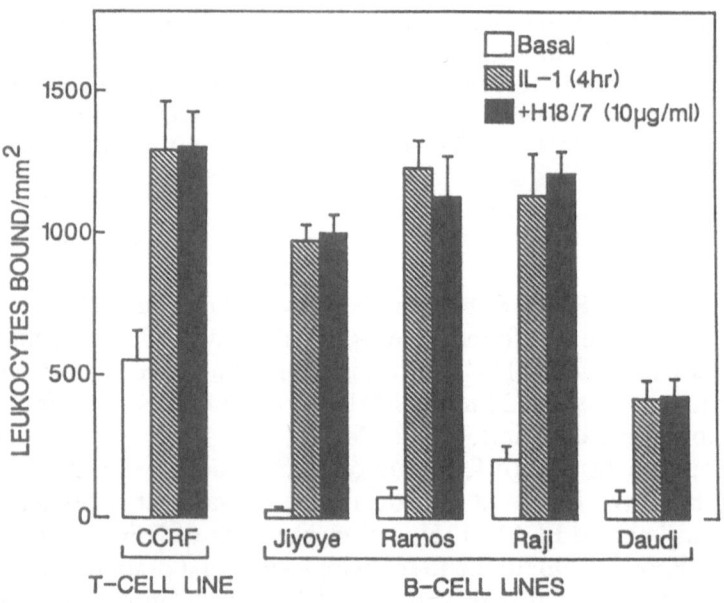

FIGURE 16.3. Adhesion of T and B cell lines to IL-1 activated HEC monolayers. Human endothelium cells monolayers were treated for 4 hours with control media (open) or with 5 U/ml of IL-1 for 4 hours (hatched). The monolayers were washed and, in certain cases (solid), anti-ELAM-1 MAb H18/7 was added 45 minutes prior to the adhesion assay.

nuclear cells. These observations suggest that ELAM-1 may act as a PMN-selective adhesion molecule on activated endothelium. Future investigations of the potential interactions of ELAM-1 and other leukocyte cell types will involve the use of cloned ELAM-1 (see below).

Expression of ELAM-1 on Activated Human Endothelium In Vivo

A series of studies on the expression of ELAM-1 in human tissues in situ has been conducted by R.S. Cotran and colleagues of our department (16–18). Briefly, ELAM-1 is not typically detected by immunohistochemical staining in the vascular endothelium or in other cellular components of normal tissues. Several pathophysiological processes, however, appear to induce ELAM-1 expression, most commonly localized to venular endothelium. Generally, this venular endothelium appears plump and "activated" and is often associated with inflammatory/immune infiltrates. ELAM-1 expression is characteristically found in sites of active inflammation, including diseases such as appendicitis and certain immunological responses such as erythema multiforme. Generally, it is not associated with chronic processes such as atherosclerosis or most cancers. However, certain lymphoid malignancies (e.g., T-cell lymphoma and Hodgkin's disease) often reveal prominent ELAM-1 expression within affected lymph nodes. This expression may occur because of high local concentrations of inflammatory/immunological cytokines released by the malignant lymphoid cells.

Molecular Cloning of ELAM-1

In collaboration with Drs. S. Stengelin and B. Seed, of the Department of Molecular Biology, Massachusetts General Hospital, Boston, Massachusetts, we have isolated a full-length cDNA encoding ELAM-1 (21) using a COS cell expression system (22,23). Sequence analysis of this cDNA has predicted a 610 amino acid polypeptide, including a 21 amino acid signal sequence (21). In our proposed structure, ELAM-1 is a Type 1 transmembrane protein with a large extracellular domain, a typical transmembrane segment, and a short cytoplasmic tail. The extracellular portion of ELAM-1 is composed of a previously undescribed domain pattern (Fig. 16.4), including an N-terminal lectin-like domain (120 amino acids) that is homologous to the asialoglycoprotein receptor (24,25) and the low affinity IgE receptor (26,27), an EGF-like domain, and six consensus repeats (60 amino acids each) that are related to those found in a recently described family of complement regulatory proteins (28,29). ELAM-1 appears to be essentially unrelated to the immunoglobulin supergene family including ICAM-1 (30,31) or to the integrin supergene family of adhesion receptors (32,33).

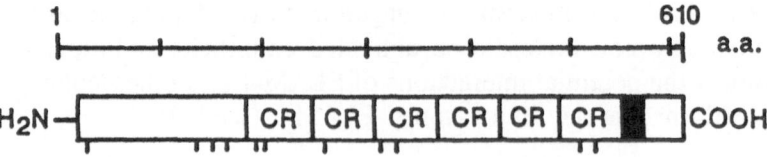

FIGURE 16.4. Schematic diagram of the ELAM-1 structure as predicted from the cDNA sequence (21).

Recent evidence suggests that the unusual mosaic structure of ELAM-1 is shared by two other molecules, the MEL-14 antigen, a lymphocyte homing receptor (34,35), and granule-membrane-protein-140 (GMP-140), a 140 KD glycoprotein that is found in the membranes of platelet and endothelial cell secretory granules (36). Like ELAM-1, both MEL-14 and GMP-140 contain an N-terminal lectin-like domain and an EGF repeat (greater than 60% identity with ELAM-1); in addition, they contain two and nine complement regulatory-like repeats, respectively. The identification of these three cell surface glycoproteins of similar overall structure defines a new gene family, the charter members of which appear to play key roles in selective adhesive interactions between circulating blood cells and the vessel wall.

Conclusions

The ELAM-1 is an inducible endothelial cell surface glycoprotein that mediates the adhesion of blood neutrophils. Its expression is relatively restricted to activated vascular endothelium in vitro and in vivo. The primary amino acid sequence of ELAM-1 suggests a complex domain structure containing a lectin-like amino terminus, an EGF-like domain and six consensus repeats similar to those found in complement regulatory proteins. The inducible expression of ELAM-1 and its complex structure suggests several potential roles in the control of the inflammatory/immune reactions at the vessel wall-blood interface.

Acknowledgments. We gratefully acknowledge our collaborators Drs. R.S. Cotran, J.S. Pober, D.L. Mendrick, M.E. Wheeler, S. Stengelin, and B. Seed who have contributed in essential ways to these studies. We also thank Dr. E. Rice for helpful discussions, A. Brock, G. Majeau, and J. Kim for expert technical assistance, as well as K. Case and V. Davis for cell culture.

References

1. Bevilacqua MP, Pober JS, Wheeler ME, Cotran RS, Gimbrone Jr. MA: Interleukin 1 acts on cultured human vascular endothelium to increase the adhesion of polymorphonuclear leukocytes, monocytes, and related leukocyte cell lines. *J Clin Invest* 76:2003–2011, 1985.

2. Bevilacqua MP, Pober JS, Wheeler ME, Cotran RS, Gimbrone Jr. MA: Interleukin-1 (IL-1) activation of vascular endothelium: Effects on procoagulant activity and leukocyte adhesion. *Am J Pathol* 121:393–403, 1985.

3. Bevilacqua MP, Wheeler ME, Pober JS, Fiers W, Mendrick DL, Cotran RS, Gimbrone Jr. MA: Endothelial-dependent mechanisms of leukocyte adhesion: Regulation by interleukin 1 and tumor necrosis factor, in Movat HZ (ed): *Leukocyte Emigration and Its Sequelae*. Basle, Karger, p 79, 1987.

4. Gamble, JR, Harlan JM, Klebanoff SJ, Vadas MA: Stimulation of the adherence of neutrophils to umbilical vein endothelium by human recombinant tumor necrosis factor. *Proc Natl Acad Sci USA* 82:8667–8761, 1985.

5. Pohlman, TH, Stanness KA, Beatty PG, Ochs HD, Harlan JM: An endothelial cell surface factor(s) induced in vitro by lipopolysaccharide, interleukin-1, and tumor necrosis factor-alpha increases neutrophil adherence by a CDw18-dependent mechanism. *J Immunol* 136:4548–4553, 1986.

6. Dunn CJ, Fleming WE: The role of interleukin-1 in the inflammatory response with particular reference to endothelial cell-leukocyte adhesion, in Kluger MJ, Oppenheim JJ, Powanda MC (eds): *The Physiologic Metabolic and Immunologic Actions of Interleukin-1*. New York, Alan R. Liss, p 45, 1985.

7. Schleimer RP, Rutledge BK: Cultured human vascular endothelial cells acquire adhesiveness for neutrophils after stimulation with interleukin 1, endotoxin and tumor-promoting phorbol diesters. *J Immunol* 136:649–654, 1986.

8. Cavender DE, Haskard DO, Joseph B, Ziff M: Interleukin 1 increases the binding of human B and T lymphocytes to endothelial cell monolayers. *J Immunol* 136:203–207, 1986.

9. Haskard D, Cavender D, Beatty PG, Springer TA, Ziff M: T Lymphocyte adhesion to endothelial cells: Mechanisms demonstrated by anti-LFA-1 monoclonal antibodies. *J Immunol* 137:2901–2906, 1986.

10. Zimmerman GA, McIntyre TM, Prescott SM: Thrombin stimulates the adherence of neutrophils to human endothelial cells in vitro. *J Clin Invest* 76:2235–2246, 1985.

11. Zimmerman GA, McIntyre TM: Neutrophil adherence to human endothelium in vitro occurs by CDw18 (Mo-1,MAC-1/LFA-1/gp150,95) glycoproteins-dependent and -independent mechanisms. *J Clin Invest* 81:531–537, 1988.

12. Dustin ML, Rothlein R, Bhan AK, Dinarello CA, Springer TA: Induction by IL-1 and interferon-gamma: Tissue distribution, biochemistry, and function of a natural adherence molecule (ICAM-1). *J Immunol* 137:245–254, 1986.

13. Dustin ML, Springer TA: Lymphocyte function-associated antigen-1 (LFA-1) interaction with intercellular adhesion molecule-1 (ICAM-1) is one of at least three mechanisms for lymphocyte adhesion to cultured endothelial cells. *J Cell Biol* 107:321–331, 1988.

14. Boyd AW, Wawryk SO, Burns GF, Fecondo JV: Intercellular adhesion molecule 1 (ICAM-1) has a central role in cell-cell contact-mediated immune mechanisms. *Proc Natl Acad Sci USA* 85:3095–3099, 1988.

15. Cybulsky MI, McComb DJ, Movat HZ: Neutrophil leukocyte emigration induced by endotoxin—mediator roles of interleukin 1 and tumor necrosis factor alpha 1. *J Immunol* 140:3144–3149, 1988.

16. Cotran RS, Gimbrone Jr. MA, Bevilacqua MP, Mendrick DL, Pober JS: Induction and detection of a human endothelial activation antigen in vivo. *J Exp Med* 164:661–666, 1986.

17. Cotran RS, Pober JS: Endothelial activation: Its role in inflammatory and immune reactions, in Endothelial cell biology. N. Simionescu, and M. Simionescu, editors. Plenum Publishing, New York. p 335, 1988.

18. Munro JM, Pober JS, Cotran RS: Tumor necrosis factor and interferon-gamma induce distinct patterns of endothelial activation and leukocyte accumulation in skin of *Papio Anubis. Am J. Pathol* 1989. In press.

19. Bevilacqua MP, Pober JS, Mendrick DL, Cotran RS, Gimbrone Jr. MA: Identification of an inducible endothelial-leukocyte adhesion molecule. *Proc Natl Acad Sci USA.* 84:9238–9242, 1987.

20. Pober JS, Bevilacqua MP, Mendrick DL, Lapierre LA, Fiers W, and Gimbrone Jr. MA: Two distinct monokines, interleukin-1 and tumor necrosis factor, each independently induce biosynthesis and transient expression of the same antigen on the surface of cultured human vascular endothelial cells. *J Immunol* 136:1680–1687, 1986.

21. Bevilacqua MP, Stengelin S, Gimbrone Jr. MA, Seed B: Endothelial-leukocyte adhesion molecule-1: An inducible receptor for neutrophils related to complement regulatory proteins and lectins. *Science* 243:1160–1165, 1989.

22. Seed B, Aruffo A: Molecular cloning of the CD2 antigen, the T cell erythrocyte receptor, by a rapid immunoselection procedure. *Proc Natl Acad Sci USA* 84:3365–3369, 1987.

23. Aruffo A, Seed B: Molecular cloning of a CD28 cDNA by a high-efficiency COS cell expression system. *Proc Natl Acad Sci USA* 84:8573–8577, 1987.

24. Spiess M, Lodish HF: Sequence of a second human asialoglycoprotein receptor: Conservation of two receptor genes during evolution. *Proc Natl Acad Sci USA* 82:6465–6469, 1985.

25. McPhaul M, Berg P: Formation of functional asialoglycoprotein receptor after transfection with cDNAs encoding the receptor proteins. *Proc Natl Acad Sci USA* 83:8863–8867, 1986.

26. Ikuta K, Takami M, Kim CW, Honjo T, Miyoshi T, Tagaya Y, Kawabe T, Yodoi J: Human lymphocyte Fc receptor for IgE: Sequence homology of its cloned cDNA with animal lectins. *Proc Natl Acad Sci USA* 84:819–823, 1987.

27. Ludin C, Hofstetter H, Sarfati M, Levy CA, Suter U, Alaimo D, Kilchherr E, Frost H, Delespesse G: Cloning and expression of the cDNA coding for a human lymphocyte IgE receptor. *EMBO J* 6:109–114, 1987.

28. Reid KBM, Bentley DR, Campbell RD, Chung LP, Sim RB, Kristensen T, Tack BF: Complement system proteins which interact with C3b or C4b. *Immunol Today* 7:230–235, 1986.

29. Klickstein LB, Wong WW, Smith JA, Weis JH, Wilson JG, Fearon DT: HUMAN C3b/C4b RECEPTOR (CRI) Demonstration of long homologous repeating domains that are composed of the short consensus repeats characteristic of C3/C4 binding proteins. *J Exp Med* 165:1095–1112, 1987.

30. Simmons D, Makgoba MM, Seed B: ICAM, an adhesion ligand of LFA-1 is homologous to the neural cell adhesion molecule NCAM. *Nature* 331:624–627, 1988.

31. Staunton DE, Marlin SD, Stratowa C, Dustin ML, Springer TA: Primary structure of ICAM-1 demonstrates interaction between members of the immunoglobulin and integrin supergene families. *Cell* 52:925–933, 1988.

32. Hynes RO: Integrins: A family of cell surface receptors. *Cell* 48:549–554, 1987.

33. Kishimoto TK, O'Connor K, Lee A, Roberts TM, Springer TA: Cloning of the beta subunit of the leukocyte adhesion proteins: Homology to an extra-

cellular matrix receptor defines a novel supergene family. *Cell* 48:681–690, 1987.

34. Siegelman MH, Van De Rijn M, Weissman IL: Mouse lymph node homing receptor cDNA clone encodes a glycoprotein revealing tandem interaction domains. *Science* 243:1165–1172, 1989.

35. Lasky LA, Singer MS, Yednock TA, Dowbenko D, Fennie C, Rodriguez H, Nguyen T, Stachel S, and Rosen SD: Cloning of a lymphocyte homing receptor reveals a lectin domain. *Cell* 56:1045–1055, 1989.

36. Johnston GI, Cook RI, and McEver RP: Cloning of GMP-140, a granule membrane protein of platelets and endothelium: sequence similarity to proteins involved in cell adhesion and inflammation. *Cell* 56:1033–1044, 1989.

Part 4
Lymphocyte Adhesion

17

Homing Receptors in Lymphocyte, Neutrophil, and Monocyte Interaction with Endothelial Cells

MARK A. JUTILA, DAVID M. LEWINSOHN,
ELLEN L. BERG, AND EUGENE C. BUTCHER

Introduction

Interactions of lymphocytes, monocytes, and granulocytes with endothelial cells in vivo exhibit striking specificity. The mechanisms involved in these interactions are important potential sites for therapeutic intervention into diverse immune and inflammatory processes. The involvement of the CD11a-c/18 family of integrin-related molecules, and the intercellular adhesion molecule-1 (ICAM-1) ligand for lymphocyte function-associated antigen-1 (LFA-1), in certain leukocyte-endothelial cell binding events has been amply demonstrated in in vitro studies well represented in this publication. Employing primarily *in vivo* and *ex vivo* approaches, we have demonstrated the existence of another class of leukocyte-endothelial cell recognition or adhesion receptors and ligands—lymphocyte homing receptors and their complementary endothelial cell ligands, the tissue-specific vascular addressins. In this chapter, we briefly summarize the important features of these molecules and emphasize that they are important not only in lymphocyte extravasation, but also in the extravasation of other leukocytes. Thus, homing receptors and vascular addressins may offer important targets for therapeutic modulation not only of tissue-specific immune responses, but also of acute and subacute inflammatory reactions characterized by granulocyte and monocyte-induced tissue damage.

Results and Discussion

In vivo studies of lymphocyte homing, in combination with studies employing an ex vivo adhesion assay developed by Woodruff and co-workers (1) in which viable lymphocytes bind to specialized lymphoid organ or inflammatory site high endothelial venules (HEV) in frozen tissue sections, have allowed the demonstration of at least three functionally distinct lymphocyte-endothelial cell recognition systems that serve to direct lymphocyte extravasation into peripheral lymph nodes, mucosal lymphoid, and extralymphoid sites, and inflamed synovium, respectively (reviewed in 2,3). Recent studies suggest that dermal vessels may express an additional specificity, and HEV in bronchus-associated lymphoid tissues may display yet another (N. Wu; and S. Rosen et al., personal communications). Indeed, there may be a large family of tissue-specific leukocyte-endothelial cell recognition systems, permitting the selective direction of particular lymphocyte subsets into diverse tissues in the body.

The 85 to 95 kD lymphocyte surface glycoprotein "homing receptors," defined by monoclonal antibody MEL-14 in the mouse (4), 1B2 in the rat (5), and by the Hermes series of antibodies in humans (6,7), are involved in tissue-specific lymphocyte/HEV adhesion. Antibodies to the Hermes antigen can inhibit lymphocyte recognition of lymph node, mucosal, or synovial HEV, and in the mouse MEL-14 selectively inhibits the in vivo extravasation of circulating lymphocytes into peripheral lymph nodes. Fluorescence energy transfer techniques have confirmed that purified gp90[Hermes] for human tonsillar lymphocytes, or from mucosal HEV-recognizing cells, interact with the 58 to 66 kD mucosal vascular addressin, a lymphocyte-binding endothelial cell molecule that defines sites of lymphocyte traffic into murine mucosal tissues (8).

cDNAs encoding gp90[Hermes] have now been isolated in the human (9,10). The sequence predicts a typical transmembrane protein with no homology with either integrin- or immunoglobulin-related adhesion molecules. The distal extracellular region of the gp90[Hermes] cDNA exhibits a striking relatedness to tandemly repeated domains of the cartilage link and proteoglycan core proteins (9,10), domains thought to be involved in interactions with the GAG hyaluronic acid. Immunologic, functional, and preliminary Northern analyses have recently allowed demonstration of related, but functionally diverse molecules on a number of cell types in humans including hematolymphoid cells (see below), fibroblasts, glial cells, and certain epithelial cells (11). Thus, we hypothesize that gp90[Hermes] is a novel family of heterotypic cell recognition/adhesion molecules. It should be mentioned, as well, that recent cross preclearing and comparative immunostaining studies demonstrate that the independently described human 80 to 95 kD lymphocyte antigens Pgp-1(12) and CD44 (13) are, in fact, gp90[Hermes] (14).

Interestingly, the MEL-14 antigen in the mouse, unlike Hermes in the human, is only expressed on lymphoid and myeloid cells. Further, recent cloning of the gp90[MEL-14] cDNA indicates that gp90[MEL-14] and gp90[Hermes] are structurally distinct (15,16). Gp90[MEL-14], but not gp90[Hermes], has a lectin binding domain, which appears to be involved in the interaction of this molecule with peripheral lymph node HEV. Thus, gp90[MEL-14] and gp90[Hermes] probably direct lymphocyte interactions with lymphoid tissue HEV through distinct and most likely complementary mechanisms.

Although additional lymphocyte surface molecules appear to contribute to lymphocyte HEV-interactions in certain settings—including LFA-1, and another integrin-related glycoprotein α/β heterodimer LPAM-1 recently shown to be involved in lymphocyte binding to mucosal HEV in the mouse (17)—the MEL-14 and Hermes antigen homing receptors are exquisitely regulated during lymphocyte differentiation, and their expression appears to be a major, if not the primary, determinant of HEV-binding ability of lymphocytes under most circumstances. It is possible that the importance of these homing receptors in leukocyte-endothelial cell interactions have eluded detection in experimental systems employing cultured endothelial cells, because the tissue-specific vascular addressins that may function as their ligands do not appear to be expressed in typical endothelial cell culture systems (E.L. Berg, and R. Hallmann, unpublished).

Homing receptors appear to be involved not only in the interaction of lymphocytes with HEV, but also in interactions of neutrophils and monocytes with endothelial cells at sites of inflammation (14, and M.A. Jutila, and E.C. Butcher, submitted). Both MEL-14 in the mouse and the Hermes series of antibodies in the human stain neutrophils, monocytes, and other leukocytes. Murine neutrophils bind to lymphoid tissue HEV in the ex vivo frozen section assay, and MEL-14 inhibits neutrophil binding to peripheral node but not to mucosal HEV, indicating as in the case of lymphocytes that antigenically distinct neutrophil cell surface molecules mediate mucosal vs. lymph node HEV recognition. Futhermore, MEL-14 inhibits the extravasation of neutrophils from the blood into an *Escherichia coli* supernatant-induced dermal site of acute inflammation, whereas control antibody anti-T200 (leukocyte common antigen) had no effect (18).

Interestingly, neutrophil binding to HEV in frozen sections of organized lymphoid tissues is constitutive, and the ability of HEV to bind applied neutrophils is not significantly altered during the course of an acute inflammatory response—for example, as measured by neutrophil binding to HEV in sections of draining lymph nodes taken at various times after footpad injection of CFA. This contrasts with the behavior of circulating neutrophils in the same lymph nodes in vivo, which although found in association with the HEV even in uninflamed nodes, demonstrate a sharp peak in high endothelial cell association roughly 3 hours after CFA injection. This observation seems to be consistent with the proposal that neutrophil activation within the vascular lumen (and not activation of the endothelium) is critical in triggering enhanced neutrophil-HEC binding acutely after an inflammatory stimulus. However, in preliminary studies using frozen sections of mouse skin neutrophils bound to venules in LPS-stimulated (Fig. 17.1) but not control tissues. The kinetics of the

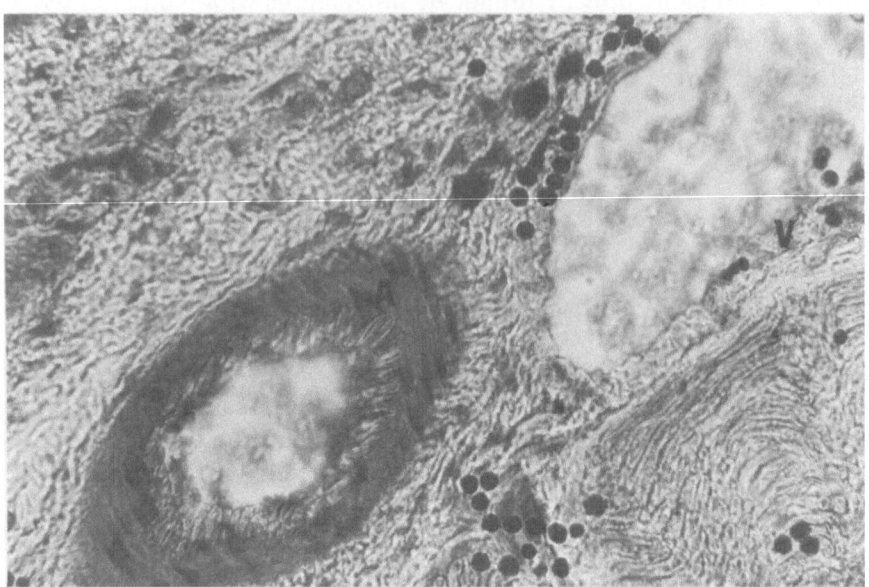

FIGURE 17.1. Bone-marrow neutrophils bind specifically to 3 hour LPS-inflamed venules in the dermis. Footpads of mice were inflamed by the subcutaneous injection of 2 to 3 ng of LPS. Three hours later the animals were injected with Leuconyl blue intravenously to visualize vessels. The inflamed tissues were harvested and frozen sections made. Bone marrow neutrophils were incubated on these sections for 15 min at 7 °C under constant rotation (80 rpm). After the assay, the sections were stained with thionine. Vessels are easily identified by the accumulation of the Leuconyl blue within their lumens. Endothelial cell binding neutrophils are in a plane above the section and preferentially take up the thionine stain. Note that neutrophils specifically bind venules (V) and not arteries (A). Mag. 200×.

increased neutrophil-binding capacity of inflamed dermal venules correlated with the extravasation of neutrophils within these venules in vivo, indicating that in this setting endothelial cell "activation" may indeed result in increased endothelial cell adhesiveness for neutrophils. These ex vivo studies constitute an important test and indeed confirmation of concepts derived from studies with cultured vascular endothelium that have suggested that both neutrophil and endothelial cell activation-dependent adhesion molecules play a role in regulating neutrophil-endothelial cell interactions (reviewed in 19,20).

The constitutive binding of neutrophils to lymph node HEV, but not to postcapillary venules (PCV) in the dermis, may reflect their ability to use gp90 homing receptors, whose ligands are expressed at high levels by HEV but are (presumably) less abundant on extralymphoid PCV early during the inflammatory response. Alternatively, vessels in lymphoid tissues may constitutively express neutrophil recognition/adhesion molecules which in the skin are only expressed on vessels during inflammation. Recently, we have characterized endothelial cell specific monoclonal antibodies that show a pattern of endothelial cell reactivity consistent with this hypothesis. The antigens recognized by these antibodies are expressed consitutively on lymphoid tissue venules, but are preferentially expressed on LPS-inflamed vs. uninflamed dermal venules (E.L. Berg and M.A. Jutila, unpublished). We are now investigating the possibility that these endothelial cell specific antigens are involved in leukocyte extravasation during inflammation.

Unlike neutrophils, monocytes fail to bind to HEV in uninflamed lymph nodes. On the other hand, they bind extremely well to HEV in lymph nodes taken 1 to 3 days after footpad injection of CFA (Fig. 17.2).

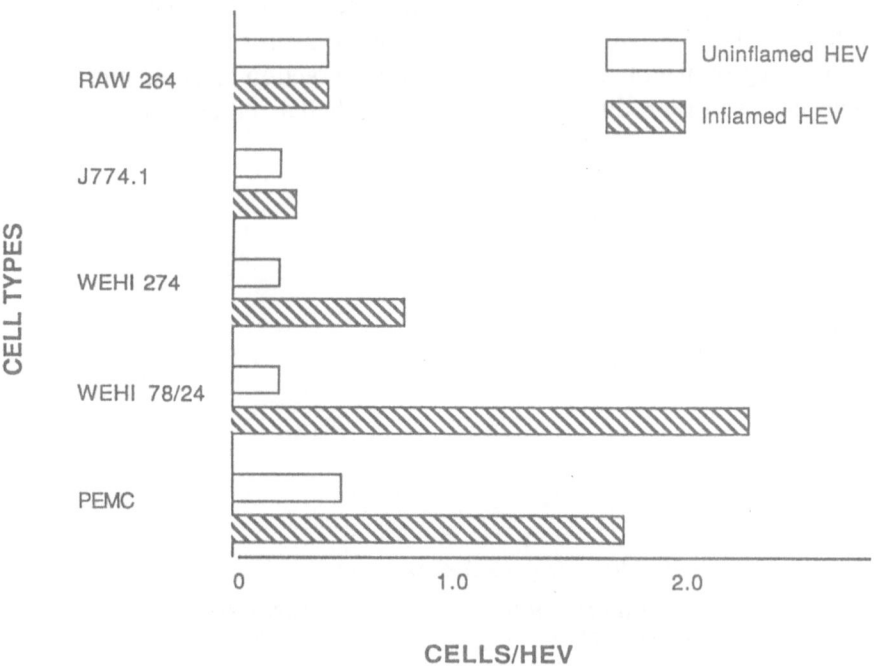

FIGURE 17.2. Thioglycollate-elicited peritoneal exudate mononuclear phagocytes (PEMC) and some monocytoid cell lines preferentially adhere to HEV in inflamed vs. uninflamed lymph nodes. Inflamed lymph nodes were collected 3 days after subcutaneous injection of CFA. Percoll separated thioglycollate-elicited PEMC, and a number of monocytoid cell lines, were incubated on frozen sections of inflamed and uninflamed lymph nodes at 7 °C under constant rotation (80 rpm). The number of cells binding/HEV was recorded. All of the PEMC and monocytoid cell lines were Mac-1 positive and phagocytic. The PEMC, WEHI 265, WEHI 274, and WEHI 78/24, but not J774.1 nor Raw 264, were MEL-14 positive.

This is the first direct demonstration of inflammation-induced alterations in the specific adhesiveness of endothelial cells for a particular leukocyte subset, the monocyte. The kinetics of altered HEV binding of monocytes, as measured in the ex vivo frozen section assay, precisely parallel the association of circulating monocytes with HEV in the same animals, as assessed immunohistologically with anti-monocyte antibodies. Thus activation of monocyte-specific endothelial cell adhesion systems probably plays a critical role in regulating monocyte extravasation during inflammatory responses and it seems unlikely that monocyte activation in the lumen is essential. (Indeed, whereas neutrophils marginate and roll slowly through venular endothelium, potentially allowing time for activation via local inflammatory mediators entering the vessel lumen from the surrounding tissue, monocytes and lymphocytes appear to pass too rapidly through local vessels to have time to be triggered significantly by inflammatory or chemotactic agents prior to their specific binding to activated endothelium.) The specificity of the inflammation-induced change in HEV for monocytes is striking, and illustrates a much more refined degree of leukocyte subset-specific interaction mechanisms in vivo than has been reproduced in in vitro endothelial cell culture systems to date.

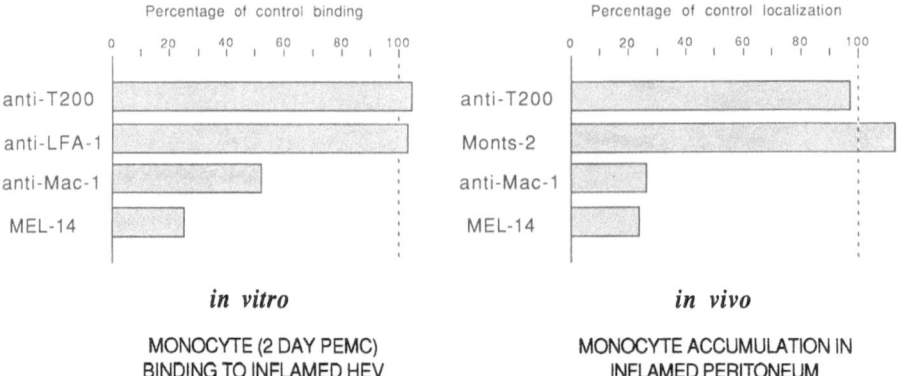

FIGURE 17.3. Antibody blocking of PEMC binding to inflamed HEV ex vivo, and blocking of monocyte accumulation into the thioglycollate inflamed peritoneal cavity in vivo. For blocking of ex vivo adhesion, PEMC were incubated with saturating levels of antibodies to the indicated antigens for 20 minutes on ice prior to the assay. For blocking monocyte extravasation in vivo, animals were injected intravenously with 500 μg of antibodies to the indicated antigens. One hour later inflammation was induced by the intraperitoneal injection of thioglycollate broth and 3 days later the number of monocytes/macrophages in the peritoneal cavity was determined. Results are presented as the percent of ex vivo binding or in vivo accumulation after treatment with medium along.

Surprisingly, monocyte binding to inflamed endothelium is inhibited by MEL-14 (Fig. 17.3). This suggests the following possibilities. (i) The MEL-14 antigen on monocytes, which has a molecular weight of approximately 110 to 120 kD vs. the 90 kD lymphocyte homing receptor, may be functionally different, incapable of recognizing the HEV determinant in uninflamed nodes and instead interacting selectively with either an inflammation modified peripheral node addressin, or with another ligand selectively induced during the inflammatory response. (ii) The homing receptor-HEV interaction may be insufficient to mediate monocyte binding, requiring the coordinate utilization of one or more accessory adhesion molecules for endothelial cell ligand(s) which are induced during inflammation. One candidate for such an accessory adhesion molecule is Mac-1, since anti-Mac-1 antibodies also significantly inhibit monocyte binding to inflamed HEV (Fig. 17.3). In this model, the Mac-1 might be necessary but not sufficient for binding, since several Mac-1$^+$ MEL-14$^-$ myelomonocytic cell lines fail to bind to inflamed HEV. Indeed, only MEL-14$^+$ Mac$^+$ monocytoid lines bind inflamed HEV. Further evidence for the involvement of both homing receptor and Mac-1 adhesion mechanisms in monocyte extravasation is confirmed by the ability of intravenously administered MEL-14 and anti-Mac-1 antibodies, but not control antibodies against other monocyte surface antigens, to substantially suppress the accumulation of monocytes into the thioglycollate-inflamed peritoneal cavity (Fig. 17.3). (iii) Finally, it is possible that the MEL-14 antigen on the monocyte might be associated with other molecules forming a complex which controls adhesion to the inflamed endothelium. In this model, blocking by MEL-14 could be due to steric hindrance or dissociation of the complex.

Therefore, based upon the information summarized here, we propose that leukocyte-endothelial cell interactions involve at least two classes of adhesion molecules in vivo, the homing receptor-related molecules that may (by analogy with the lymphocyte homing receptor) function not only to provide adhesion but also to direct tissue specificity; and the integrin-related LFA-1/Mac-1/p150,95 family, which may provide accessory adhesive functions in certain settings, but in others may actually determine sites of leukocyte extravasation by mediating interactions with induced endothelial cell ligands. In some situations, for example lymphocyte interactions with well-developed HEV, homing receptors may themselves be sufficient to orchestrate leukocyte extravasation; this would be consistent with the ability of lymphocytes to enter lymph nodes and sites of delayed-type hypersensitivity in CD18-deficient patients. On the other hand, in other situations such as leukocyte interactions with cultured endothelium (which have not been shown to express tissue-specific homing receptor ligands), or perhaps in early acute tissue inflammation with selective neutrophil extravasation, the CD11a-c/18 (especially Mac-1)-mediated adhesion may predominate and be sufficient. In most inflammatory settings, however, it seems likely that both molecular classes are involved coordinately, indeed both may be required to mediate efficient leukocyte extravasation except in unusual circumstances. If this is the case, both homing receptors and CD11a-c/18 molecules, and their endothelial cell ligands, must be considered equally appropriate targets for therapeutic intervention; and in fact, it seems likely that even more dramatic anti-inflammatory effects will be possible when we are able to interfere with both classes of adhesion molecules simultaneously.

Finally, as a caveat, the molecules defined to date seem to us insufficient to allow a simple hypothesis to explain the truly exquisite regulation of leukocyte-endothelial cell interactions observed in vivo. Additional classes of adhesion molecules, and possibly leukocyte subset-specific chemoattractants that could selectively influence the diapedesis of neutrophils vs. monocytes or lymphocytes following initial endothelial cell attachment, most likely await discovery in the future.

References

1. Stamper HB, Woodruff JJ: Lymphocyte homing in lymph nodes: *In vitro* demonstration of the selective affinity of recirculating lymphocytes for high endothelial venules. J *Exp Med* 144:828, 1976.
2. Gallatin WM, St. John TP, Siegelman M, Reichert R, Butcher EC, Weissman IL: Lymphocyte homing receptors. *Cell* 44:673, 1986.
3. Jalkanen S, Reichert R, Gallatin WMG, Weissman IL, EC Butcher: Homing receptors and the control of lymphocyte migration. *Immunol Rev* 91:39, 1986.
4. Gallatin WM, Weissman IL, Butcher EC: A cell surface molecule involved in organ-specific homing of lymphocytes. *Nature* 304:30, 1983.
5. Chin YH, Rasmussen RA, Woodruff JJ, Easton TG: A monoclonal anti-HEBFPP antibody with specificity for lymphocyte surface molecules mediating adhesion to Peyer's patch high endothelium of the rat. *J Immunol* 136:256, 1986.
6. Jalkanen S, Bargatze RF, Herron L, Butcher EC: A lymphoid cell surface glycoprotein involved in endothelial cell recognition and lymphocyte homing in man. *Eur J Immunol* 16:1195, 1986.
7. Jalkanen S, Bargatze RF, delos Toyos J, Butcher EC: Lymphocyte recognition of high endothelium: Antibodies to distinct epitopes of an 85-95 kD glycoprotein antigen differentially inhibit lymphocyte binding to lymph node, mucosal, or synovial endothelial cells. *J Cell Biol* 105:983, 1987.
8. Streeter PS, Lakey Berg E, Rouse BTN, Bargatze RF, Butcher EC: Tissue specific endothelial cell molecule involved in lymphocyte homing. *Nature* 331:41, 1988.
9. Goldstein, LA, Zhou DFH, Picker LJ, Minty CN, Bargatze RF, Ding JF and Butcher EC: A human lymphocyte homing receptor, the Hermes antigen, is related to cartilage proteoglycan core and link proteins. *Cell* 56:1063, 1989.
10. Stamenkovic, I. Amiot M, Pesando JM, Seed B: A lymphocyte molecule implicated in lymph node homing is a member of the cartilage link protein family. *Cell* 56:1057, 1989.

11. Picker, LJ, Nakache M, Butcher EC: Monoclonal antibodies to human lymphocyte homing receptors define a novel class of adhesion molecules on diverse cell types. *J Cell Biol*, in press.
12. Haynes BF, Harden EA, Telen MJ, Hemler ME, Strominger JL, Palker TJ, Scearce RM, Eisenbarth GS: Differentiation of human T lymphocytes. I. Acquisition of a novel human cell protein (p80) during intrathymic T cell maturation. *J Immunol* 131:1195, 1983.
13. Cobbold S, Hale G, Waldman H: Non-lineage, LFA-1 family and leukocyte common antigens: New and previously defined clusters, in McMichael AJ (ed.): *Leukocyte Typing III: White Cell Differentiation Antigens* Oxford, Oxford University Press, 1987.
14. Picker LJ, de los Toyos J, Telen MJ, Haynes BF, Butcher EC: Identity of CD44 (In (Lu)-related p80), Pgp-1, and the Hermes class of lymphocyte homing receptors. *J Immunol* 142:2046, 1989.
15. Siegelman, MH, van de Rijn M, Weissman IL: Mouse lymph node homing receptor cDNA clone encodes a glycoprotein revealing tandem interaction domains. *Science* 243:1165, 1989.
16. Laskey LA, Singer MS, Yednock TA, Dowbenko D, Fennie C, Rodriquez H, Nguyen T, Stachel S, Rosen SD: Cloning of a lymphocyte homing receptor reveals a lectin domain. *Cell* 56: 1045, 1989.
17. Holzmann B, McIntyre BW, Weissman IL: Identification of a murine Peyer's patch-specific lymphocyte homing receptor as an integrin molecule with an α chain homologous to human VLA-4α. *Cell* 56:37, 1989.
18. Lewinsohn DM, Bargatze RF, Butcher EC: A common endothelial cell recognition system shared by neutrophils, lymphocytes, and other leukocytes. *J Immunol* 138:4313, 1987.
19. Cybulsky MI, Chan MK, Movat HZ: Biology of disease. Acute inflammation and microthrombosis induced by endotoxin, interleukin-1, and tumor necrosis factor and their implication in gram-negative infection. *Lab Invest* 58:365, 1988.
20. Cotran RS: New roles for the endothelium in inflammation and immunity. *Am J Pathol* 129:407, 1988.

18

Antigen-Independent Adhesion: A Critical Process in Human Cytotoxic T Cell Recognition

STEPHEN SHAW, M. WILLIAM MAKGOBA, AND
YOJI SHIMIZU

Introduction

Unlike solid organs, in which cell-cell contacts are largely established during morphogenesis, cells of the immune system are mobile and their contacts with other cells are transient ones. The ability to establish transient contacts with other cells is critical to the function of essentially all leukocytes. Cell-cell contact is critical to the function of mature circulating T lymphocytes both in the induction of their responses (e.g., by T cell interaction with an accessory cell) and in the effector arm of their responses (e.g., T cells providing help for B cells or effecting cell-mediated cytotoxicity). Our interest in T lymphocyte adhesion derives from commitment to understanding cytotoxic T lymphocyte (CTL) recognition and to a growing certainty that T cell adhesion is a complex and poorly understood process which is critical to CTL recognition. Although our studies emphasize the relevance of adhesion to CTL, it is apparent that other T cell responses must be similarly dependent on cell-cell adhesion. It is pointless to duplicate in detail recent reviews dealing with many of the important issues in T cell adhesion [1–5]. Instead, we will focus primarily on discussion of the importance of antigen-independent adhesion in T cell recognition.

Results

Antigen-specificity is a hallmark of T cell responses; however, an important element of T cell adhesion is antigen-nonspecific. This contrast between the antigen-specificity of cell-mediated lysis (CML) and the antigen-independence of adhesion of the cells which mediate that lysis is illustrated in Figure 18.1. A typical cytotoxic T cell clone specific for the DPw2 alloantigen kills lymphoblastoid B cell line (LCL) targets which

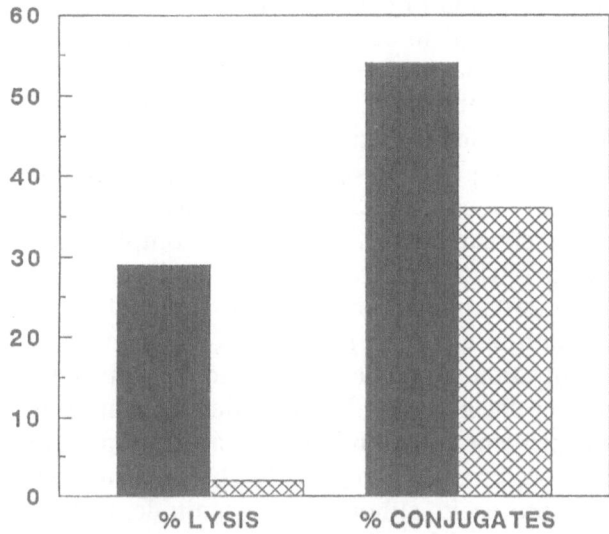

FIGURE 18.1. Contrast between the antigen-specificity of CML and relative antigen-independence of adhesion. A DPw2-specific CTL clone was assayed by a standard 4 hour ^{51}Cr release assay at a 4:1 E/T ratio for its ability to kill an LCL target expressing the DPw2 antigen (filled bars) and a similar cell lacking the DPw2 antigen (hatched bars). In a parallel assay the same effector and targets were assayed by a two color flow cytometric assay [29] for formation of stable effector target pairs (conjugates) following a 6 minute incubation at 37 °C in a pellet at an E/T ratio of 4:1.

express the relevant antigen but fails to kill similar targets which lack the antigen. In contrast, the T cell clone adheres well to each "target," forming a large number of conjugates with each target under conditions similar to that in which the lysis is measured. Thus, T cells adhere to other cells in the absence of expression of the relevant antigen.

The inference that antigen-independent T cell adhesion is a critical early step in T cell recognition emerged independently from our laboratory and from the laboratory of Spits and co-workers [6,7]. It emerged at a time when studies of murine cytotoxic T cell interactions had highlighted the importance of antigen-specific adhesion as the first step in CTL recognition (reviewed in [1]); consequently, the proposition that antigen-independent adhesion is an important biological process for T cells has been subjected to vigorous discussion. A variety of lines of evidence, which have been further elucidated since the original reports, provide credibility.

1. Theoretical considerations, which we have outlined previously [8], predict that a component of antigen-independent adhesion would be essential to achieving sensitivity of T cell recognition. Cells are mutually

repulsive [9]. Initiation of contact between a T cell and an opposing cell would be expected to occur readily only if it is mediated by receptors and ligands that are present at sizable concentrations and which interact with reasonable affinity. The T cell receptor (TCR) is an improbable candidate for initiation of adhesion. The difficulty is illustrated by considering T cell recognition under conditions of low antigen concentration in which the TCR is unlikely to encounter, during a brief repulsive intercellular interaction, the rare complexes of the relevant antigen together with histocompatibility antigen. (If the affinity of TCR/antigen/MHC interaction is low, this would further restrict its role in initiation of adhesion.) Therefore, other molecules present at higher concentration would be expected to participate in initial adhesion which then facilitates TCR interaction with antigen.

2. Antigen-independent adhesion is not an aberration unique to our experimental procedure; it is commonly observed in a wide variety of experimental protocols, although its magnitude varies considerably. Although antigen-specific adhesion is more prominent than antigen-independent adhesion in many murine studies, an element of antigen-nonspecific adhesion has usually been observed. For example, it has been noted in studies of murine T-helper cell interactions with B cells [10,11]. This predictability of occurrence of antigen-independent adhesion is consistent with the concept that it is critical to recognition.

3. The occurrence of antigen-independent adhesion correlates closely with the efficiency of CML in several informative contexts. Spits and co-workers observed that the capacity of human CTL to lyse transfected murine targets correlated with the capacity of those CTL to engage in antigen-independent conjugates with those targets [6]. We have noted that the capacity of many monoclonal antibodies (MAbs) to inhibit CML is predictable based on their effect on antigen-independent adhesion [8].

4. Antigen-independent adhesion is clearly *not* nonspecific adhesion but rather it is mediated by specific interactions. Figure 18.2 illustrates one aspect of its specificity. The experimental design is comparison of adhesion of a human T cell clone and a murine T cell clone on target cells from both species. The data demonstrate a strong preferential interaction between T cells and target cells of the same species. This is consistent with the interpretation that there are specific receptor ligand interactions which are mediating this adhesion and that these molecules involved have diverged sufficiently between species that the interaction is much less efficient between species than within a species.

5. Finally, the molecules which mediate antigen-independent adhesion (LFA-1, CD2, LFA-3, and ICAM-1) are ones which are involved in the overall process of T cell recognition, as illustrated by responses such as CML or antigen-specific recognition. The molecules LFA-1, CD2, and LFA-3 were found to be important in T cell responses such

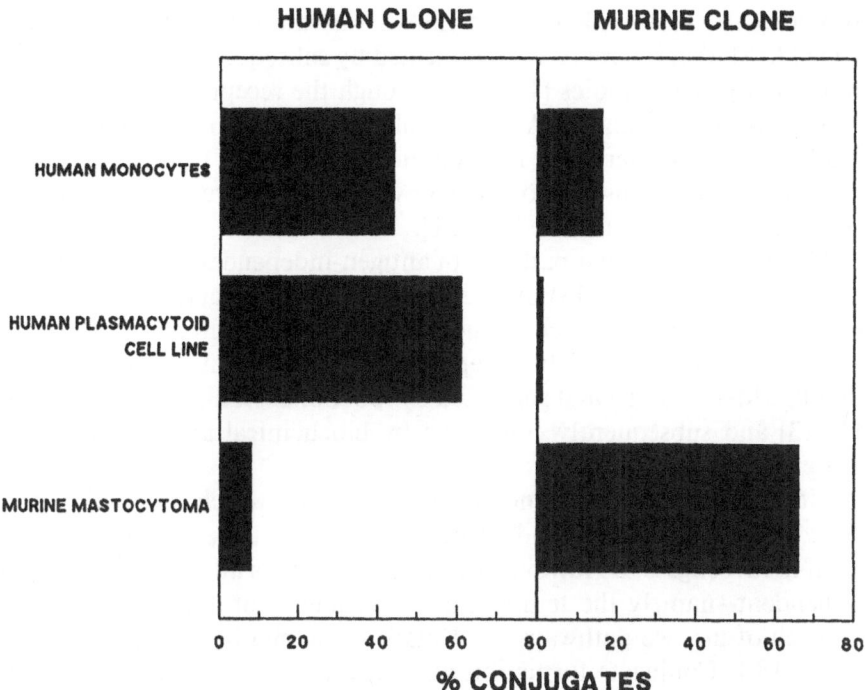

FIGURE 18.2. Species specificity of antigen-independent adhesion. A human T cell clone and a murine T cell clone are assayed for antigen-independent adhesion on three target cell lines: human peripheral blood monocytes, a human plasmacytoid cell line, and a murine mastocytoma. In a parallel assay no killing is observed in any of the combinations tested, so this is not specificity conferred by the antigen-specific receptor.

as proliferation and CML (reviewed in [1,2]) before appreciation of their role in antigen-dependent adhesion [6–8]. The converse was true of ICAM-1, which was identified by screening for a MAb which was able to inhibit adhesion [12,13]. Appreciation of the role of ICAM-1 in adhesion prompted investigation of its broader role in T cell recognition; an important role of ICAM-1 was confirmed in studies of CML [14,15] and antigen-specific proliferation [15,16]. The simplest interpretation of the involvement of these molecules in both antigen-independent adhesion and in CML is that they function in CML by mediating antigen-independent adhesion. This inference by no means precludes the possibility that these molecules have additional functions in CML, for example by transducing other regulatory signals to either cell during the interaction.

Studies of antigen-independent adhesion demonstrate involvement of two molecular pathways [3,8]. One involves the interaction of T cell CD2 with LFA-3 on the opposing cell. Studies of antigen-independent adhesion provided the first evidence that LFA-3 was the cell surface ligand for CD2 [7,17]; that inference was confirmed by subsequent biochemical and molecular genetic studies [18,19]. Although the receptor-ligand relationship between CD2 and LFA-3 is established, studies by Bernard and coworkers indicate that there may be more complexity in this interaction than is currently appreciated. They have identified three more molecules which appear to be involved [20,21].

The second molecular pathway of antigen-independent adhesion is mediated by LFA-1 (CD11a) on the T cell surface interacting with ligands on the opposing side which include ICAM-1 and apparently other as yet unidentified molecules. The receptor ligand relationship between LFA-1 and ICAM-1 was also first suggested by functional studies of cell adhesion [12,13] and subsequently confirmed by biochemical and molecular genetic studies [22,23].

Most aspects of antigen-independent adhesion which we have studied have proved to be independent of the details of experimental protocol used to investigate it. However, one of our previous findings is technique-dependent—namely the temperature dependence of adhesion mediated by each of the two pathways [7]. Details of the findings are presented in Table 18.1. Conjugate formation was measured between a human cytotoxic T cell clone and an antigen-negative target cell; formation was measured in the presence of blocking MAb used to isolate the LFA-1 pathway (using αLFA-3) and the CD2 pathway (using αLFA-1). The experimental variables are: (i) whether the T cells and target cells are centrifuged together (columns 2 and 4) or allowed to settle together (columns 1 and 3); and (ii) whether the T cell/target cell pellets are maintained at 4 °C continuously (column 1 and 2) or incubated subsequently at 37 °C for 6 minutes (columns 3 and 4). The results demonstrate: (i) that the LFA-1

TABLE 18.1. Effect of Centrifugation and Temperature on Two Pathways of Antigen-Independent Adhesion

| Monoclonal Antibodies | Pathways Remaining | % Conjugate Formation Following Incubation at: | | | |
| | | 4 °C only | | 4 °C then 37 °C | |
		No spin	Spin	No spin	Spin
None	Both			72	79
αLFA-1	CD2	6	30	34	57
αLFA-3	LFA-1	2	1	23	30
αLFA-1+αLFA-3	Neither			1	2

Antigen-independent adhesion between a CTL clone and an antigen-negative plasmacytoid cell line at a 4:1 E:T ratio. The T cell and target were allowed to settle together at 1 g at 4 °C for 1 hour. Where indicated, the tubes were then centrifuged for 2 seconds in a microfuge (microfuge B, Beckman, Palo Alto, CA) (estimated 9,000 g) at 4 °C. Where indicated, the incubation mixture was then rapidly warmed to 37 °C in a water bath for 6 minutes.

pathway mediates significant adhesion only when the incubation temperature is raised: (ii) the CD2 pathway can mediate adhesion not only at 37 °C but also at 4 °C. However, the 4 °C adhesion occurs only when the cells are forced into tight approximation by centrifugation. Thus, centrifugation appears to bypass some steps of adhesion, as has been previously reported by Grinnell for fibroblast adhesion to substrate [24]. Since appreciating this affect, we modified our standard experimental protocol so that cells settle into a pellet [8].

Either one of these two molecular pathways appears capable, by itself, of mediating adhesion between T cells and other cells. For example, T cell rosetting is mediated by the CD2/LFA-3 pathway [18,19,25,26] whereas T cell adhesion to some other cells, such as the myeloid cell U937, is mediated almost exclusively by LFA-1/ICAM-1 interaction [23,27]. Although either pathway can operate in isolation, CTL interactions with a wide variety of target cells usually involve both pathways, albeit in varying proportions [8]. Functional studies indicate a more than additive interaction between these two pathways [8,14], suggesting that the efficiency of T cell adhesion is augmented by the two pathways operating together. Definitive studies with mixes of purified ligand will be required to characterize details of potential synergistic interactions of these two pathways.

These two pathways both mediate antigen-independent adhesion, but several of their many differences warrant mention. The adhesion receptor in one pathway (CD2) is virtually restricted to the T cell lineage, while the other's (LFA-1) is present and mediates adhesion on a variety of hematopoietic cells [2]. For the CD2/LFA-3 pathway there appears to be a single ligand which is expressed constituitively on most other cell types and, therefore, well suited to serve as a "handle" on virtually all cells. For the LFA-1 pathway, the ligand story is more complex. There is probably more than a single ligand; the known ligand, ICAM-1, is absent from most cells but dramatically upregulated under conditions found at inflammatory sites and, therefore, well suited to enhancing leukocyte involvement at sites of inflammation [28]. Our interpretation of these two pathways is that the function of adhesion is sufficiently critical to T cells that it is served by two complementary systems which can be regulated independently to optimize adhesion under the varied circumstances which the T cell encounters.

Conclusions

There are at least two pathways of antigen-independent adhesion which facilitate T cell recognition. Detailed understanding of the structure and function of the molecules involved, as well as the cell biology of the adhesion process, will be essential to understanding T cell recognition.

References

1. Martz E: LFA-1 and other accessory molecules functioning in adhesions of T and B lymphocytes. *Hum Immunol* 18:3–37, 1987.
2. Springer, TA, Dustin ML, Kishimoto TK, Marlin SD: The lymphocyte function-associated LFA-1, CD2, and LFA-3 molecules: Cell adhesion receptors of the immune system. *Annu Rev Immunol* 5:223–252, 1987.
3. Shaw S, Shimizu Y: Two molecular pathways of human T cell adhesion: Establishment of receptor-ligand relationship. *Curr Opin Immunol* 1:92–97, 1988.
4. Bierer BE, Burakoff SJ: T cell adhesion molecules. *FASEB J.* 2:2584–2590, 1988.
5. Sanders ME, Makgoba MW, Shaw S: Human naive and memory T cells: Reinterpretation of helper-inducer and suppressor-inducer subsets. *Immunol Today* 9:195–199, 1988.
6. Spits H, Van Schooten W, Keizer H, Van Seventer G, Van de Rijn M, Terhorst C, De Vries JE: Alloantigen recognition is preceded by nonspecific adhesion of cytotoxic T cells and target cells. *Science* 232:403–405, 1986.
7. Shaw S, Luce GEG, Quinones R, Gress RE, Springer TA, Sanders ME: Two antigen-independent adhesion pathways used by human cytotoxic T cell clones. *Nature* 323:262–264, 1986.
8. Shaw S, Luce GG: The lymphocyte function-associated antigen (LFA)-1 and CD2/LFA-3 pathways of antigen-independent human T cell adhesion. *J Immunol* 139:1037–1045, 1987.
9. Bell GI: Cell-cell adhesion in the immune system. *Immunol Today* 4:237–240, 1983.
10. Sanders VM, Snyder JM, Uhr JW, Vitetta ES: Characterization of the physical interaction between antigen-specific B and T cells. *J Immunol* 137:2395–2404, 1986.
11. Kupfer A, Swain SL, Janeway, Jr. CA, Singer SJ: The specific direct interaction of helper T cells and antigen-presenting B cells. *Proc Natl Acad Sci USA* 83:6080–6083, 1986.
12. Rothlein R, Dustin ML, Marlin SD, Springer TA: A human intercellular adhesion molecule (ICAM-1) distinct from LFA-1. *J Immunol* 137:1270–1274, 1986.
13. Dustin ML, Rothlein R, Bhan AK, Dinarello CA, Springer TA: Induction by IL 1 and Interferon-gamma: Tissue distribution, biochemistry, and function of a natural adherence molecule (ICAM-1). *J Immunol* 137:245–254, 1986.
14. Makgoba MW, Sanders ME, Luce GE, Gugel EA, Dustin ML, Springer TA, Shaw S: Functional evidence that intercellular adhesion molecule-1 (ICAM-1) is a ligand for LFA-1 in cytotoxic T cell recognition. *Eur J Immunol* 18:637–640, 1988.
15. Boyd, AW, Wawryk SO, Burns GF, Fecondo JV: Intercellular adhesion molecule 1 (ICAM-1) has a central role in cell-cell contact-mediated immune mechanisms. *Proc Natl Acad Sci USA* 85:3095–3099, 1988.
16. Dougherty, GJ, Murdoch S, Hogg N: The function of human intercellular adhesion molecule-1 (ICAM-1) in the generation of immune response. *Eur J Immunol* 18:35–40, 1988.

17. Vollger LW, Tuck DT, Springer TA, Haynes BF, Singer KH: Thymocyte binding to human thymic epithelial cells is inhibited by monoclonal antibodies to CD-2 and LFA-3 antigens. *J Immunol* 138:358–363, 1987.
18. Plunkett ML, Sanders ME, Selvaraj P, Dustin M, Springer TA: Rosetting of activated T lymphocytes with autologous erythrocytes: Definition of the receptor and ligand molecules as CD2 and lymphocyte function-associated antigen-3 (LFA-3). *J Exp Med* 165:664–676, 1987.
19. Seed B, Aruffo A: Molecular cloning of the CD2 antigen, the T-cell erythrocyte receptor, by a rapid immunoselection procedure. *Proc Natl Acad Sci USA* 84:3365–3369, 1987.
20. Bernard A, Aubrit F, Raynal B, Pham D, Boumsell L: A T-cell surface molecule different from CD2 is involved in spontaneous rosette formation with erythrocytes. *J Immunol* 140:1802–1807, 1988.
21. Bernard A, Tran HC, Boumsell L: Three different erythrocyte surface molecules are required for spontaneous T-cell rosette formation. *J Immunol* 139:18–23, 1987.
22. Marlin, SD, Springer TA: Purified intercellular adhesion molecule-1 (ICAM-1) is a ligand for lymphocyte function-associated antigen-1 (LFA-1). *Cell* 51:813–819, 1987.
23. Makgoba MW, Sanders ME, Luce GEG, Dustin ML, Springer TA, Clark EA, Mannoni P, Shaw S: ICAM-1 a ligand for LFA-1-dependent adhesion of B, T and myeloid cells. *Nature* 331:86–88, 1988.
24. Grinnell F: Studies on the mechanism of cell attachment to a substratum with serum in the medium: Further evidence supporting a requirement for two biochemically distinct processes. *Arch Biochem Biophys* 165:524–530, 1974.
25. Makgoba MW, Shaw S, Gugel EA, Sanders ME: Human T cell rosetting is mediated by LFA-3 on autologous erythrocytes. *J Immunol* 138:3587–3589, 1987.
26. Sayre PH, Chang HC, Hussey RE, Brown NR, Richardson NE, Spagnoli G, Clayton LK, Reinherz EL: Molecular cloning and expression of T11 cDNAs reveal a receptor-like structure on human T lymphocytes. *Proc Natl Acad Sci USA* 84:2941–2945, 1987.
27. Makgoba MW, Sanders ME, Luce GE, Gugel EA, Dustin ML, Springer TA, Shaw S: Intercellular adhesion molecule-1 (ICAM-1) monoclonal antibody inhibits cytotoxic T lymphocyte recognition. *Ann NY Acad Sci* 532:427–428, 1987.
28. Dustin ML, Singer KH, Tuck DT, Springer TA: Adhesion of T lymphoblasts to epidermal keratinocytes is regulated by interferon γ and is mediated by intercellular adhesion molecule 1 (ICAM-1). *J Exp Med* 167:1323–1340, 1988.
29. Luce GG, Gallop PM, Sharrow SO, Shaw S: Enumeration of cytotoxic cell-target cell conjugates by flow cytometry using internal fluorescent stains. *BioTechniques* 3:270–272, 1985.

19

Inhibition of Human Mixed Lymphocyte Reactions by Monoclonal Antibodies to Intercellular Adhesion Molecule-1 (ICAM-1)

Vincent J. Merluzzi, Robert Rothlein, Chester Wood, Carol D. Stearns, Ronald B. Faanes, and Kathleen Last-Barney

Introduction

Cell-cell adhesion is believed to be an integral event in the development of effective immunological responses. A family of molecules has been described which mediate leukocyte cell-cell and cell-substrate adhesions and are necessary for cell migration through endothelium. This family has been designated the lymphocyte function-associated antigen-1 (LFA-1)[1] family or the CD18 complex of adhesion molecules (1–10).

Recently, it has been shown that a ligand for LFA-1 is the intercellular adhesion molecule designated ICAM-1 (11–13). The ICAM-1 is necessary for effective cellular interactions during an immune response mediated through LFA-1 dependent cell adhesion.

It has been shown that ICAM-1 is induced by pro-inflammatory cytokines such as interleukin-1 (IL-1), gamma interferon (IFN-g), and tumor necrosis factor-α (TNF) (9,10). The induction of ICAM-1 during immune responses or inflammatory disease allows for the interaction of leukocytes with each other and with endothelial cells.

An inherited immunodeficiency disease characterized by progressive soft tissue infections, diminished pus formation, granulocytosis, and delayed umbilical cord separation has been described. This immune deficiency disease called *leukocyte adhesion deficiency disorder* (LADD), in which patients are unable to mount an inflammatory response, is directly associated with a deficiency in the LFA-1, Mac-1 adhesion molecules (6,7). Deficiency in ICAM-1 or other possible ligands for LFA-1, Mac-1 adhesion molecules may also interfere with physiologic immune responses.

The experiments described here have shown that ICAM-1 becomes upregulated on human peripheral blood monocytes during in vitro culture. The addition of monoclonal antibodies (MAbs) to ICAM-1 and LFA-1 to the mixed lymphocyte reaction (MLR) significantly inhibited cell proliferation. When suboptimal concentrations of these antibodies were co-cultured, additive suppressive effects were observed. When suboptimal concentrations of cyclosporin A (CyA), dexamethasone (Dex), or imuran were combined with suboptimal concentrations of anti-ICAM-1, the inhibition of the MLR was additive.

Additional experiments examined the actions of anti-ICAM-1 and immunosuppressive agents and anti-LFA-1 antibodies in a secondary MLR. In these studies, anti-ICAM-1 was not as effective in inhibiting the secondary MLR when compared to the primary MLR. However, combinations of anti-ICAM-1 with anti-LFA-1 antibodies seem to synergize in the secondary MLR reaction.

Results

Initial experiments were designed to test the effect of anti-ICAM-1 monoclonal antibodies on the human MLR. Table 19.1 shows that purified anti-ICAM-1 MAbs inhibited the MLR in a dose dependent manner with significant suppression apparent at 20 ng/ml. Purified mouse IgG had little or no suppressive effect. Suppression of the MLR by the anti-ICAM-

TABLE 19.1. Effect of Anti-ICAM-1 Antibody on the One-Way Mixed Lymphocyte Reaction

Responder Cells[a]	Stimulator Cells[b]	Antibody[c] (μg/ml)	^{3}HT Incorporation (CPM)
−	−	−	445[d] ± 286
−	+	−	148 ± 38
+	−	−	698 ± 177
+	+	−	42,626 ± 3,538
+	+	mIgG (10.0)	36,882 ± 4,084 (14%)
+	+	mIgG (0.4)	35,550 ± 3,097 (17%)
+	+	mIgG (0.02)	42,815 ± 2,791 (0%)
+	+	R6.5 (10.0)	8,250 ± 1,165 (81%)
+	+	R6.5 (0.4)	16,142 ± 1,923 (62%)
+	+	R6.5 (0.02)	28,844 ± 4,362 (32%)

[a]Responder cells (6.25 × 10⁵/ml).
[b]Stimulator cells (6.25 × 10⁵ml, irradiated at 1000R).
[c]Purified MAb to ICAM-1 (R6.5) or purified mouse IgG (mIgG) at final concentrations (μg/ml).
[d]Mean ± SD of four to six cultures, numbers in parentheses indicate percent inhibition of MLR.

1 MAb occurs when the antibody is added within the first 24 hours of culture (Table 19.2). Since ICAM-1 is an inducible molecule, it was desirable to test whether ICAM-1 becomes upregulated on cultured cells alone and during a MLR. Flow cytometry was performed on both fresh and cultured human peripheral blood mononuclear cells and during the MLR (Days 1–4 after culture initiation). Figure 19.1A shows the results of ICAM-1 expression on fresh, uncultured gated monocytes. It can be seen that ICAM-1 expression is not significantly greater than the control samples ("no antibody") or the mouse IgG control. This was also true when these same cells were mixed with noncultured, irradiated cells from an unrelated donor (Fig. 19.1B). In the following panels (Fig. 19.1C and 19.1D), it can be seen that cultured monocytes express increased amounts of ICAM-1 when cultured alone (Fig. 191C) or in the presence of stimulator cells in the MLR (Fig. 19.1D). Also of note was that ICAM-1 was not expressed to a large degree in the lymphocyte gate at any time point (data not shown). The increased expression of ICAM-1 during the first 24 hours after culture on monocytes corresponds with the ability of anti-ICAM-1 antibody to inhibit the MLR when added during the first 24 hours after culture initiation.

It was important to examine whether the responder to stimulator (R/S) ratio affected the ability of anti-ICAM-1 to inhibit the MLR. In Table 19.3, a 1-to-1 R/S ratio is shown to be optimum. Addition of 1 μg/ml (a dose effective for inhibition) of anti-ICAM-1 antibody to the cultures was shown to be effective for inhibition at all R/S combinations (Table 19.3). The generation of a maximum MLR response in this experimental system does not favor or disfavor the ability of anti-ICAM-1 antibody to effectively inhibit the reaction.

TABLE 19.2. Time of Addition of Anti-ICAM-1

			[3]HT Incorporation (CPM)		
			Time of Addition of Medium or Antibody		
R[a]	S[b]	Additions[c]	Day 0	Day 1	Day 2
−	−	Medium	205[d] ± 57	476 ± 323	247 ± 168
−	+	Medium	189 ± 40	ND[e]	ND
+	−	Medium	1,860 ± 1,507	ND	ND
+	+	Medium	41,063 ± 7,204	45,955 ± 7,219	50,943 ± 7,527
+	+	R6.5	17,781 ± 3,168 (57%)[f]	38,409 ± 4,118 (16%)	47,308 ± 5,118 (7%)

[a]Responder cells (6.25 × 10[5]/ml).
[b]Stimulator cells (6.25 × 10[5]/ml, irradiated at 1000R).
[c]Culture medium or purified MAb to ICAM-1 (R6.5) at 10μg/ml were added on Day 0 and at 24-hour intervals.
[d]Mean ± SD of four to six cultures.
[e]ND = not done.
[f]Percent inhibition.

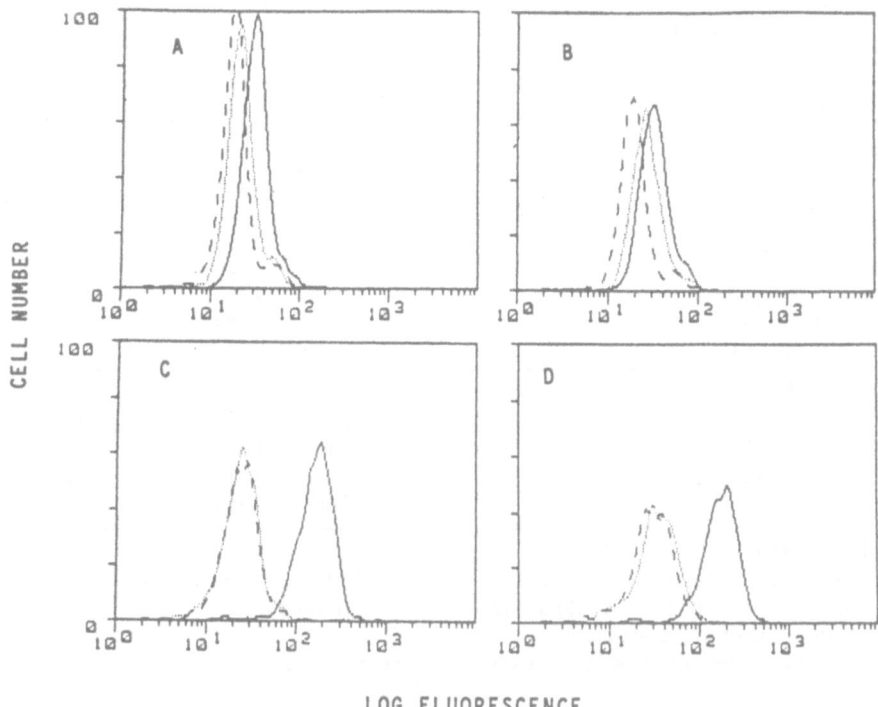

LOG FLUORESCENCE

FIGURE 19.1. Expression of ICAM-1 on human peripheral monocytes. (A) Responder cells not cultured; (B) Responder × Stimulator cells not cultured; (C) Responder cells cultured for 24 hours; and (D) Responder × Stimulator cells cultured for 24 hours. - - - - No monoclonal antibody; .... Mouse Immunoglobulin; ------- Mouse anti-ICAM-1.

The effects of anti-LFA-1 antibody on the MLR were also performed. LFA-1 is a molecule constitutively expressed on leukocytes as a heterodimer consisting of one α subunit of 180 kD and one noncovalently associated β subunit of 95 kD (5). Purified MAb to either subunit of the LFA-1 molecule inhibited the MLR. In Table 19.4, the effects of MAbs to the α subunit can be seen. Significant inhibition of the MLR can be seen as low as 20 ng/ml. In Table 19.5, MAbs to the β subunit are also effective in inhibiting the MLR. Effective inhibition of the MLR by the anti-LFA-1β antibody is apparent at 200 ng/ml.

A ligand for LFA-1 is ICAM-1 which is necessary for effective cellular interactions during an immune response mediated through LFA-1 dependent cell adhesion (11–13). It can be seen in Tables 19.4 and 19.5 that when suboptimal concentrations of MAbs to ICAM-1 (20 ng/ml, Table 19.4; 200 ng/ml, Table 19.5) and either LFA-1α (20 ng/ml, Table 19.4) or LFA-1β (200 or 40 ng/ml, Table 19.5) were co-cultured in the MLR, additive suppressive effects were observed.

TABLE 19.3. Effect of Anti-ICAM-1 on Human One-Way MLR (Dose Curve of Irradiated Stimulators)

Group	R6.5[c]	[3]HT Incorporation[d] (CPM)	Percent Inhibition[e]
Media	—	92 ± 20	—
Responders[a] (R)	—	2,365 ± 461	—
Stimulators[b] (S)	—	135 ± 32	—
S × R (5:1)	—	29,121 ± 3,103	—
S × R (1:1)	—	60,708 ± 3.148	—
S × R (0.5:1)	—	34,076 ± 5,694	—
S × R (0.25:1)	—	23,028 ± 4,183	—
S × R (0.125:1)	—	14,799 ± 1,828	—
S × R (5:1)	+	15,222 ± 1,686	48
S × R (1:1)	+	30,870 ± 2,327	49
S × R (0.5:1)	+	18,198 ± 1,501	47
S × R (0.25:1)	+	12,352 ± 2,214	46
s × R (0.125:1)	+	6,685 ± 358	55

[a]Responder cells (6.25 × 10[5]/ml).
[b]Stimulator cells (multiple concentrations of 6.25 × 10[5]/ml, irradiated at 1000R).
[c]Purified MAb to ICAM-1 (R6.5) at 1 μg/ml final concentration.
[d]Mean ± SD of five to six cultures.
[e]Percent inhibition of MLR based upon appropriate S × R ratio.

TABLE 19.4. Effect of Anti-ICAM-1 and Anti-LFA-1 Alpha on the Human MLR

Group[a]	Antibody[b] (ng/ml)	[3]HT Incorporation[c] (CPM)	Percent Inhibition[d]
Media	—	82 ± 33	—
Stimulators (S)	—	195 ± 42	—
Responders (R)	—	420 ± 135	—
R × S	—	16,080 ± 1,759	—
R × S	R3.1 (500)	1,085 ± 219	93
R × S	R3.1 (100)	1,208 ± 201	93
R × S	R3.1 (20)	11,387 ± 675	29
R × S	R3.1 (4)	16,367 ± 1,033	0
R × S	R6.5 (20)	11,164 ± 831	31
R × S	R6.5 (20)+R3.1 (20)	5,772 ± 552	64

[a]Responder cells and irradiated stimulator (1000R) cells at 6.25 × 10[5]/ml.
[b]Purified MAb to ICAM-1 (R6.5) or LFA-1-α (R3.1) at final concentrations (ng/ml).
[c]Mean ± SD of five to six cultures.
[d]Percent inhibitation of MLR.

Immunosuppressive agents such as CyA, DeX, and imuran significantly inhibit the human MLR. Combinations of suboptimal concentrations of a single immunosuppressant with suboptimal concentrations of

TABLE 19.5. Effect of Anti-ICAM-1 and Anti-LFA-1 Beta on the Human MLR

Group[a]	Antibody[b] (ng/ml)	[3]HT Incorporation[c] (CPM)	Percent Inhibition[d]
Media	—	92 ± 20	—
Stimulators (S)	—	132 ± 32	—
Responders (R)	—	2,365 ± 461	—
R × S	—	53,701 ± 8,408	—
R × S	R.15 (200)	29,210 ± 3,784	46
R × S	R.15 (40)	51,514 ± 2,378	4
R × S	R.15 (8)	60,940 ± 5,130	0
R × S	R6.5 (200)	39,567 ± 2,559	26
R × S	R6.5 (200)+R.15 (200)	12,492 ± 2,076	77
R × S	R6.5 (200)+R.15 (40)	23,923 ± 3.064	56

[a]Responder cells and irradiated stimulator (1000R) cells at 6.25×10^5/ml.
[b]Purified MAb to ICAM-1 (R6.5) or LFA-1β (R.15) at final concentrations (ng/ml).
[c]Mean ± SD of five to six cultures.
[d]Percent inhibition MLR.

TABLE 19.6. Effect of Anti-ICAM-1 and Dexamethasone on the Human MLR

Group[a]	Reagent[b] (ng/ml)	[3]HT Incorporation[c] (CPM)	Percent Inhibition[d]
Media	—	156 ± 81	—
Stimulators (S)	—	101 ± 20	—
Responders (R)	—	4,461 ± 2,289	—
R × S	—	34,199 ± 2,924	—
R × S	R6.5 (200)	5,329 ± 1,045	84
R × S	R6.5 (40)	14,563 ± 3,588	57
R × S	R6.5 (8)	26,224 ± 8,568	23
R × S	Dex (5000)	5,349 ± 1,326	84
R × S	Dex (500)	8,854 ± 7,205	74
R × S	Dex (50)	14,158 ± 4,323	59
R × S	R6.5 (8) + Dex (50)	7,759 ± 5,130	77

[a]Responder cells and irradiated stimulator (1000R) cells at 6.25×10^5/ml.
[b]Purified MAb to ICAM-1 (R6.5); Dexamethasone (Dex); at final concentrations (ng/ml).
[c]Mean ± SD three to four cultures.
[d]Percent inhibition of MLR.

anti-ICAM-1 result in additive suppressive effects on the MLR (Tables 19.6, 19.7, and 19.8).

The effectiveness of anti-ICAM-1 to inhibit the proliferative response resulting in a secondary MLR was also been examined. Anti-ICAM-1 was compared to the more classical therapeutic agents such as Dex and CyA. In Table 19.9, anti-ICAM-1 was not seen to be as effective in inhibiting a secondary MLR as the primary reaction. Cyclosporin A and

TABLE 19.7. Effect of Anti-ICAM-1 and Cyclosporin A on the Human MLR

Group[a]	Reagent[b] (ng/ml)	³HT Incorporation[c] (CPM)	Percent Inhibition[d]
Media	—	87 ± 22	—
Stimulators (S)	—	206 ± 64	—
Responders (R)	—	987 ± 461	—
R × S	—	31,640 ± 3,613	—
R × S	R6.5 (1000)	12,081 ± 4,810	62
R × S	R6.5 (200)	12,316 ± 4,507	61
R × S	R6.5 (40)	17,307 ± 6,814	45
R × S	R6.5 (8)	26,282 ± 7,450	17
R × S	CyA (100)	7,246 ± 1,478	77
R × S	CyA (10)	23,617 ± 1,725	25
R × S	CyA (1)	32,267 ± 2,773	0
R × S	R6.5 (8)+CyA (10)	19,204 ± 1,077	39

[a]Responder cells and irradiated stimulator (1000R) cells at 6.25 × 10⁵/ml.
[b]Purified MAb to ICAM-1 (R6.5); Cyclosporin A (CyA); at final concentrations (ng/ml).
[c]Mean ± SD of five to six cultures.
[d]Percent inhibition of MLR.

TABLE 19.8. Effect of Anti-ICAM-1 and Imuran on the Human MLR

Group[a]	Reagent[b] (ng/ml)	³HT Incorporation[c] (CPM)	Percent Inhibition[d]
Media	—	78 ± 16	—
Stimulators (S)	—	174 ± 126	—
Responders (R)	—	3,419 ± 2,108	—
R × S	—	49,570 ± 5,435	—
R × S	R6.5 (200)	18,350 ± 3,732	63
R × S	R6.5 (40)	33,583 ± 3,184	32
R × S	R6.5 (8)	44,374 ± 4,271	11
R × S	Imuran (100)	6,542 ± 767	86
R × S	Imuran (10)	24,237 ± 1,526	51
R × S	Imuran (1)	42,710 ± 4,487	14
R × S	R6.5 (8)+Imuran (1)	34,246 ± 6,455	31

[a]Responder cells and irradiated stimulator (1000R) cells at 6.5 × 10⁵/ml.
[b]Purified MAb to ICAM-1 (R6.5); Imuran; at final concentrations (ng/ml).
[c]Mean ± SD of five to six cultures.
[d]Percent inhibition of MLR.

Dex, however, were equally effective in inhibiting the primary and secondary MLR. The effectiveness of anti-ICAM-1 alone and in combination with anti-LFA-1 antibodies to inhibit a secondary MLR was also examined (Table 19.10). In the primary MLR response, combinations of the antibodies at suboptimal concentrations result in additive suppressive

TABLE 19.9. Effect of Anti-ICAM-1 and Immunosuppressants on the
Secondary MLR

Group[a]	Reagent[b] (μg/ml)	^{3}HT Incorporation (CPM)[c] I[d]	II[d]	Percent Inhibition[e] I[d]	II[d]
Media	—	92	310	—	—
Responders (R)	—	1,076	516	—	—
Stimulators (S)	—	115	425	—	—
R × S	—	7,100	16,438	—	—
R × S	R6.5 (5)	1,219	10,731	83	35
R × S	R6.5 (0.2)	1,971	13,725	72	17
R × S	R6.5 (0.008)	6,675	15,591	6	5
R × S	Dex (50)	276	181	96	99
R × S	Dex (0.05)	971	646	86	97
R × S	CyA (1)	502	155	93	99
R × S	CyA (0.01)	1,152	860	84	95
R × S	CyA (0.001)	3,817	3,287	46	80

[a]Responder cells and irradiated stimulator (1000R) cells at 6.25 × 10^5/ml.
[b]Purified MAb to ICAM-1 (R6.5); Dexamethasone (Dex); Cyclosporin A (CyA) at final
concentrations (μg/ml).
[c]Mean of five to six cultures after 3-day incubation.
[d]Primary MLR (I); secondary MLR (II).
[e]Percent inhibition of MLR.

effects (Tables 19.4, 19.5, and 19.10). Antibodies to ICAM-1 are not
effective in inhibiting a secondary MLR (Tables 19.9 and 19.10). Combinations, however, of suboptimal concentrations of anti-ICAM-1 with
anti-LFA-1 antibodies in the secondary reaction (Table 19.10) synergize
for the inhibition of the MLR.

Discussion

These data suggest that the antibody against ICAM-1 may have therapeutic and diagnostic potential in immune-mediated disorders regulated
by LFA-1, Mac-1 cell to cell interactions. Therapeutically, the in vivo
use of an antibody against an inducible cell surface molecule (ICAM-1)
may allow selective downregulation of an immune response. This is in
counter distinction to antibodies against constitutively expressed molecules that may inhibit the immune response in a less restricted manner.

Immune mediated disorders dependent on cell to cell interactions may
be promising targets for ICAM-1 MAb therapy. For example, prophylactic
use of antibodies against LFA-1 α in haploidentical individuals for bone-
marrow transplantation has resulted in improved engraftment, long-term

TABLE 19.10. Effect of Combinations of Anti-ICAM-1 and Ant-LFA-1 on the Secondary MLR

Group[a]	Antibody[b] (μg/ml)	^{3}HT Incorporation (CPM)[c]		Percent Inhibition[e]	
		I[d]	II[d]	I[d]	II[d]
Media	—	145	143	—	—
Responders (R)	—	252	216	—	—
Stimulators (S)	—	145	164	—	—
R × S	—	5,785	39,866	—	—
R × S	R6.5 (.04)	3,327	37,760	42	5
R × S	R3.1 (.04)	1,674	24,302	71	39
R × S	R6.5 (.04)+ R3.1 (.04)	555	10,174	90	74
R × S	R6.5 (.04)	3,327	37,760	42	5
R × S	R.15 (.2)	3,114	26,847	46	33
R × S	R6.5 (0.4)+ R.15 (.2)	1,764	13,655	70	66

[a]Responder cells and irradiated stimulator (1000R) cells at 6.25 × 10^5/ml.
[b]Purified MAbs to ICAM-1 (R6.5); LFA-1α (R3.1); LFA-1β (R.15) at final concentrations (μg/ml).
[c]Mean of five to six cultures after 3-day incubation.
[d]Primary MLR (I); secondary MLR (II).
[e]Percent inhibition of MLR.

survival, and development of tolerance to the donor histocompatibility antigens (14). The ICAM-1 is a ligand for LFA-1, and theoretically antibodies to ICAM-1 may perform in a similar manner. Diagnostically, antibodies against ICAM may be important in differentiating an immune from a nonimmune-induced rejection response. In individuals with nickel-induced contact dermatitis, ICAM-1 is expressed in the first few hours and peaks at 24 hours. Non-immune stimuli which evoke toxic dermatological responses (e.g., croton oil) do not induce ICAM-1 on Langerhan cells (Lange Wantzin et al., submitted). During acute renal rejection, it is also difficult to determine whether rejection is because of vascular, infectious, or immune-triggered events. Renal biopsy and assessment of upregulation of ICAM-1 expression may allow one to differentiate immune from non-immune mediated reactions.

In summary, the upregulation of ICAM-1 expression on monocytes and the ability of antibody against ICAM-1 to inhibit the MLR suggests that ICAM-1 MAbs may have diagnostic and therapeutic potential in acute graft rejection. The ICAM-1 MAbs may also have utility in related immune-mediated disorders dependent on LFA-1, Mac-1 regulated cell-to-cell interactions.

References

1. Sanchez-Madrid F, Nagy J, Robbins E, Simon P, Springer TA: A human leukocyte differentiation antigen family with distinct alpha subunits and a common beta subunit: The lymphocyte function-associated antigen (LFA-1), the C3bi complement receptor (OKM1/Mac-1), and the p150,95 molecule. *J Exp Med* 158:1785, 1983.
2. Mentzer SJ, Gromkowski SH, Krensky AM, Burakoff SJ, Martz E: LFA-1 membrane molecule in the regulation of homotypic adhesions of human B lymphocytes. *J Immunol* 135:9, 1985.
3. Rothlein R, Springer TA: The requirement for function-associated antigen 1 in homotypic leukocyte adhesion stimulated by phorbol ester. *J Exp Med* 163:1132, 1986.
4. Martz E: LFA-1 and other accessory molecules functioning in adhesions of T and B lymphocytes. *Human Immunol* 18:3, 1986.
5. Springer TA, Dustin ML, Kishimoto TK, Marlin SD: The lymphocyte function-associated LFA-1 CD2, and LFA-3 molecules: Adhesion receptors of the immune system. *Ann Rev Immunol* 5:223, 1987.
6. Anderson DC, Schmalsteig FC, Finegold MJ, Hughes BJ, Rothlein RR, Miller LJ, Kohl S, Tosi MF, Jacobs RL, Waldrop TC, Goldman AS, Shearer WT, Springer TA: The severe and moderate phenotypes of heritable Mac-1, LFA-1 deficiency: Their quantitative definition and relation to leukocyte dysfunction and clinical features. *J Inf Dis* 152:668, 1985.
7. Anderson DC, Springer TA: Leukocyte adhesion deficiency: An inherited defect in the Mac-1, LFA-1, and p150, 95 glycoproteins. *Ann Rev Med* 38:175, 1987.
8. Schleimer RP, Rutledge BK: Cultured human vascular endothelial cells acquire adhesiveness for neutrophils after stimulation with interleukin 1, endotoxin and tumor promoting phorbol diesters. *J Immunol* 136:649, 1986.
9. Dustin ML, Rothlein R, Bhan AK, Dinarello CD, Springer TA: Induction by IL 1 and interferon-y: Tissue distribution, biochemistry and function of a natural adherence molecule (ICAM-1). *J Immunol* 137:245, 1986.
10. Pober JS, Gimbrone Jr MA, Lapierre LA, Mendrick DL, Fiers W, Rothlein R, Springer TA: Activation of human endothelial cells by lymphokines: Overlapping patterns of antigenic modulation by interleukin 1, tumor necrosis factor and immune interferon. *J Immunol* 137:1893, 1986.
11. Rothlein RR, Dustin ML, Marlin SD, Springer TA: A human intercellular adhesion molecule (ICAM-1) distinct from LFA-1. *J Immunol* 137:1270, 1986.
12. Marlin SD, Springer TA: Purified intercellular adhesion molecule-1 (ICAM-1) is a ligand for lymphocyte function-associated antigen 1 (LFA-1). *Cell* 51:813, 1987.
13. Makgoba MW, Sanders ME, Luce GEG, Dustin ML, Springer TA, Clarke EA, Mannoni P, Shaw S: ICAM-1 a ligand for LFA-1 dependent adhesion of B, T and myeloid cells. *Nature* 331:86, 1988.
14. Fischer A, Blanche S, Veber F, Delaage M, Mawas C, Griscelli C, LeDeist F, Lopez M, Olive D, Janossy G: Prevention of graft failure by an anti-HLFA-1 monoclonal antibody in HLA-mismatched bone-marrow transplantation. *Lancet* 1:1058, 1986.

20

Lymphocyte Function-Associated Antigen-1 (LFA-1) Is a Signaling Molecule for Human T Lymphocytes

MARY C. WACHOLTZ, SUNIL S. PATEL, AND
PETER E. LIPSKY

Introduction

The CD18 family of integrin molecules defines a group of structurally related glycoproteins found on hematopoietic cells (1). The different members of this group (LFA-1, Mac-1, p150,95) all share a common β chain (95 kD; CD18), but have different α chains (LFA-1-180 kD, CD11a; Mac-1-165kD, CD11b; p150,95-150kD, CD11c). In a number of different circumstances, these proteins have been found to play an important role in cell to cell interactions and adhesion (2). Consistent with this, the LFA-1 molecule has been found to play a role in a number of the functional activities of lymphocytes. Much of the understanding of the involvement of LFA-1 molecules in lymphocyte function has been defined by use of anti-LFA-1 monoclonal antibodies (MAbs) and includes mediation of the adhesion of the target and effector cells in cytotoxic T lymphocyte and NK cell mediated killing. LFA-1 also plays a role in the induction of T cell proliferation by accessory cell-dependent stimuli by facilitating the induction of required physical interactions between responding T cells and accessory cells. Similarly, LFA-1 plays a role in the development of cell to cell contact required for some T-cell dependent B cell responses (2,3). The role of LFA-1 in these various cell contact dependent immune responses has been confirmed by analyzing the functional responsiveness of cells from children with a genetic deficiency in LFA-1 expression (2).

The bulk of the data has supported the conclusion that LFA-1 might serve as an adhesion molecule. Some evidence, however, has also suggested the possibility that LFA-1 might function as a signaling molecule. In studies designed to examine the role of LFA-1 in mediating accessory cell-T cell interactions, an anti-CD18 MAb was found to inhibit monocyte-supported anti-CD3 stimulated proliferation of both CD4(+) and

CD8(+) T cells (4). The effect of the anti-CD18 MAb could not be explained by a nonspecific inhibitory effect on all T cell functions since it did not inhibit PMA and ionomycin or PMA and anti-CD3 induced T cell proliferation in the presence or absence of accessory cells. Consistent with an effect on initial T cell-accessory cell interactions, the MAb was effective only when added early in culture. When immobilized anti-CD3 MAb was used as a mitogenic stimulus to study the effect of LFA-1 in the absence of accessory cells, soluble anti-CD18 MAb did not affect the anti-CD3 induced proliferative response (5). However, co-stimulation with an immobilized anti-CD18 MAb and a suboptimal concentration of anti-CD3 MAb immobilized to the same surface markedly enhanced T cell proliferation (5). This suggested the possibility that engagement of LFA-1 molecules might transmit information to the cell that resulted in enhanced activation. It remained possible, however, that co-stimulation might have resulted from increased binding of the responding lymphocytes to the anti-CD3 coated surface or from stabilization of cellular-solid phase interactions.

Additional experimental results tended to support the view that engagement of LFA-1 might deliver a signal to the cell. In studies using suboptimal immobilized OKT3 as an accessory cell-independent stimulus (6), addition of soluble anti-CD18 MAb was found to inhibit the subsequent proliferative response, whereas addition of soluble anti-CD11a MAb enhanced proliferation. Thus, independent binding of MAb to different chains of the LFA-1 molecule appeared to regulate the capacity of T cells to proliferate in response to MAb to CD3. More recent data have suggested that purified T cells might proliferate in response to immobilized anti-CD18 MAb directly when co-stimulation with PMA was provided (7). This observation suggests that MAb to LFA-1 can stimulate T cells in the absence of anti-CD3 mediated activation and, therefore, implies that immobilized anti-LFA-1 might provide an appropriate co-mitogenic signal, comparable to the increase in intracellular free calcium ($[Ca^{2+}]_i$) usually induced by anti-CD3 MAb or calcium ionophore in PMA stimulated T cells. The possibility that MAb to LFA-1 might induce an increase in $[Ca^{2+}]_i$ in a small number of T cells has been suggested (8), although not confirmed (9). Finally, in the murine system, a MAb to CD11a has been shown to induce B cell proliferation directly in the absence of any co-stimulatory signals (10).

All of these results are consistent with the conclusion that LFA-1 can function as a signaling molecule on lymphocytes. There is no clear-cut evidence, however, that such apparent signaling might not actually result from alterations in cellular interactions occurring in multicellular cultures. Moreover, if LFA-1 functions as a signal-transducing molecule, the exact nature of the signal has not been delineated. To address these issues, functional studies using normal human T cell clones were undertaken. As normal human T cell clones can be activated in culture by cross-linking certain surface antigens in the absence of accessory cells (11,12), these cells were used to examine the effect of anti-LFA-1 MAb on T cell activation without the need for accessory cells or solid-phase immobilization of the mitogenic stimulus. The results of these studies clearly demonstrate that although MAb to LFA-1 do not functionally activate human T cell clones alone, cross-linking the LFA-1 molecule conveys a unique signal that can result in enhanced activation of T cells co-stimulated by engagement of the CD3 molecular complex.

Results

Initial experiments were carried out to determine whether MAb to LFA-1 could generate immediate biochemical changes in T cells associated with cellular activation. Therefore, the effect of the anti-CD18 MAb, 60.3, on intracellular calcium ($[Ca^{2+}]_i$) was examined, as an increase in $[Ca^{2+}]_i$ is one of the earliest biochemical changes that occurs after stimulation of the T cell and may be an obligatory component of mitogenic signaling to the cell (13). Flow cytometry was used to measure $[Ca^{2+}]_i$, as previously described (9), because this technique allows analysis of the percentage of cells which respond and the mean $[Ca^{2+}]_i$ of the cells as a function of time. Addition of the anti-CD18 MAb alone (data not shown) or addition of a cross-linking secondary antibody, goat anti-mouse immunoglobulin (GaMIg), to cells pretreated with 60.3 did not result in an increase in $[Ca^{2+}]_i$ (Fig. 20.1).These results confirm our previous findings using fresh T cells that binding or cross-linking CD18 did not lead to an increase in $[Ca^{2+}]_i$ (9).

The effect of concomitant cross-linking of CD3 and LFA-1 or HLA-A,B,C and LFA-1 was examined next. As shown in Figure 20.1, cross-linking CD3 alone resulted in a rapid increase in $[Ca^{2+}]_i$ followed by a rapid decline toward baseline. When cells were also pretreated with the MAb 60.3, the response was more sustained. This enhanced response was specific for the combination of MAb to CD3 and LFA-1 as cross-linking HLA-A,B,C and LFA-1 resulted in a diminished and delayed signal compared to the calcium response typically elicited by cross-linking HLA-A,B,C alone (9). This co-stimulatory effect was not unique to the anti-CD18 MAb, 60.3, as both the anti-CD11a MAb, TS1/22, and a second anti-CD18 MAb, TS1/18 (14), also sustained the CD3 response and diminished the HLA-A,B,C response after cross-linking (Fig. 20.2). Thus, although cross-linking LFA-1 alone did not result in a detectable change in $[Ca^{2+}]_i$, concomitant cross-linking with other surface antigens modified the resultant calcium signal, and, in particular, sustained the CD3 induced increase in $[Ca^{2+}]_i$.

Perturbation of the LFA-1 surface antigen also modified cellular activation as measured by IL2 production or DNA synthesis. As shown in

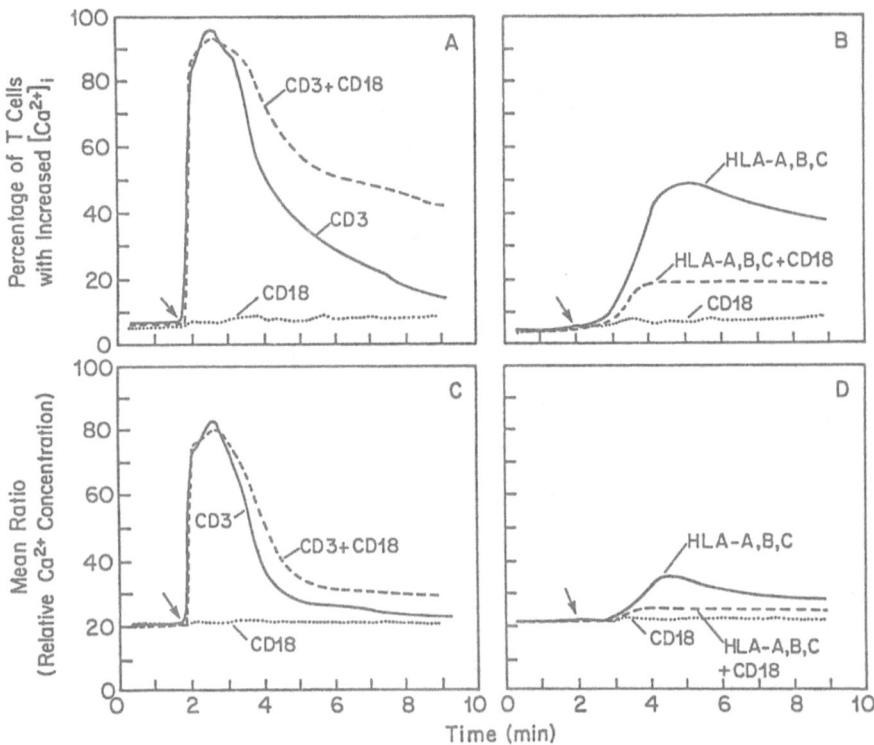

FIGURE 20.1. Cross-linking CD18 and CD3 prolongs the [Ca²⁺]ᵢ response. Clone NP18 (CD8+) cells were pretreated at 4 °C with the indicated MAb (CD3-OKT3; CD18-60.3) and washed. Cells warmed to 37 °C were analyzed with the flow cytometer for 2 minutes to establish the baseline [Ca²⁺]ᵢ. At 2 minutes (indicated by the arrow), the cross-linking secondary antibody, GaMIg, was added, and the measurement continued for up to 8 minutes. The data were analyzed in terms of the percentage of T cells which increased [Ca²⁺]ᵢ over threshold, which was defined as the mean ratio which included 95% of the cells before addition of GaMIg (A,B), and in terms of the mean fluorescence ratio (violet to blue) which reflects the mean [Ca²⁺]ᵢ of the population (C,D). In the absence of GaMIg, none of the MAb induced an increase in [Ca²⁺]ᵢ, whereas GaMIg induced no increase in [Ca²⁺]ᵢ, in cells stained with an irrelevant control MAb.

Table 20.1, pretreating T cell clones with an anti-CD18 MAb followed by culture with or without a cross-linking GaMIg did not stimulate IL2 production. Pretreating the CD4(+) clone, but not the CD8(+) clone, with an anti-CD3 MAb alone resulted in IL2 production, and culture with GaMIg resulted in a modest increase in IL2 production by either clone above that caused by the anti-CD3 MAb alone. Pretreatment with both an anti-CD3 and an anti-CD18 MAb followed by culture with GaMIg resulted in a three- to sevenfold increase in IL2 production com-

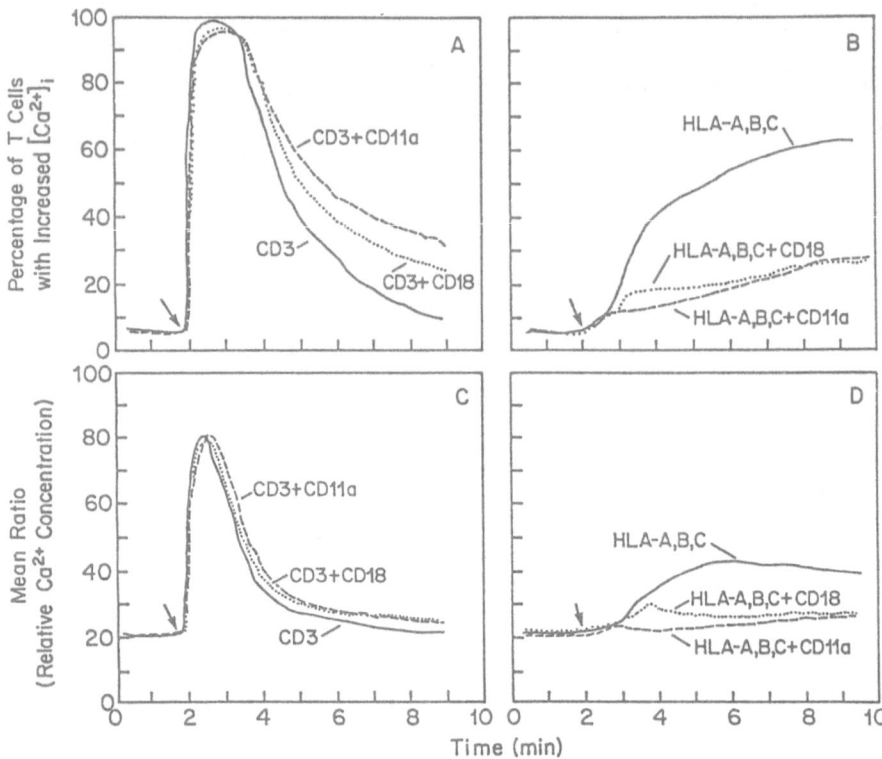

FIGURE 20.2. Both anti-CD11a or anti-CD18 MAb prolong the $[Ca^{2+}]_i$ response induced by anti-CD3. Clone NP1 (CD4+) cells were pretreated at 4 °C with the indicated MAb (CD3-OKT3; CD11a-TS1/22; CD18-TS1/18) and washed. Subsequent flow cytometric analysis was done as described in Figure 20.1.

pared to the amount of IL2 produced after cross-linking CD3 alone. Cross-linking HLA-A,B,C alone consistently resulted in a small increase in IL2 production as previously observed (9). Cross-linking both HLA-A,B,C and LFA-1 did not enhance IL2 production in any experiment and sometimes decreased IL2 secretion as demonstrated with the CD8(+) clone.

The enhancement of the response to cross-linking CD3 was also observed when DNA synthesis was examined (Table 20.2). Pretreating cells with an anti-CD18, CD11a, or CD3 MAb alone with or without GaMIg did not result in increased ³H-thymidine incorporation. However, pretreating cells with both an anti-CD3 and an anti-CD18 MAb followed by culture with GaMIg stimulated DNA synthesis. The response was not specific to cross-linking the β-chain of LFA-1 as pretreatment with an anti-CD11a MAb was also effective at co-stimulation of DNA synthesis. Cross-linking HLA-A,B,C alone stimulated DNA synthesis which was

TABLE 20.1. Effect of Cross-linking CD3 and CD18 on IL2 Production by Human T Cell Clones

Monoclonal Antibody	CD4+		CD8+	
	Control	GaMIg	Control	GaMIg
	IL2 Production (Units/10⁶ cells/24 hrs)			
Control	0.1	0.1	0.0	0.0
CD18	0.1	0.1	0.0	0.0
CD3	3.4	4.6	0.0	0.4
CD3+CD18	0.6	13.8	0.4	2.8
HLA-A,B,C	0.1	2.6	0.0	1.2
HLA-A,B,C+CD18	0.1	2.0	0.0	0.4

Cloned T cells were prepared as previously described (19). Clones were pretreated with saturating concentrations of the indicated MAb (CD18-60.3; CD3-OKT3; HLA-A,B,C-W6/32), washed and incubated (1 × 10⁵ cells/well) with or without GaMIg. Supernatants were harvested at 24 hours and assayed for IL2 production using CTLL cells as described (4).

TABLE 20.2. Cross-linking Either the α or β Chain of the LFA-1 Molecule Enhances anti-CD3 Induced T Cell DNA Synthesis

Monoclonal Antibody	DNA Synthesis	
	Control	GaMIg
	(cpm × 10⁻³ ± SEM)	
Control	1.6 ± 0.1	1.4 ± 0.1
CD11a	1.6 ± 0.1	1.7 ± 0.0
CD18	1.3 ± 0.2	1.3 ± 0.2
CD3	1.6 ± 0.1	1.2 ± 0.1
CD3+CD11a	2.3 ± 0.1	8.1 ± 0.3
CD3+CD18	2.1 ± 0.0	5.2 ± 0.6
HLA-A,B,C	1.8 ± 0.0	12.6 ± 0.6
HLA-A,B,C+CD11a	1.8 + 0.2	12.3 + 0.8

Clone NP7 (CD4+) was pretreated with saturating concentrations of the indicated MAb (Control-P1.17; CD3-OKT3; CD11a-TS1/22; CD18-TS1/18; HLA-A,B,C-W6/32), washed, and incubated (1 × 10⁵ cells/well) either with or without GaMIg. Cells were cultured for 36 hours. Eighteen hours prior to harvesting, 1 μCi of ³H-thymidine was added. All data are expressed as mean cpm ± SEM of triplicate determinations.

not increased by the presence of an anti-CD11a MAb. When IL2 production was measured in this CD4(+) clone (data not shown), cross-linking CD3 alone, but not the α or β chain of the LFA-1 molecule generated a small amount of IL2 release. When cells were pretreated with both an anti-CD3 and an anti-CD11a or an anti-CD18 MAb, a 3.5- to four-fold enhancement of IL2 release was observed after cross-linking. Cross-linking HLA-A,B,C alone stimulated IL2 release by this clone which was not changed by concomitant cross-linking of LFA-1. Thus, cross-linking CD3 and LFA-1 at the same time prolonged the CD3 in-duced increases in $[Ca^{2+}]_i$ and enhanced functional activation of T cell clones.

The next experiments examined whether co-stimulation required that CD3 and LFA-1 molecules be cross-linked together. Since GaMIg binds both the anti-CD3 and the anti-LFA-1 MAb, cross-linking of CD3 to CD3, LFA-1 to LFA-1, or CD3 to LFA-1 could all be taking place. To determine whether cross-linking LFA-1 separately from cross-linking CD3 was an effective stimulus or whether cross-linking LFA-1 was re-quired, isotype specific secondary antibodies were utilized (Table 20.3). In these experiments, cells were stained with an IgG1 anti-CD3 MAb or an IgG2a anti-LFA-1 MAb, either alone or in combination. Pretreatment with the anti-CD3 MAb alone induced IL2 release by this clone that was slightly increased by cross-linking with the appropriate isotype specific antibody or GaMIg. Cross-linking LFA-1 alone did not generate detect-able IL2 production. When cells were pretreated with both an anti-CD3 and an anti-LFA-1 MAb, cross-linking CD3 alone was no more effective at inducing IL2 production than cross-linking cells reacted only with an anti-CD3 MAb. However, when LFA-1 was cross-linked alone on a cell also carrying an anti-CD3 MAb, IL2 release was increased 5.5-fold com-pared to cross-linking CD3 on a cell pretreated with both MAb and was equal to the response induced by cross-linking both with GaMIg. Cross-linking CD3 and LFA-1 separately (IgG1 + IgG2a) was less effective at inducing IL2 production than cross-linking LFA-1 alone or using GaMIg to cross-link, although it induced a greater amount of IL2 production than cross-linking CD3 only on cells stained with both MAb. Similar results were noted when proliferation was examined (Table 20.4). In par-ticular, cross-linking only LFA-1 when cells were stained with both an anti-CD3 and an anti-LFA-1 MAb induced a greater degree of prolifer-ation than cross-linking CD3 alone or CD3 + LFA-1 together.

TABLE 20.3. Co-stimulation of IL2 Production Does Not Require Cross-linking of CD3 to CD18

Monoclonal Antibody		IL2 Production				
Specificity	Isotype	Control	Anti-IgG1	Anti-IgG2a	IgG1 + IgG2a	GaMIg
		(Units/10^6 cells/24 hours)				
Control	IgG1/G2a	0.0	0.0	0.0	0.0	0.0
CD3	IgG1	0.6	1.2	0.8	1.2	1.0
CD18	IgG2a	0.0	0.0	0.0	0.0	0.0
CD3 + CD18	IgG1 + G2a	0.4	1.6	8.8	3.2	8.8

Clone LM2 (CD4+) was pretreated with the indicated MAb (Control-P1.17[IgG2a]/ MOPC-[IgG1]; CD3-Leu4; CD18-60.3), washed, and incubated (1 × 10⁵ cells/well) with isotype specific secondary antibody or GaMIg. After 24 hours, supernatants were harvested and assayed for IL2 production (4).

TABLE 20.4. Separately Cross-linking LFA-1 increases DNA Synthesis by Anti-CD3 Stimulated T Cell Clones

Monoclonal Antibody		DNA Synthesis				
Specificity	Isotype	Control	Anti-IgG1	Anti-IgG2a	IgG1 + IgG2a	GaMIg
		(^{3}H-Thymidine Incorporation, cpm × 10^{-3} ± SEM)				
Control	IgG1/G2a	0.5 ± 0.1	1.1 ± 0.4	0.7 ± 0.0	0.7 ± 0.1	0.7 ± 0.0
CD3	IgG1	0.9 ± 0.1	5.3 ± 0.3	2.0 ± 0.2	5.0 ± 0.2	1.9 ± 0.1
CD18	IgG2a	0.7 ± 0.1	1.1 ± 0.1	0.5 ± 0.2	0.5 ± 0.2	0.8 ± 0.0
CD3 + CD18	IgG1 + G2a	0.7 ± 0.1	4.3 ± 0.1	8.0 ± 0.3	4.2 ± 0.1	5.0 ± 0.3

Clone LM2 (CD4+) was pretreated with the MAb described in Table 20.3, washed, and incubated (1 × 10^5 cells/well) with isotype specific antibody or GaMIg for 36 hours. Eighteen hours prior to harvesting, ^{3}H-thymidine was added. All data are expressed as mean cpm ± SEM of triplicate determinations.

Discussion and Conclusions

The results presented here indicate that LFA-1 can function as more than an adhesion molecule. In this experimental system, T cell activation required only cross-linking of certain surface antigens, thus eliminating the requirement for T cell-accessory cell or T cell-solid phase interactions. Use of this system made it possible to examine the capacity of LFA-1 molecules to deliver signals directly to T cells. When $[Ca^{2+}]_i$ was measured to assess the induction of an early activation signal, perturbation of LFA-1 alone was not found to generate a detectable change in $[Ca^{2+}]_i$. This was similar to our previous results using freshly prepared T4 cells, in which MAb to LFA-1 did not induce an increase in $[Ca^{2+}]_i$ with or without cross-linking (9). Although another study (8) did suggest that cross-linking LFA-1 molecules in peripheral blood mononuclear cells might induce a change in $[Ca^{2+}]_i$, the cells were not purified and the modest increase seen was limited to a small number of cells. The current studies demonstrate that MAb to LFA-1, although not directly inducing a change in $[Ca^{2+}]_i$, can affect the calcium signal. Thus, cross-linking LFA-1 in conjunction with MAb to other surface antigens, including CD3 or HLA-A,B,C, resulted in a new pattern of increase in $[Ca^{2+}]_i$. Detection of this response required the sensitivity of flow cytometric analysis as it was most apparent when the percentage of cells responding was measured. The change in average $[Ca^{2+}]_i$ (mean ratio) resulting from co-stimulation with anti-CD3 and anti-LFA-1 was small, corresponding to approximately an increase of 150 nM. The change in the calcium signal strongly indicates that the LFA-1 molecule may play a signaling role in early T cell activation. The modification of the $[Ca^{2+}]_i$ signal induced by MAb to LFA-1 translated into enhanced activation of T cell clones when CD3 was the second molecule engaged. This was manifested both by enhanced IL2 production and proliferation. Unlike the results of van Noesel et al. (6) in which addition of soluble anti-CD18 MAb inhibited and soluble anti-CD11a MAb increased anti-CD3-stimulated proliferation, we observed that the prolongation of the calcium signal and increased CD3 mediated cell activation occurred whether anti-CD18 or anti-CD11a MAb were used. However, cross-linking of the MAb to either the α or β chain of LFA-1 was essential for co-stimulation. Moreover, the synergistic co-stimulation exhibited specificity for CD3 as cross-linking HLA-A,B,C and LFA-1 somewhat diminished the calcium signal and did not increase IL2 production and proliferation. Although the reason for this difference between the anti-CD3 and anti-HLA-A,B,C MAb induced response is not known, it may be important that the response induced by MAb to class I MHC molecules is absolutely dependent on cross-linking, whereas there is often a small response in the clones to anti-CD3 alone, and addition of GaMIg frequently is inhibitory of the subsequent response (9,11).

The mechanism by which MAb to LFA-1 modify cellular activation is unknown, but includes several possibilities. It is possible that LFA-1 physically interacts with CD3 to generate an enhanced activation signal. There is precedent for such co-stimulation in that MAb to CD4 or CD8 can magnify the response to anti-CD3 MAb when co-immobilized, when presented together bound to Sepharose beads, when heteroconjugate MAb to CD3 are linked to MAb specific for CD4 or CD8, or when cells pretreated with both MAb are cross-linked with a secondary antibody (5,12,15–17). It has also been suggested that 5% of surface CD3 is complexed with CD4 on T cells (18). Thus, it is possible that LFA-1 may also physically interact with CD3 to amplify the signal. Cross-linking could encourage the interaction. The current data make this unlikely, however, since cross-linking of CD3 with LFA-1 was no more effective at delivering a co-stimulatory signal than cross-linking LFA-1 independently of CD3. It remains possible that association of CD3 and LFA-1 may occur physiologically, although there is no evidence for such an interaction. It would appear to be more likely that cross-linking of LFA-1 could generate an independent signal that is too small to be readily detected, at least as a calcium signal, but which synergistically amplifies signal generation via CD3. The findings that cross-linking LFA-1 alone is sufficient to co-stimulate anti-CD3 activated cells supports this conclusion. Moreover, such experiments are highly suggestive that LFA-1 alone can convey independent information to the cell under the proper circumstances. Cross-linking LFA-1 with a secondary antibody may mimic the cross-linking that could occur when LFA-1 molecules were engaged by their natural ligand, ICAM-1, during accessory cell-T cell interactions. Such receptor-ligand engagement may not only serve to facilitate cellular adhesion, but also to deliver co-stimulatory signals that amplify T cell activation.

Acknowledgments. The authors would like to thank Drs. P.G. Beatty and T.A. Springer for the various monoclonal antibodies employed in this study.

References

1. McMichael AJ et al. (ed): *Leucocyte Typing III. White Cell Differentiation Antigens.* New York. Oxford University Press, 1987.
2. Springer TA, Dustin ML, Kishimoto TK, Martin SD: The lymphocyte function-associated LFA-1, CD2, and LFA-3 molecules: Cell adhesion receptors of the immune system. *Ann Rev Immunol* 5:223, 1987.
3. Beatty, PG, Ledbetter JA, Martin PJ, Price TH, Hansen JA: Definition of a common leukocyte cell-surface antigen (Lp95-150) associated with diverse cell-mediated immune functions. *J Immunol* 131:2913, 1983.
4. Geppert TD, Lipsky PE: Accessory cell-T cell interactions involved in anti-CD3-induced T4 and T8 cell proliferation: Analysis with monoclonal antibodies. *J Immunol* 137:3065, 1986.
5. Geppert TD, Lipsky PE: Activation of T lymphocytes by immobilized monoclonal antibodies to CD3. Regulatory influences of monoclonal antibodies to additional T cell surface determinants. *J Clin Invest* 81:1497, 1988.
6. van Noesel C, Miedeman F, Brouwer M, de Rie MA, Aarden LA, van Lier RAW: Regulatory properties of LFA-1 α and β chains in human T-lymphocyte activation. *Nature* 333:850, 1988.
7. Carrera AC, Rincon M, Sanchez-Madrid F, Lopez-Botet M, de Landazuri MO: Triggering of co-mitogenic signals in T cell proliferation by anti-LFA-1 (CD18, CD11a), LFA-3, and CD7 monoclonal antibodies. *J Immunol* 141:1919, 1988.
8. Ledbetter JA, June CH, Grosmaire LS, Rabinovitch PS: Crosslinking of surface antigens causes mobilization of intracellular ionized calcium in T lymphocytes. *Proc Natl Acad Sci USA* 84:1384, 1987.
9. Geppert TD, Wacholtz MC, Davis LS, Lipsky PE: Activation of human T4 cells by cross-linking class I MHC molecules. *J Immunol* 140:2155, 1988.
10. Mishra GC, Berton MT, Oliver KG, Krammer PH, Uhr JW, Vitetta ES: A monoclonal anti-mouse LFA-1α antibody mimics the biological effects of B cell stimulatory factor-1 (BSF-1). *J Immunol* 137:1590, 1986.
11. Geppert TD, Wacholtz MC, Patel SS, Lightfoot E, Lipsky PE: Activation of human T cell clones and Jurkat cells by cross-linking class I major histocompatibility complex molecules. *J Immunol* 142:3763, 1989.
12. Wacholtz MC, Patel SS, Lipsky PE.: Patterns of costimulation of T cell clones by cross-linking CD3, CD4/CD8, and class I MHC molecules. *J Immunol*, 142:4201, 1989.
13. Weiss A, Imboden J, Hardy K, Manger B, Terhorst C, Stobo J: The role of the T3/antigen receptor complex in T-cell activation. *Ann Rev Immunol* 4:619, 1986.
14. Ware CF, Sanchez-Madrid F, Krensky AM, Burakoff SJ, Strominger JL, Springer TA: Human lymphocyte function associated antigen-1 (LFA-1): Identification of multiple antigenic epitopes and their relationship to CTL-mediated cytotoxicity. *J Immunol* 131:1182.

15. Emmrich F, Kanz L, Eichmann K: Cross-linking of the T cell receptor complex with the subset-specific differentiation antigen stimulates interleukin 2 receptor expression in human CD4 and CD8 T cells. *Eur J Immunol* 17:529, 1987.
16. Anderson P, Blue ML, Morimoto C, Schlossman SF: Coaggregation of the T-cell receptor with CD4 and other T-cell surface molecules enhances T-cell activation. *Proc Natl Acad Sci* 84:9209, 1987.
17. Emmrich F, Rieber P, Jurrie R, Eichmann K: Selective stimulation of human T lymphocyte subsets by heteroconjugates of antibodies to the T cell receptor and to subset-specific differentiation antigens. *Eur J Immunol* 18:645, 1988.
18. Anderson P, Blue ML, Schlossman SF: Comodulation of CD3 and CD4. Evidence for a specific association between CD4 and approximately 5% of the CD3:T cell receptor complexes on helper T lymphocytes. *J Immunol* 140:1732, 1988.
19. Patel SS, Thiele DL, Lipsky PE: Major histocompatibility complex-unrestricted cytolytic activity of human T cells. Analysis of precursor frequency and effector phenotype. *J Immunol* 139:3886, 1987.

21

Involvement of Lymphocyte Function-Associated Antigen-1 and Intercellular Adhesion Molecule-1 in Monocyte and T-Cell Antigen-Specific Interactions

Nancy Hogg, Graeme Dougherty, and Anne-Marie Buckle

Introduction

The cell surface molecules directly involved in the specific interaction between T cells and monocytes have been well described—MHC class II molecules and processed antigen on the monocyte side and the antigen specific receptor on the T cell side. These minimal requirements for antigen presentation have been simulated by using T cells and as accessory cell, "null" cells such as the murine L cell, transfected with human MHC class II products (1). Most of these studies have been successful only with cloned T cell lines and not with resting T cell populations such as would be found in peripheral blood. This has led to the suggestion that other molecular interactions are required to activate such cells (2,3). T cells, particularly CTL, use the adhesion molecule lymphocyte function-associated antigen-1 (LFA-1) to interact with target cells (4). The molecule to which LFA-1 binds in many but not all, of these interactions is intercellular adhesion molecule-1 (ICAM-1) (5,6). We have examined the role of the leukocyte integrin family, namely LFA-1, CR3/Mac-1 and p150,95 in the early interaction between resting T cells and monocytes which is required for the generation of antigen specific T cell proliferation. These molecules are all well expressed by monocytes, whereas T cells express LFA-1 with CR3 and p150,95 found only on a subset of CD8+ T cells (7,8).

Materials and Methods

Monoclonal Antibodies, Antigens, and Mitogens

The monoclonal antibodies (MAbs) used were MHM-23 and H-52 (β chain/CD18) (9), MHM-24 (LFA-1 α chain/CD11a) (9), 44 (CR3/Mac-1/CD11b) (10), p150,95 (CD11c), 52 (MHC class II) (12), and RR1/1 (anti-ICAM-1) (13).

Purified protein derivative (PPD) used at 10 μg/ml, was a gift from Dr V. Barnard, Ministry of Agriculture, Fisheries and Food, Weybridge, Surrey, UK. Pokeweed mitogen (PWM) was used at 2 μg/ml (Sigma Chemical Co.).

Cell Populations

Peripheral blood mononuclear cells (PBMC) were prepared from fresh heparinized blood by centrifugation through Ficoll-Hypaque (Pharmacia, $\delta = 1.08$ kg/l). Monocytes were purified from PBMC by adherence to microexudate-coated petri dishes (95% CD14+) and the T cell population fractionated from the microexudate nonadherent cells by passage over nylon wool columns (>95% CD3+) as previously reported (11,14). Alternatively, T cells were fractionated directly from PBMC using negative antibody selection and magnetic beads (Dynabeads M450, Dynal A.S.).

Proliferation Assays

Two $\times 10^5$ PBMC or 1×10^5 T cells and 2×10^4 monocytes were cultured in 200 μl of RPMI-1640 plus 5% AB serum in the presence of antigen, mitogen, and MAb which was usually used as ascitic fluid at a final dilution of 1:100. In the experiments in which PWM-pulsed monocytes were used as stimulators of T cell proliferation, monocytes were resuspended in medium containing 10 μg/ml of PWM and incubated at 37 °C for 1 hour then thoroughly washed. In addition, T cells and monocytes were preincubated with MAb (ascites,1:50) for 1 hour on ice before removing unbound MAb by washing. Cultures were pulsed with 1 μ Ci/well-tritiated thymidine for 8 hours on Day 2 or 3 for mitogen responses or Day 5 for antigen responses.

FACS Analysis

Briefly, 10 μl of cell suspension containing $2 \times 10^7 - 4 \times 10^7$ cells/ml were incubated on ice for 40 minutes with 50 μl of MAb tissue culture supernatant. The cells were washed three times with cold medium to remove unbound antibody, resuspended in 50 μl of fluorescein-conjugated F(ab')2 rabbit anti-mouse IgG (Dakopatts, 1:10) and incubated for a further 30 minutes on ice. After an additional three washes, the percentage of positively staining cells was determined using a FACScan (Becton Dickinson).

Results

The participation of the various members of the LFA-1 family (LFA-1, CR3/Mac-1, and p150,95) in the early stages of an immune response was investigated by determining the ability of specific MAb to inhibit T cell proliferation in vitro (15). A large number of MAb specific for monocyte or T cell molecules had no effect. Inhibition was achieved, however, with MAbs directed against LFA-1 α and β chains but not with MAbs specific for CR3/Mac-1 nor p150,95 even though these latter MAbs have been shown to have functional effects by their interference with binding of iC3b-coated erythrocytes to myeloid cells (10,16). Thus, not unexpectedly, it could be concluded that LFA-1 was important in the interaction between monocytes and T cells.

To discover the cell type on which LFA-1 was functional, fractionated peripheral blood T cells and monocytes were separately incubated with either anti-β chain or anti-LFA-1 α chain MAbs before the initiation of the experiments. Figure 21.1 shows that preincubation of neither T cells nor monocytes with MAb achieved blocking effects. To substantially in-

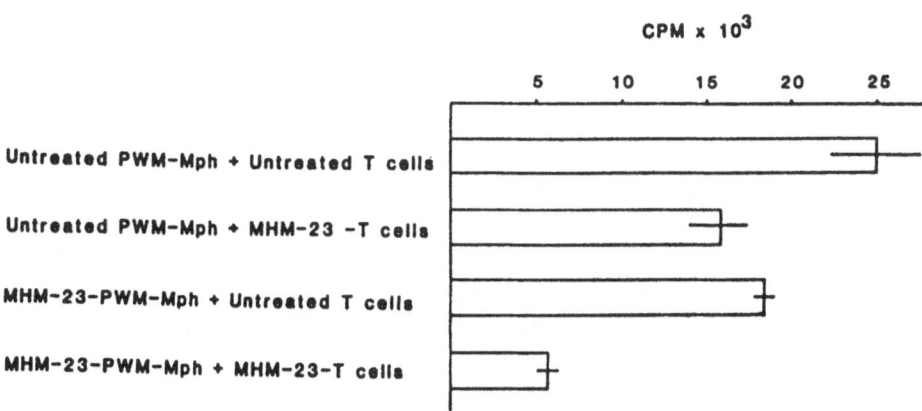

FIGURE 21.1. Involvement of monocyte and T cell LFA-1 in the generation of a proliferation response. Cultures contained 2×10^4 PWM-pulsed monocytes pretreated (1 hour at 0 °C) with medium or a 1:50 dilution of MAb MHM-23 (anti-LFA-1/CR3/p150,95 β subunit) and 1×10^5 T lymphocytes pretreated (1 hour at 0 °C) with medium or a 1:50 dilution of MAb MHM-23. Each bar represents the mean cpm ± SD of quadruplicate cultures. (Reprinted with permission of the *European Journal of Immunology*.)

hibit T cell proliferation, blocking of both T cell and monocyte LFA-1 was required.

This result suggested the requirement for a reciprocal interaction of LFA-1 with ligand on both monocyte and T cell surface. As ICAM-1 is the only ligand which has been so far characterized for LFA-1, we next investigated whether an anti-ICAM-1 MAb would interfere with T cell proliferation (17). Figure 21.2 shows that MAb RR1/1 specific for ICAM-1, as well as MAbs to LFA-1 α chain (MHM-24), common β chain (H-52) and MHC class II molecules (52) blocked the reaction whereas MAbs used as controls, (MAbs 44 (CR3), 3.9 (p150,95) and 10.1 (FcRI)) did not inhibit T cell proliferation. Therefore, as blocking of both T cell and monocyte LFA-1 is required to inhibit responsiveness and as anti-ICAM-1 MAb also blocks the response, these findings can be most simply interpreted by suggesting that there is a reciprocal reaction between LFA-1 and ICAM-1 on monocytes and T cells.

Therefore, it was interesting to investigate the levels of ICAM-1 expressed by these cells. Whereas the majority of monocytes express ICAM-1 (Fig. 21.3a), it has been reported that peripheral blood T cells lack ICAM-1 (17–19). However, Figure 21.3b shows that a proportion (20–40%) of freshly isolated T cells do express ICAM-1, although at a lower level than monocytes. ICAM-1 is expressed on the T cells which also

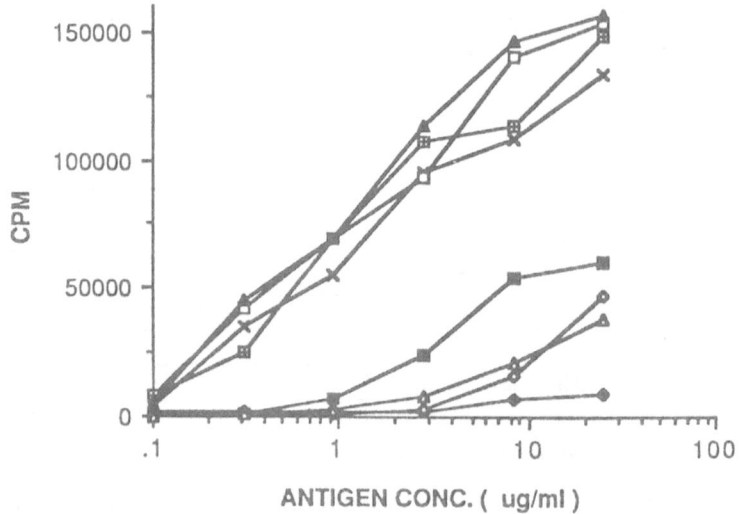

FIGURE 21.2. Effect of various MAbs on antigen-induced proliferation in vitro. Cultures contained 2×10^5 PBMC, PPD (10 μg/ml) and medium (×), or a 1:100 dilution of MAb 3.9 (⊞), 44 (□), 10.1 (▲), RR 1/1 (■), MHM-24 (◊), 52 (△) or H-52 (♦) ascites. Each point represents the mean cpm of quadruplicate cultures. Standard deviations were <10%. (Reprinted with permission of the *European Journal of Immunology*.)

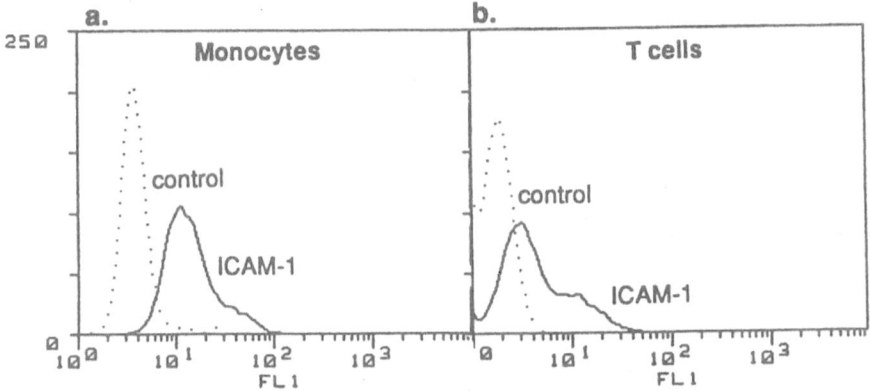

FIGURE 21.3. Expression of ICAM-1 on freshly isolated peripheral blood monocytes and T cells. Using FACScan, levels of ICAM-1 were assessed by MAb RR1/1 binding to (*a*) monocytes (*b*) T cells. A nonreactive IgG1 MAb was used as an isotype control.

express LFA-3 and high levels of LFA-1 and UCHL1, a cell profile which is the characteristic phenotype of memory T cells (Buckle and Hogg, submitted; 20). As expected functional testing of ICAM-1⁺ and ICAM-1⁻ T cells fractionated by flow cytometry has revealed that the response to recall antigens (PPD, influenza virus) is found in the ICAM-1⁺ subpopulation (Buckle and Hogg, submitted). This result suggests that those T cells expressing ICAM-1 precisely define the responding cells of the memory T cell pool.

Discussion

Interaction between T cells and monocytes during the initial phases of an immune response requires the adhesion molecules LFA-1 and ICAM-1. Monocyte LFA-1 and T cell LFA-1 perform a role in these early events that is essentially over by 8 hours (15). Because anti-ICAM-1 can also block T cell proliferation, the simplest hypothesis is that ICAM-1 is the molecule with which LFA-1 interacts in a reciprocal manner on both T cell and monocyte surface (17; see Fig. 21.4). Monocyte ICAM-1 is readily detected and T cells that express ICAM-1 at low levels are the T cells which bear the memory T cell phenotype (Buckle and Hogg, submitted). These molecules, such as LFA-3 plus heightened levels of LFA-1 and UCHL1, which distinguish memory from naive T cells, can be regarded as hallmarks of previous activation events. Several of these molecules including ICAM-1 are implicated in cell adhesion events, which suggests that facilitated cell interactions resulting in clustering, might be partially

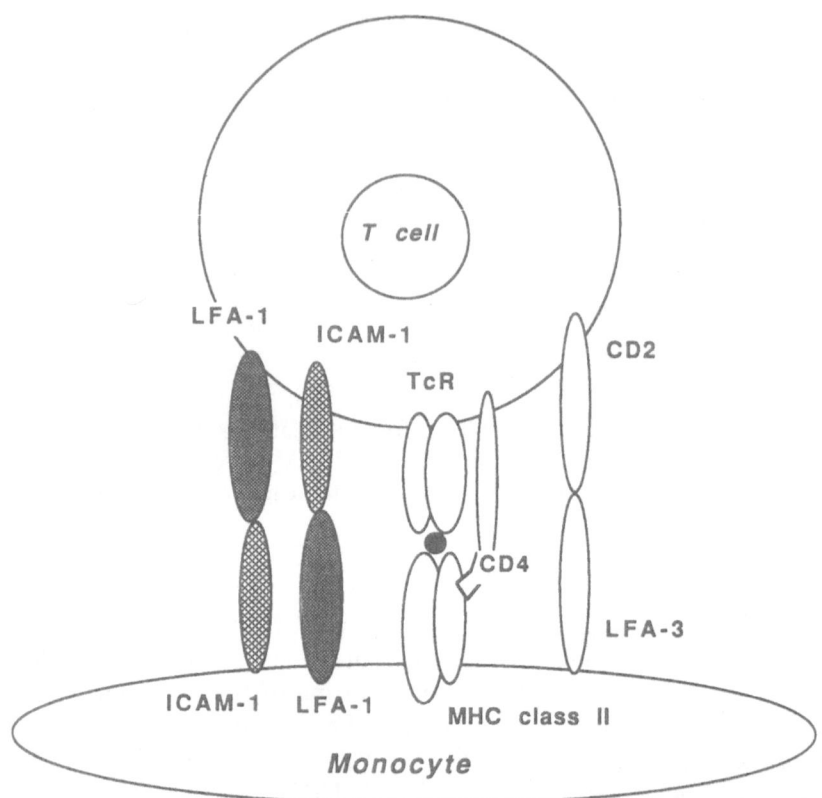

FIGURE 21.4. Schematic representation of the molecular interactions that occur between T cells and monocytes at the early stages of immune responsiveness.

responsible for the more rapid secondary response of memory cells (20,21). It can be predicted that T cell clones would display a similar phenotype both of adhesion molecules and other features of permanently activated cells. Several clones which we have investigated are ICAM-1+(17). Such an "activation" phenotype could explain why they are so readily triggered by MHC class II transfectants.

The precise role(s) of LFA-1 and ICAM-1 have not been clearly defined. They may perform a "nonspecific" adhesive role in bringing into contact T cell and monocyte in order that more specific receptors are able to interact. Alternatively, they may strengthen weak initial conjugate formation as has been demonstrated for LFA-1 in the macrophage/tumor cell interaction in the mouse (22). We do not know where the two LFA-1/ICAM-1 interactions described here occur in the sequence of early events. Whether they occur simultaneously or at different times in the T cell/monocyte interaction is equally unknown. The in vitro clustering reactions in which LFA-1 participates frequently require cell activation (23). The rapid phosphorylation of the LFA-1 β chain after phorbol ester treatment also suggests that LFA-1 functioning is altered after activation (24,25). By analogy, LFA-1 interactions, in vivo, must therefore be secondary to other activating events such as would follow on from specific interactions of the TcR with ligand. This would have the desirable effect of restricting otherwise unfavorable LFA-1/ligand binding which could occur with ease between circulating T cells, monocytes, and other cells. However, it is possible that weak binding, provided by one or the other of the LFA-1/ ICAM-1 pair in the nonactivated state, could provide a sufficient stimulus to permit the interaction of the specific receptors prior to the signalling of cell activation. For example, low affinity binding of LFA-1/ICAM-1 could precede cell activation followed by a high affinity interaction between the same pair of molecules. Further uncertainty about this sequence of events comes from the information that other ligand(s) for LFA-1 are expressed on both T cells (13) and on monocytes (26). Much remains to be learned about the role which adhesion molecules perform in the early communication between T cells and monocytes which precedes the proliferation of T cells.

Acknowledgment. This work was supported by the Imperial Cancer Research Fund.

References

1. Germain RN, Malissen B: Analysis of the expression and function of class II major histocompatibility complex-encoded molecules by DNA-mediated gene transfer. *Ann Rev Immunol* 4:281, 1986.
2. Gougeon ML, Bismuth G, Theze J: Differential effects of monoclonal antibodies anti-L3T4 and anti-LFA1 on the antigen-induced proliferation of T-helper-cell clones: Correlation between their susceptibility to inhibition and their affinity for antigen. *Cell Immunol* 95:75, 1985.
3. Lechler RI, Norcross MA, Germain RN: Qualitative and quantitative studies of antigen-presenting cell function by using I-A-expressing L cells. *J Immunol* 135:2914, 1985.
4. Krensky AM, Sanchez-Madrid F, Robbins E, Nagy JA, Springer TA, Burakoff SJ: The functional significance, distribution, and structure of LFA-1, LFA-2, and LFA-3: Cell surface antigens associated with CTL-target interactions. *J Immunol* 131:611, 1983.
5. Simmons D, Makgoba MW, Seed B: ICAM, an adhesion ligand of LFA-1, is homologous to the neural cell adhesion molecule NCAM. *Nature* 331:624, 1988.
6. Marlin SD, Springer TA: Purified intercellular adhesion molecule-1 (ICAM-1) is a ligand for lymphocyte function-associated antigen 1 (LFA-1). *Cell* 51:813, 1987.
7. Keizer GD, Borst J, Visser W, Schwarting R, de Vries JE, Figdor CG: Membrane glycoprotein p150,95 of human cytotoxic T cell clones is involved in conjugate formation with target cells. *J Immunol* 138:3130, 1987.

8. Takeuchi T, DiMaggio M, Levine H, Schlossman SF, Morimoto C: CD11 molecule defines two types of suppressor cells within the T8⁺ population. *Cell Immunol* 111:398, 1988.

9. Hildreth JEK, Gotch FM, Hildreth PDK, McMichael AJ: A human lymphocyte-associated antigen involved in cell-mediated lympholysis. *Eur J Immunol* 13:202, 1983.

10. Malhotra V, Hogg N, Sim RB: Ligand binding by the p150,95 antigen of U937 monocytic cells: Properties in common with complement receptor type 3 (CR3). *Eur J Immunol* 16:1117, 1986.

11. Hogg N, Takacs L, Plamer DG, Selvendran Y, Allen C: The p150,95 molecule is a marker of human mononuclear phagocytes: Comparison with expression of class II molecules. *Eur J Immunol* 16:240, 1986.

12. Allen CA, Hogg N: Association of colorectal tumor epithelium expressing HLA-D/DR with CD8-positive T-cells and mononuclear phagocytes. *Cancer Res* 47:2919, 1987.

13. Rothlein R, Dustin ML, Marlin SB, Springer TA: A human intercellular adhesion molecule (ICAM-1) distinct from LFA-1. *J Immunol* 137:1270, 1986.

14. Julius MH, Simpson E, Herzenberg LA: A rapid method for the isolation of functional thymus-derived murine lymphocytes *Eur J Immunol* 3:645, 1973.

15. Dougherty GJ, Hogg N: The role of monocyte lymphocyte function-associated antigen-1 (LFA-1) in accessory cell function. *Eur J Immunol* 17:943, 1987.

16. Myones BL, Dalzell JG, Hogg N, Ross GD: Neutrophil and monocyte cell surface p150,95 has iC3b-receptor (CR₄) activity resembling CR₃. *J Clin Invest* 82:640, 1988.

17. Dougherty GJ, Murdoch S, Hogg N: The function of human intercellular adhesion molecule-1 (ICAM-1) in the generation of an immune response. *Eur J Immunol* 18:35, 1988.

18. Dustin ML, Rothlein R, Bhan AK, Dinarello CA, Springer TA: Induction by IL 1 and interferon-γ: Tissue distribution biochemistry, and function of a natural adherence molecule (ICAM-1). *J Immunol* 137:245, 1986.

19. Hogg N, Horton MA: Myeloid antigens: New and previously defined clusters, in McMichael AJ et al (ed): *Leucocyte Typing III*. Oxford, Oxford University Press, p 576, 1987.

20. Sanders ME, Makgoba MW, Sharrow SO, Stephany D, Springer TA, Young HA, Shaw S: Human memory T lymphocytes express increased levels of three adhesion molecules (LFA-3, CD2 and LFA-1) and three other molecules (UCHL1, CDw29 and PGP-1) and have enhanced IFN-g production *J Immunol* 140:1401, 1988.

21. Sanders ME, Makgoba MW, Shaw S: Human naive and memory T cells: Reinterpretation of helper-inducer and suppressor-inducer subsets. *Immunol Today* 9:195, 1988.

22. Strassmann G, Springer TA, Somers SD, Adams DO: Mechanisms of tumor cell capture by activated macrophages: Evidence for involvement of lymphocyte function-associated (LFA)-1 antigen. *J Immunol* 136:4328, 1986.

23. Springer TA, Dustin ML, Kishimoto TK, Marlin SD: The lymphocyte function-associated LFA-1, CD2 and LFA-3 molecules: Cell adhesion receptors of the immune system. *Ann Rev Immunol* 5:223, 1987.

24. Hara T, Fu SM: Phosphorylation of α, β subunits of 180/100-kD polypeptides (LFA-1) and related antigens, in Reinherz EL et al. (eds): New York, *Leukocyte*

Typing II, Vol.3. Human Myeloid and Hematopoietic Cells. Springer-Verlag, p 77, 1985.

25. Chatila TA, Geha RS: Phosphorylation of T cell membrane proteins by activators of protein kinase C. *J Immunol* 140:4308, 1988.

26. Makgoba MW, Sanders ME, Luce GEG, Guge EA, Dustin ML, Springer TA, Shaw S: Functional evidence that intercellular adhesion molecule-1(ICAM-1) is a ligand for LFA-1-dependent adhesion in T cell-mediated cytotoxicity. *Eur J Immunol* 18:637, 1988.

22

Immunosuppression of Cynomolgus Recipients of Renal Allografts by R6.5, a Monoclonal Antibody to Intercellular Adhesion Molecule-1

A. Benedict Cosimi, Cheryl Geoffrion,
Terri Anderson, David Conti,
Robert Rothlein, and
Robert B. Colvin

Introduction

Immunosuppression with heterologous antibodies directed against human T lymphocytes has been investigated in human allograft recipients for more than 20 years. Polyclonal preparations (e.g., ATG and ALG) were initially the only agents available. Because of their unquestioned efficacy, these agents remain an important part of the clinical armamentarium in spite of the continuing difficulties encountered when trying to produce consistent batches of antisera with predictable immunosuppressive potency. The more recent development of monoclonal antibodies (MAbs) reactive with human T-cell-specific surface determinants has provided a more refined approach to the prevention or treatment of allograft rejection episodes (1). The most widely studied MAb in clinical protocols has been the pan-T-cell reactive OKT3. As with most heterologous antisera that have been clinically effective, the immunosuppression achieved following administration has been attributed to T cell depletion, and, in the case of OKT3, modulation of the CD3 molecule.

Of continuing concern is the nonspecific T-cell depletion/suppression resulting from such therapy which obviously increases the risk of infection in these recipients and may not be necessary for adequate rejection control. Thus, manipulation of the immune response by MAbs with more selective reactivity continues to be evaluated. One such approach might be to pursue limited cell depletion by directing therapy toward interruption of the attachment of cytotoxic cells to their targets via antigen-nonspecific accessory molecules, such as lymphocyte function-associated antigen-1 (LFA-1) or intercellular adhesion molecule-1 (ICAM-1).

The LFA-1 (CD11a) is a member of the integrin family of adhesion receptors and shares a common β chain (CD18) with the C3bi receptors Mac-1 (CD11b) and p150,95 (CD11c). The LFA-1 is expressed primarily by cells of bone-marrow origin, including circulating T cells. The ICAM-1 is a 90 kD glycoprotein of the immunoglobulin gene superfamily, with closest sequence homology to the neural adhesion molecules, NCAM, and myelin-associated glycoprotein (2–4). The ICAM-1 is expressed on human vascular endothelium, activated T and B lymphocytes, macrophages, germinal center dendritic cells, thymic epithelial cells, and certain other epithelial cells in vivo (2). Little or no ICAM-1 is detectable on peripheral blood leukocytes (2), although blood monocytes express variable amounts (5). Interaction of LFA-1 and ICAM-1 is required for optimal T cell function in vitro, as judged by the inhibitory effects in vitro of antibodies to each of these components on T cell lysis (6), B cell help (7), or antigen-induced mitogenesis (5).

We have begun to assess the potential clinical applicability of this concept using nonhuman primate transplantation models. In this report, we describe briefly in vivo and in vitro observations of the immunosuppressive capacity and toxicity of R6.5, a murine IgG2a MAb (Boehringer-Ingelheim) reactive with human ICAM-1. A detailed report will appear elsewhere.

Results

Sixteen male Cynomolgus monkeys weighing 3.75 to 5.75 kg received heterotopic renal allografts (8). In most recipients, simultaneous bilateral native nephrectomy was performed, and subsequent uremic death was considered the endpoint of allograft survival. In selected recipients, unilateral native nephrectomy and contralateral ureteral ligation were performed at the time of transplantation. This was done to evaluate any histopathologic changes that might develop in the autologous kidney following administration of R6.5. Renal and hepatic function and hematologic parameters were monitored biweekly. Frequent open allograft biopsies for histopathologic evaluation were performed as were complete autopsies on nonsurviving recipients.

Prophylactic Regimen

"Prophylactic" immunosuppression was employed in 10 recipients. R6.5 was administered intravenously as the only immunosuppression beginning 48 hours prior to the kidney allograft procedure. The MAb was administered as a single daily dosage of 0.5 to 2 mg/kg per day for 12 consecutive days after which no further immunosuppression was given. In these recipients, survival ranged from 7 to 34 days (mean 22.9 day).

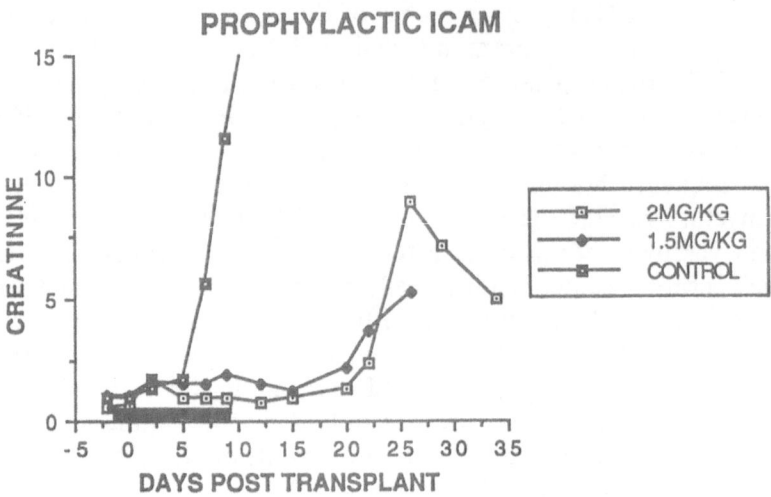

FIGURE 22.1. Course of cynomolgus monkeys with renal allografts treated prophylactically with anti-ICAM-1 compared with control untreated monkeys.

This is in contrast to untreated recipients in whom survival ranged from 8 to 13 days. A comparison of the clinical course of treated and untreated recipients is depicted in Figure 22.1. Survival in eight of the R6.5-treated recipients (20–34 days) was prolonged well beyond that of the longest surviving control animal.

Allografts biopsied during the prophylactic course of R6.5 contained mononuclear cells that infiltrated the interstitium to a moderate degree, focal edema, and little or no hemorrhage. Mononuclear accumulation along the arterial endothelium was absent or focal. Glomeruli were normal.

The allograft kidneys at autopsy showed marked interstitial infiltrates of mononuclear cells and involvement of arteries. One graft was infarcted and lost during therapy. This loss was attributed to a technical failure, since little infiltrate was present in the graft. One animal died on the seventh day with histologic evidence of cellular rejection, but a Cr of 2.7. The only other findings at the autopsy was the presence of pigment in the liver and spleen consistent with old malaria. One animal developed severe acute pyelonephritis that extended to the peritoneum and was probably the cause of death at 22 days, since rejection was not severe.

Therapeutic Regimen

In six recipients, a therapeutic protocol was used. These animals were begun on intramuscular cyclosporine (CsA) injection, 15 mg/kg per day at the time of transplantation. The dose was then tapered biweekly by

2.5 mg/kg per day decrements until subtherapeutic serum levels resulted in an acute rejection episode. At this point, R6.5, 1 to 2 mg/kg per day, was added as the only therapeutic measure for 10 days. Cyclosporine was continued at the dosage level reached when rejection occurred. In previous studies, it was shown that recipients treated with tapering dosages of CsA develop rapidly progressive renal failure once a subtherapeutic threshold is reached (9). Uremic death occurs within 5 to 14 days if no further anti-rejection measures are instituted. Of the six monkeys immunosuppressed initially with CsA in this study, five demonstrated prompt and remarkable improvement in renal function following institution of R6.5 therapy. Mean serum creatinine in these animals peaked at 4.7 mg % during rejection and fell to 1.6 mg % following the 10-day course of monoclonal antibody treatment. With the addition of no further immunosuppression, chronic rejection leading to death developed in four recipients at 38, 45, 54, and 60 days after the initially treated rejection episode. The fifth animal died after developing seizures 11 days following the initial rejection episode. This presumably resulted from CsA neurotoxicity since no other etiology was apparent at postmortem examination. The clinical course of one of these animals is detailed in Figure 22.2.

Prior to treatment, the grafts showed intense mononuclear infiltrates in the interstitium with edema and sometimes hemorrhage; tubules and arterial intima were often invaded by mononuclear cells. No detectable histologic changes were detected 1 hour after the first dose of R6.5. Subsequently, the grafts showed decreased interstitial edema and hemor-

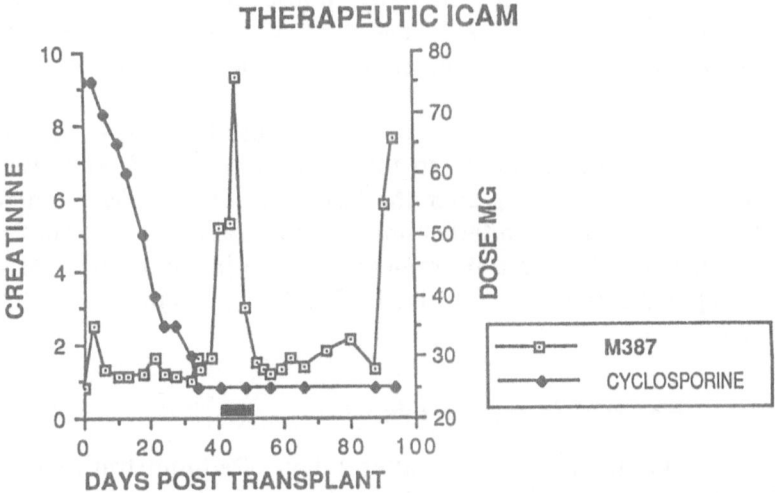

FIGURE 22.2. Course of a monkey treated for rejection that developed during taper of CsA.

rhage, but a variable effect on the mononuclear infiltrate. Two animals showed improvement of renal function and a decreased interstitial infiltrate. Three others, however, showed improvement of renal function without an obvious decrease in the infiltrate. The sixth recipient developed ventricular fibrillation and died while receiving a bolus injection of unfiltered R6.5 antibody on Day 5 of therapy. Serum creatinine had stabilized at 1.6 mg % at the time of death. Autopsy revealed a focal chronic myocarditis was present, which predated the R6.5 treatment, as judged histologically.

Localization of R6.5 After In Vivo Administration

Allografts that were treated prophylactically showed accumulation of mouse IgG along the vascular and glomerular endothelium and on infiltrating cells in samples taken after several days of therapy, as judged by immunoperoxidase staining. In most instances the addition of R6.5 in vitro increased the staining intensity, most notably on the infiltrating cells and the tubules. Samples taken five or more days after completion of the course showed no residual mouse IgG and somewhat increased reactivity with R6.5 compared with grafts sampled during therapy.

Samples of renal allografts on the therapeutic protocol taken 1 hour after the first dose of R6.5 showed uniformly intense accumulation of mouse IgG on the endothelium of the glomeruli, peritubular capillaries, and arteries. Addition of further R6.5 to the tissue sections increased the staining of the endothelium, indicating that saturation of R6.5 epitopes had not been reached. At 1 hour, mouse IgG was not detectable on the infiltrating mononuclear cells, although these cells reacted with R6.5 applied to sections in vitro. Subsequently, mouse IgG was detectable in the interstitial infiltrate as well as the endothelium. Tubules were stained in vitro with R6.5 in rejection, but mouse IgG did not reach this site in vivo.

After completion of the course of R6.5, mouse IgG was cleared from the endothelium and infiltrate in the graft within days; none was detectable four or more days after the last dose. The ICAM-1 remained detectable on the graft endothelium and infiltrating cells. Samples of chronic rejection showed a prominent decrease in the number of peritubular capillaries that were positive for ICAM-1, even in areas with little infiltrate.

R6.5 in the Autologous Kidney

The normal kidney was sampled after 2 days of administration of R6.5 (before the transplantation procedure). Mouse IgG was present on the endothelium of glomeruli, arteries, and capillaries. None was detectable in other renal structures. Addition of R6.5 in vitro increased the staining intensity of the endothelium in all samples. The intensity and distribution of R6.5 reactivity was similar to that in untreated control kidneys. The autologous kidneys removed after 2 days of R6.5 treatment were normal by light microscopy, despite the presence of mouse IgG on the endothelium.

Four autologous kidneys were left in place with their ureters ligated. These kidneys showed mild interstitial edema, dilation of tubules, and a focal mononuclear infiltrate, changes typical of acute obstruction. Mouse IgG was detected in the same distribution and about the same intensity as in the normal kidneys sampled after 2 days of R6.5. In most samples, addition of R6.5 in vitro increased the endothelial-staining intensity over that of anti-mouse IgG alone. In one kidney sampled 5 days after the last dose of R6.5, no mouse IgG was detectable and the endothelium had decreased R6.5 reactivity.

R6.5 in Other Organs

Mouse IgG deposited transiently on the endothelium of the heart, lung, and spleen without eliciting an inflammatory response or histologic alteration. R6.5 accumulated in germinal centers of the spleen and lymph nodes and remained detectable, along with ICAM-1 antigen, for more than 30 days after the last dose. During therapy, mouse IgG deposited on scattered mononuclear cells in the sinusoids of the spleen and in lymph nodes.

Conclusions

With the introduction of MAbs into clinical immunotherapeutic protocols, it has become possible to suppress only selected elements of the immune response. The ultimate goal, of course, would be the elimination or blocking of just the cells responding to an allograft while leaving intact all other reactivity. Most previous approaches have been directed against T-lymphocytes (e.g., OKT3) or subpopulations of T-cells including activated cells (e.g., anti IL-2R MAbs). The current study was undertaken to determine whether allograft rejection could be blocked by a novel approach, namely interference with LFA-1/ICAM-1 adhesion between cytotoxic cells and their targets.

The most important new information is that ICAM-1 is a critical molecule in the pathogenesis of graft rejection. In vitro studies had indicated that antibodies to ICAM-1 or its receptor (LFA-1) block leukocyte adhesion and function; however, several ICAM-1/LFA-1 independent adhesive mechanisms can be demonstrated in vitro. It was not known which, if any, were relevant to T cell mediated injury in vivo. These studies demonstrate that an antibody to the leukocyte adhesion molecule ICAM-

1 can delay the onset of acute cellular allograft rejection and reverse established acute rejection. The immunosuppressive efficacy of R6.5 in this model compares favorably with that produced by MAbs directed against T-cells surface antigens (10). R6.5 may act simply by decreasing T cell adhesion or by preventing other functional consequences of LFA-1/ICAM-1 interaction.

The most striking site of deposition of R6.5 was on the vascular endothelium of the graft, especially the arteries, arterioles, capillaries, and glomeruli. Notable in the R6.5 prophylactic group was the almost complete lack of vascular rejection (11) during treatment and the lack of progression of vascular rejection in the therapeutic group. In the therapeutic protocol, recovery of renal function was accompanied by decreased interstitial edema and hemorrhage, which are believed to result form peritubular capillary injury (11). This suggests that blockage of the T-mediated vascular component of rejection may be the major mechanism of action of R6.5.

These studies were designed to test the anti-rejection activity of R6.5, rather than its toxicity, because these animals ultimately develop uremia from the rejection process. Some comment should be made, however, on the extra-graft pathologic findings. No vascular or glomerular lesions were found in the autologous kidneys. No vascular lesions were found in the other organs, with the exception of two animals with microscopic pulmonary emboli. Whether these were related to the R6.5 treatment is uncertain. The lack of vascular injury induced by R6.5 contrasts dramatically with the acute endothelial inflammatory reaction mediated by other antibodies (11–13). The intriguing possibility that R6.5 itself prevents antibody-dependent cell-mediated cytotoxicity (ADCC) of the endothelial cells is supported by the observation that antibodies to LFA-1 block ADCC by human mononuclear cells (14).

These observations suggest further possible strategies of intervention. T cell lysis of some targets is LFA-1 dependent and ICAM-1 independent and escapes inhibition by anti-ICAM-1 alone. The addition of antibodies to the ICAM-1 receptor, LFA-1, should be synergistic or additive with anti-ICAM-1, according to experiments in vitro (Merluzzi, et al., chapter 19, this volume). Monoclonal antibodies to the β chain of the LFA-1 molecule inhibit reflow damage and extension of myocardial infarction in myocardial ischemia models in dogs (Todd et al., chapter 9; Seewaldt et al., chapter 10, this volume). It has not been established whether these antibodies act at the leukocyte or endothelial level. Anti-ICAM-1 however, might also have clinical application during perfusion of organs prior to transplant to prevent ischemic injury.

Acknowledgments. This work was supported by NIH grant PO1-HL18646 and Boehringer-Ingelheim Pharmaceuticals, Inc.

References

1. Cosimi AB, Colvin RB, Burton RC, Rubin RH, Goldstein G, Kung PC, Hansen WP, Delmonico FL, Russell PS: Use of monoclonal antibodies to T cell subsets for immunologic monitoring and treatment in recipients of renal allografts. *N Engl J Med* 305:308–314, 1981.
2. Dustin ML, Rothlein R, Bhan AK, Dinarello CA, Springer TA: Induction by IL 1 and interferon gamma: Tissue distribution, biochemistry, and function of a natural adherence molecule (ICAM-1). *J Immunol* 137:245–254, 1986.
3. Staunton DE, Marlin SD, Stratowa C, Dustin ML, Springer TA: Primary structure of ICAM-1 demonstrates interaction between members of the immunoglobulin and integrin supergene families. *Cell* 52:925–933, 1988.
4. Simmons D, Makgoba MW, Seed B: ICAM, an adhesion ligand of LFA-1, is homologous to the neural cell adhesion molecule NCAM. *Nature* 331:624–627, 1988.
5. Dougherty GJ, Murdoch S, Hogg N: The function of human intercellular adhesion molecule 1 (ICAM-1) in the generation of an immune response. *Eur J Immunol* 18:35–39, 1988.
6. Dustin ML, Springer TA: Lymphocyte function associated antigen 1 (LFA-1) interaction with intercellular adhesion molecule 1 (ICAM-1) is one of at least three mechanisms for lymphocyte adhesion to cultured endothelial cells. *J Cell Biol* 107:321–331, 1988.
7. Boyd AW, Wawryk SO, Burns GF, Fecondo JV: Intercellular adhesion molecule 1 (ICAM-1) has a central role in cell cell contact mediated immune mechanisms. *Proc Natl Acad Sci USA* 85:3095–3099, 1988.
8. Cosimi AB, Burton RC, Kung PC, Colvin RB, Goldstein G, Lifter J, Rhodes W, Russell PS: Evaluation in primate renal allograft recipients of monoclonal antibody to human T-cell subclasses. *Transplant Proc* 13:499–503, 1981.
9. Wright JK Jr, Barrett LV, Delmonico FL, Cosimi AB: Preclinical evaluation of immunosuppression selective for T cells recognizing class I histocompatibility antigens. *Transplant Proc* 19:1106–1109, 1987.
10. Cosimi AB, Colvin RB, Jaffers GJ, Giorgi JV, Delmonico FL, Fuller TC, Russell PS: Immunologic monitoring of monoclonal antibody therapy: Comparison of five antibodies as immunosuppressants of renal allograft rejection. *Transplant Proc.* 16:1459–1461, 1984.
11. Colvin RB: Renal allografts, in Colvin RB, Bhan AK, McCluskey RT (eds): *Diagnostic Immunopathology,* New York, Raven Press, pp. 151–197, 1988.
12. Barba LM, Caldwell PR, Downie GH, Camussi G, Brentjens JR, Andres G: Lung injury mediated by antibodies to endothelium. I. In the rabbit a repeated interaction of heterologous anti-angiotensin-converting enzyme antibodies with alveolar endothelium results in resistance to immune injury through antigenic modulation. *J Exp Med* 158:2141–2158, 1983.
13. Ahern AT, Artruc SB, Della Pelle P, Cosimi AB, Russell PS, Colvin RB, Fuller TC: Hyperacute rejection of HLA-AB identical renal allografts associated with B lymphocyte and endothelial reactive antibodies. *Transplant* 33:103–106, 1982.
14. Kohl S, Loo LS. Schmalstieg FS, Anderson DC: The genetic deficiency of leukocyte surface glycoprotein Mac-1, LFA-1, p150, 95 in humans is associated with defective antibody-dependent cellular cytotoxicity in vitro and defective protection against herpes simplex virus infection in vivo. *J Immunol* 137:1688–1694, 1986.

Index